21,20
Sm 4
21 March 1997

IDEALS AND REALITIES

IDEALS AND REALITIES
Selected Essays of Abdus Salam
(Third Edition)

Editors: C H Lai
and
Azim Kidwai

World Scientific
Singapore • New Jersey • London • Hong Kong

Published by

World Scientific Publishing Co. Pte. Ltd.,
P O Box 128, Farrer Road, Singapore 9128
USA office: 687 Hartwell Street, Teaneck, NJ 07666
UK office: 73 Lynton Mead, Totteridge, London N20 8DH

The editors and the Publisher are grateful for the permission granted to reproduce the articles collected in this volume.
The choice of the title of this volume has been inspired by Seyyed Hossein Nasr's book "Ideals and Realities in Islam", George Allen and Unwin Ltd., London 1966.

IDEALS AND REALITIES (THIRD EDITION)

ISBN 981-02-0080-3
     981-02-0081-1 (pbk)

Library of Congress Cataloging-in-Publication Data

Salam, Abdus, 1926—
    Ideals and realities.

    1. Physics--Developing countries. 2. Science-Developing countries.
3. Physicists--Pakistan--Biography.   I. Lai, C. H.   II. Kidwai, Azim.
III. Title.
QC71.S28   1989   530   89-24808
ISBN 981-02-0080-3
ISBN 981-02-0081-1 (pbk.)

Printed by Singapore National Printers Ltd

Contents

Editorial Note

The first two editions of *Ideals and Realities* got a response from the readers that showed the book to be a *tour de force*. In the meantime, Professor Abdus Salam had made some more valuable presentations like "Notes on Science, Technology and Science Education in the Development of the South" — a document he had prepared for the 4th and 5th meetings of the South Commission. Professor Abdus Salam is one of the thought leaders in the world today and his views coming through such researched presentations need be given wider dissemination.

To enlarge the second edition of *Ideals and Realities* thus turned into an imperative. It also provides an opportunity to improve upon the collection in selection and presentation.

In its latest form, the volume provides a greater fund of thoughts of a Nobel Laureate whose first love and main passion is physics, but whose concern for the Third World and promotion of science in the developing countries knows no bounds and matches his main passion.

Most of the new material added in the third edition concerns the problems of the Third World and his ideals about the shape of things to be.

We would like to thank Professor Abdus Salam for the advice and guidance we received from him in the selection of the material for this volume.

Azim Kidwai and *C. H. Lai*

Editorial Notes to First and Second Edition

An attempt is made here to collect in one volume some representative non-technical writings of Professor Abdus Salam. The essays touch on many different themes, and in particular discuss the social and economic dimensions of science. As a scientist from Pakistan, Professor Salam has a personal understanding of the various difficulties that scientists in developing countries encounter. His concern in this aspect is quite manifest in many of his writings in this volume, which also offer some insightful analyses. A few other essays record his effort in the establishment of the International Centre for Theoretical Physics at Trieste, and his excitement in nurturing its growth.

Perhaps of particular interest is Professor Salam's view on the development and the international nature of science. His insistence that there can be no permanent dominance in science by a single nation and that "scientific thought and its creation is the common and shared heritage of mankind" deserves much thought.

It should be pointed out that the essays collected here are written in Professor Salam's personal style, on subjects clearly dear to his heart. A number of them have overlapping portions, and we have decided to leave them intact. It is felt that any substantial editing would reduce much of the flavour and emphasis of the original articles.

We have also included in this volume a few accounts of Professor Salam himself and of the International Centre for Theoretical Physics at Trieste. They provide glimpses of the man Professor Salam as he appears to scientists and friends who have more than a superficial acquaintance with him, as well as of his efforts towards the internationalisation of science, epitomised by the International Centre for Theoretical Physics at Trieste.

It has been a difficult task making the selection for this volume — we can only wish that the collection has not suffered too badly from our lack of experience and insight.

April 1983 C. H. Lai & Zafar Hassan

Note added:

The publication of the first edition of Ideals and Realities was a remarkable success, judging from the response we received. It was also clear that some improvements could be made to this collection of Professor Salam's essays, both in its selection and presentation. This second edition affords a chance to do just that. I particularly want to thank Professor Salam for his constant interest in this book, to which my editorial efforts are simply no match.

November, 1986
Cambridge, Massachusetts, USA C. H. Lai

The Less-Developed World: How Can We be Optimists?

This article was written in 1964 by Abdus Salam and is reprinted from **The World in 1984** *(Vol. 1), edited by Nigel Calder, Penguin Books Ltd., England, 1964 © New Scientist.*

Between the frontiers of the three super-states Eurasia, Oceania, and Eastasia, and not permanently in possession of any of them, there lies a rough quadrilateral with its corners at Tangier, Brazzaville, Darwin, and Hongkong. These territories contain a bottomless reserve of cheap labour. Whichever power controls equatorial Africa, or the Middle East or Southern India or the Indonesian Archipelago, disposes also of the bodies of hundreds of millions of ill-paid and hardworking coolies, expended by their conquerors like so much coal or oil in the race to turn out more armaments, to capture more territory, to control more labour, to turn out more armaments, to capture more territory, to control . . .

Thus George Orwell — in his only reference to the less-developed world. I wish I could disagree with him. Orwell may have erred in not anticipating the withering of direct colonial controls within the 'quadrilateral' he speaks about; he may not quite have gauged the vehemence of urges to political self-assertion. Nor, dare I hope, was he right in the sombre picture of conscious and heartless exploitation he has painted. But he did not err in predicting persisting poverty and hunger and overcrowding in 1984 among the less privileged nations.

I would like to live to regret my words but twenty years from now, I am positive, the less-developed world will be as hungry, as relatively undeveloped, and as desperately poor, as today. And this, despite the fact that we know the world has enough resources — technical, scientific, and material — to eliminate poverty, disease, and early death, for the whole human race.

The visible portents of 1984 are there for all to see. Notwithstanding every physical and ideological exhortation, the agricultural production of all but the richest countries is static. It would seem that the industry of food production is as investment-intensive as any other. We are only just beginning to speak, in not too muted tones, of high birth-rates. There are none among the rich nations willing enough to sponsor a fair price structure for the commodity market — the one major resource the poorer countries possess for financing their meagre development plans. There are likely to be higher and still higher tariff walls against their cheap manufactures. And every year the battle to keep the trickle of foreign aid programmes flowing becomes fiercer and fiercer. The United Nations Development Decade from all indications is likely to end with a whimper.

But this is not what makes me pessimistic. Never in the history of mankind has a change happened all at once. The one great change of the first half of this century — the passing of the Colonial Age — was the culmination of fifty years of crusading. In most places it all started with a few men, whose passionate fury first overhelmed their own peoples and then succeeded in rousing the liberal conscience of their captors, bringing home to them also the utter economic futility of holding down an unwilling people. This is the normal process of change. What makes me worried is that no such thing has yet happened in the underdeveloped world so far as the harder crusade against poverty is concerned. And in the few places where realisation has come, it has not been purposeful enough yet to bring down the internal, social and the organisational barriers, nor to be able to defy external pressures. In the next twenty years I trust this crusade will come to be preached with the fury it deserves within the poorer countries. I can only hope it remains inward-turning — that it does not become a destructive wave of antagonism against those fortunate few among the nations of the world who somehow inherited most of this earth's resources and do not quite need them all.

But this will take time. For 1984 itself, I am resigned to Orwell's grim picture, with want and misery unchecked — unless, of course, there rises earlier somewhere a new Messiah, the one who can preach that in this age when technological miracles are indeed possible, the raising of living standards everywhere to a decent human level is first and foremost a moral problem, and a collective world responsibility.

IDEALS AND REALITIES

I

SCIENCE AND THE WORLD

Ideals and Realities

Lecture given by Abdus Salam at the University of Stockholm, 23 September 1975. Published in a shorter version in Bulletin of the Atomic Scientists, September 1976.

I am deeply honoured and much appreciate the opportunity to give the first lecture in this series on Human, Global and Universal Problems, particularly just after the conclusion of one of the most momentous of special sessions of the United Nations General Assembly dealing with this subject. This session, as you all know, was convened to discuss the global crisis in the human family's continuing and near-permanent polarisation between the rich and the desperately poor and the latter's demand for a *New International Economic Order.* I have looked forward to the opportunity of speaking to you today because I know that Sweden is one of the few countries of the world which has understood the issues; it is the ONLY country at present which is fulfilling the United Nations targets of aid. Its youths led the world in 1972 so far as global concerns go. My purpose today is to have a dialogue with you and to explore what are the ways in which the almost total incomprehension among the rich nations of what the poor are really demanding can be removed — and the urgency of the crisis mankind is facing brought home to developed societies.

The short-term crisis the world faces is simply this: The developing world — some nine-tenths of humanity — is bankrupt. We — the poor — owe the rich — one-tenth of mankind — some 50 billion dollars. The poorest amongst us cannot even pay the interest on our borrowings — far less find the 10 billion dollars we collectively need to import 10 million tons of cereals every year to feed ourselves. My own country, Pakistan, owes some 6 billion dollars — roughly equal to Pakistan's GNP for one year, roughly equal to Pakistan's six years' export earnings. Last week's London's prestigious *Economist* magazine starkly said, "The poorest among the poor who can neither borrow more nor draw on reserves will cut on their imports — their people will simply starve."

But this short-term crisis is only a part of a longer-term crisis. Our world is terribly unbalanced in income and in consumption. At least three-quarters of the world's income, three-quarters of its investment, its services and almost all of the world's research are concentrated in the hands of a quarter of its people. They consume 78% of its major minerals, and for armaments alone, as much as the rest of the world combined. In 1970, the world's richest one billion earned an income of $3,000 per person per year; the world's poorest one billion, no more than a $100 each. And the awful part of it is that there is absolutely nothing in sight — no mechanism whatsoever — which can stop this disparity. Development on the traditional pattern — the market economics — is expected to increase the one hundred dollars per capita of the poor to all of one hundred and three dollars by 1980, while the $3,000 earned by the rich will grow to $4,000 — that is an increase of $3 against $1,000 over an entire decade.

No wonder the poor nations consider visions of any growth and development on the traditional economic system a vicious fraud. This is the system which in the last 20 years created liquidity and credits of 120 billion dollars allocating just 5% of these to the poor nations. This is the system which pays 200 billion dollars for world commodities, but only one-sixth of this reaches the primary producer himself — the rest, five-sixths, going to the distributor and the middleman in rich countries — this is the system which gave 7 billion dollars of aid last year and took away almost exactly the same amount from the poor in depressed commodity prices. No wonder they are demanding, in Omar Khayyam's words, "Ah love! could thou and I with fate conspire, to grasp this sorry scheme of things entire, would not we shatter it to bits — and then remould it nearer to the heart's desire."

Over the past three to four years, some of the brighter young economists of the Third World countries, Brazil, Mexico, Algeria, Pakistan and others — helped by some of the most distinguished figures in World Economics — have been groping towards a new synthesis of development and outer limits of growth. I am ashamed for my own profession, for there were no scientists or technologists associated with them. It is this new synthesis — embodied in the so called Cocoyoc and Rio declarations — which formed the basis of the Resolution on a Declaration on Establishment of a New International Economic Order, adopted in 1974 by the sixth special session of the United Nations Assembly. The present session — just concluded, the follow-up from the last — was convened to put some teeth into the Charter of Economic Rights promulgated by the United Nations Assembly in 1974.

Among the poor these Declarations have been likened to the great Declarations of Rights of Man in the 18th century by Tom Paine and the Communist Manifesto of the 19th century. What the establishments in the richer countries really think of the International Economic Order is hard to fathom. During 1974, the reaction might have been typified by the words of one of the richer nations' delegates to the United Nations who referred to the "shadow world of rhetoric" and "the drawback of so many short-lived resolutions, each longer than the last, one a repetition of the other, virtually unreadable . . ." This year, though the response was still not outright commitment, Dr. Kissinger presented to the Assembly, on behalf of the richer nations, a welcome package of co-operative funds, joint institutes and aid initiative. I shall speak of these later, but in any case what is needed is not just that the Foreign and Finance Ministries of the developed countries should respond to the demands of the poor, but that the intellectuals and the general public become aware of these and truly comprehend them.

In this spirit, I shall therefore try to convey to you how a humble natural scientist from a developing country — who is not an economist, but one who passionately loves the United Nations and its work — views the global crisis of the disparity of the rich and the poor.

To get behind the psychological thinking of the poorer humanity, you must understand how recent in our view this disparity — which makes untermenschen of us today — is. It is good to recall that three centuries ago, around the year 1660, two of the greatest monuments of modern history were erected, one in the West and one in the East: St. Paul's Cathedral in London and the Taj Mahal in Agra. Between them, the two symbolize, perhaps better than words can describe, the comparative level of architectural technology, the comparative level of craftsmanship and the comparative level of affluence and sophistication the two cultures had attained at that epoch of history.

But about the same time there was also created — and this time only in the West — a third monument, a monument still greater in its eventual import for humanity. This was Newton's Principia, published in 1687. Newton's work had no counterpart in the India of the Mughuls. I would like to describe the fate of the technology which built the Taj Mahal when it came into contact with the culture and technology symbolized by the Principia of Newton.

The first impact came in 1757. Some one hundred years after the building of the Taj Mahal, the superior firepower of Clive's small arms had inflicted a humiliating defeat on the descendants of Shah Jahan. A hundred years later still — in 1857 — the last of the Mughuls had been forced to relinquish the Crown of Delhi to Queen Victoria. With him there passed away not only an empire, but also a whole tradition in art, technology, culture and learning. By 1857, English had supplanted Persian as the language of Indian state and learning. Shakespeare and Milton had replaced the love lyrics of Hafiz and Omar Khayyam in school curricula, the medical canons of Avicenna had been forgotten and the art of muslin making in Dacca had been destroyed making way for the cotton prints of Lancashire.

The next hundred years of India's history were a chronicle of a more subtly benevolent exploitation. I shall not speak of this, but only of the scientific and technological milieu I was brought up in as a young man in British India. The British set up something like 31 liberal High Schools and Arts Colleges in what is now Pakistan, but for a population then approaching 40 million people, just one College of Engineering and one College of Agriculture. The results of these policies could have been foreseen. The chemical revolution of fertilisers and pesticides in agriculture touched us not. The manufacturing crafts went into complete oblivion. Even a steel plough had to be imported from England. It was in this milieu that I started research and teaching in modern physics some 25 years ago at Lahore, in the University of the Panjab.

Pakistan had then just won its independence after one hundred years of British rule. We then had a per capita income of $80 a year, a literacy rate of 20%, a population growing 3% a year and an irrigation system for agriculture which was breaking down. There was no social security and there was high child mortality — only five children out of twelve lived beyond one year. A child — a male child — was the only social security for old age one could budget for, making high birth rate imperative.

Pakistan — very willingly — accepted to become part of the free world economic bloc. We were relieved of worries of increasing population needing growing of more food. The US surpluses of wheat — under P. L. 480 — gratefully came, at first, in such abundance that one of our Finance Ministers spoke of curtailing wheat cropping in Pakistan by law to grow tobacco instead. We imported highly talented Development Planners from Harvard University. They told us we did not need to put up a steel industry. We could in any case buy any amount from Pittsburgh. We leased out our oil imports and even the distribution of petroleum products within the country to multinations who conducted — in that age of oil surpluses — a half-hearted search for it.

Pakistan was thus a classic case of a post-colonial economy; political tutelage was interchanged for an economic tutelage. In the scheme of things, we were to provide cheap commodities — principally jute, tea, cotton, raw unprocessed leather. It was in 1956 that I remember hearing for the first time of the scandal of commodity prices — of a continuous downward trend in the prices of what we produced, with violent fluctuations superposed, while industrial prices of goods we imported went equally inexorably up as a consequence of the welfare and security policies the developed countries had instituted within their own societies. All this was called Market Economics. And when we did build up manufacturing industries with expensively imported machinery — for example, cotton cloth — stiff tariff barriers were raised against these imports from us. With our cheaper labour, we were accused of unfair practices.

To give you an idea of these tariffs — suppose Pakistan exported cotton seeds, these would attract only $100 a ton as tariff. But woe-betide if the seed was crushed into oil. The oil fell into the category of manufactures, and the tariff shot up to $600. We were to be markets for steel, for machinery, for fertilizer, for armaments. We must not export anything remotely resembling manufacture. No wonder we have been bankrupted.

Of indigenous science and technology — or indeed of any technological manpower development — there was neither need, nor appreciation, nor any role for it. Any technology we needed, we bought. It came hedged with all types of restrictions. For example, no product which used this technology could be exported. And in any case, not all technology was for sale. Pakistan, for example, could not buy the technology of penicillin manufacture in 1955. My brother, together with a few other young chemists from Pakistan, re-invented the process, producing as a result of their inexperience penicillin at 16 times the world price. In the early 1950's, I looked upon my future as contributing to Pakistan's advance to technology and development as non-existent. I could help my country in only one way — as a good teacher — and that was to produce more physicists, who for lack of any industry, would in their turn, become teachers themselves, or leave the country.

But soon it became clear to me that even this role — that of a good teacher — would increasingly become impossible for me to maintain. In that extreme isolation in Lahore, where no physics literature ever penetrated, with no international contacts whatsoever and with no other physicists around in the whole country, I was a total misfit. I knew that all alone, I had no hope pf changing Pakistan policies, so far as valuing science and technology were concerned. There was but one recourse, to make a call on the international scientific community to help in preserving one's professional integrity. My hope lay with the United Nations organisation and its agencies. And thus in 1954 started my involvement with these.

It is now two decades since I have been engaged in a very humble way with science and international affairs. I can divide this period into two distinct decades — the first decade from 1954 to 1964 — the decade of innocence and hope — and the second decade 1964 to 1974, of growing frustration and a feeling of hopelessness of it all. The third decade is beginning for me now. Perhaps this decade will bring more hope.

But to go back to my personal story; the first opportunity I got of playing a minor role in public affairs came in 1955 with the Atoms for Peace Conference held in Geneva. You may recall that this was the first scientific conference held under the United Nations auspices, the first conference when East-West secrecy, which extended till then to even trivial scientific information like neutron scattering cross-sections, was partly lifted. At this conference was promised atomic plenty to the world for energy, for isotope applications, for new and revolutionary genetic varieties of crops.

For me personally, this conference was important, for this was my first introduction to the United Nations. I remember entering that Holy Edifice in New York in June 1955 and falling in love with all that the organization represented — the Family of Man, in all its hues, its diversity, brought together for Peace and Betterment. I did not then realise how weak an organisation it was, how fragile and how frustrating in its inaction, but I shall speak of this later. It seemed to me then that any ideas I may have of helping Pakistan physics — and developing countries' physics — must be implemented through United Nations action.

The second occasion when I came in contact with the organisation came in 1958, at the Second Atoms for Peace Conference. This conference was similar to the one in 1955; its major achievement was a furthering of the process of declassification of nuclear fusion. For me, the greatest gain was that I had the privilege of working as a secretary under one of the greatest Swedes in international affairs — Dr. Sigvard Eklund — now Director-General of the International Atomic Energy Agency. From that date, started a most cherished personal friendship and one which transformed my life.

One consequence of the 1958 conference was that the Pakistan Government became interested in atomic energy. Pakistan has no oil, little gas, some hydro-potential. Pakistan needed atomic energy. In 1958, President Ayub Khan assumed power; I was recalled to Pakistan and asked to help with the creation of an Atomic Energy Commission.

We decided that in the absence of any other scientific organisation in the country, it was our mandate to create research teams and research institutes in all fields of national endeavour — agriculture, health, besides an atomic industry. For this and to fulfil the needs of Pakistan universities, we must train in the great institutions of the world, mathematicians, chemists, physicists and agriculturists.

We instituted a training programme for scientific manpower within our meagre resources. I say meagre, because at its peak, the total research expenditure in all universities and all research establishments in Pakistan never exceeded 4 million dollars — a sum which you in Sweden spend on one department of physics in one university. With these meagre outlays, it was clearly impossible for Pakistan science to achieve any semblance of excellence. For ending the isolation of Pakistan science — the problem I had faced — we would still depend on international help.

For mobilising this help, an opportunity came in 1960 when I was fortunate to represent Pakistan at the General Conference of the International Atomic Energy Agency in Vienna. I suggested at this conference that the international scientific community represented through the scientific agencies of the United Nations organisation should accept as one of its responsibilities the caring for its deprived members — that there should be set up a network of first-rate international centres — in various pure and applied disciplines of science and technology which should offer their facilities principally to short-term senior visitors from developing countries. I envisaged a system of associateships available at these centres — through which top scholars from developing countries would be given long-term — five years — appointments enabling them to spend three months of their summer vacation working together with their peers from developed countries at these centres, recharging their batteries and taking back with them newer ideas, newer techniques, newer impetus. This would end the isolation which, for example, I had suffered and which, in my view, was the principal cause of brain drain of scientists — in contrast to the brain drain of doctors or engineers.

In 1961, the values of high level scientific and technological contacts were rather strikingly brought home to us in Pakistan. Pakistan had inherited from the 19th century one of the most extensive networks of irrigation canals — some 10,000 miles long — irrigating 23 million acres of land. Some of these canals were as large as the Colorado River. They were carefully designed as to width, depth and slope in such a way that silty water moved just fast enough so that it neither eroded their beds nor choked them by depositing sediment.

But in 1961 something had gone grievously wrong with the system. After a few decades of operation, the canal network slowly began to stifle the very fertility it was meant to create by spreading the blight of water-logging and salinity in areas through which the canals passed. One million acres of land were passing out of cultivation every year during 1950 to 1960.

In 1961, Professor J. Wiesner, President Kennedy's Science Adviser, assembled a team of university scientists, hydrologists, agriculturists and engineers, led by Roger Revelle, to advise on this problem of water-logging and salinity. This team suggested continuous pumping out of saline water to lower the water table, but with the important caveat that the pumping operation must be simultaneous over a contiguous area as large as a million acres — otherwise the quantity of water seeping in from the periphery would exceed the quantities pumped out. Pumping had been tried on parcels of land smaller than one million acres, but proved ineffective. Some of you may recall that Blackett during the last war was called upon to suggest to the British Admiralty whether merchant ships should cross the Atlantic in a few large convoys or many small ones — given that the number of available destroyers protecting against enemy submarines was fixed. Noting that ratio of area to circumference maximises with large radius, Blackett had suggested fewer large convoys rather than many small ones. Revelle's team's suggestion for Pakistan was a remark equally simple and it equally simply worked.

My next involvement with the United Nations system — and also the first disenchantment with the establishments representing their countries at this forum — came in 1962 when Dag Hammarskjöld projected a United Nations Conference to be held the following year, on Science and Technology. He had the vision of transforming the developing world — through technological projects like the one I have just mentioned. I had the privilege of a long interview with Dag — the only time I met him — and to share his semimystical reverence for what science and technology — if applied meaningfully — could achieve for the poor. He recognised clearly that this needed first and foremost investment, even if relevant technology was available. Much more even than the leaders of the developing world, he recognised that it was important to establish an indigenous scientific capacity in developing countries for research and development. This was needed, at the very least, to achieve an awareness of the significant development of world science and technology, an awareness which would enable a country to select and negotiate the purchase and ensure the effective assimilation of technology which its economic and social objectives required. He recognised that it is not just know-how which the developing countries need; it is also the *know-why*, if technological development was to be a graft which should take in the poorer world.

The conference proposed by Hammarskjöld was held in 1963, unfortunately after his tragic death. We, from the developing countries, proposed the creation of a World Science and Technology Agency — a Technical Development Authority, backed by an international bank for technological development. Besides strengthening indigenous science in developing countries, the Authority would have acted as a planning and programming body which would carry out feasibility studies, devise programmes and arrange their implementation. Being a United Nations organisation it would associate with its work and give maturity to local scientific and technological organisations and talent, giving them training and intimate knowledge of the complex new techniques. Its very existence would have emphasised what the planning economist so often forgets — that the modern world and its problems are a creation of modern science and technology.

We proposed this; we lobbied for this, but we met with a complete blank wall of incomprehension — or worse — from the delegates from the industrialised countries — who, by and large, opposed the idea of any such Science and Technology Agency. It seemed that they preferred the scientific and technological effort of the United Nations to remain weak and fragmented within the system. There appeared no desire on their part to share technology with the developing world except through the existing system of licensing, operating in the manner I have described earlier in the context of Pakistan and the story of penicillin manufacture. The net legacy of this conference was the creation of an eighteen-man Advisory Committee on Science and Technology. We met for eleven years — twice a year; after eleven years labour, we have recommended yet another United Nations Conference on Science and Technology to be held in 1978, this to meet and create the same Science and Technology Development Agency we proposed fifteen years ago. This time we are likely to get this because Dr. Kissinger gave the proposed Conference his blessing three weeks ago.

I was meeting the same incomprehension in respect of my second suggestion at the forum of the IAEA regarding the idea of the creation of a Theoretical Physics Centre, particularly from some of the countries where theoretical physics in fact flourishes. One delegate went as far as to say; "Theoretical physics is the Rolls-Royce of sciences — what the developing countries want is nothing more than bullock-carts." To him a community of 25 physicists and 15 mathematicians, all told, trained at a high level for a country like Pakistan with a population of 60 million, was simply 40 men wasted. That these were the men responsible for all norms and all standards in the entire spectrum of Pakistan's education in physics and mathematics was totally irrelevant. He was himself an economist, who had wandered into a scientific organisation like the IAEA. He could fully understand that we needed more high level economists, but physicists and mathematicians — that was wasteful luxury.

For the first time, it also began to be borne in upon me, how weak the United Nations system really was in terms of resources. Even today, 12 years later, the United Nations family has miniscule resources. Let me give you the figures (Table I).

The total of the funds within the United Nations system for development do not add up to the funds available, for example, to the Ford Foundation — and this for service to 140 nations of whom 82 are

Table I
Budget Figures (in millions US$)

	1975	1976
UN	540	620
UNEP	6	6
UNIDO	31	45
IAEA	32	37
WHO	115	125
UNESCO	255 (including 100 from UNDP)	
ILO	94	135
FAO	117	unavailable
ICAO	12	13
IMCO	unavailable	11

desperately poor. The United Nations was created as a community of equal nations — but some were more equal than others. It was financially weak because the rich nations would not contribute to its revenues; it was functionally weak because the powerful nations respected its resolutions only when they were extensions of the decisions of their own foreign policies.

In 1964 when IAEA did agree to the physics centre, its Board voted as a sum of $55,000 to create an international centre. Fortunately, the Government of Italy came through with an annual grant of $350,000 and the Centre was set up in Trieste.

To complete the story of the Centre, it started operating in 1964. It is now co-sponsored by IAEA and UNESCO, together with UNDP, who both contribute around a quarter million dollars each; plus the Italian Government with a grant of $350,000 and SIDA with a grant of $100,000. During the 11 years of its existence so far, it has received some 6,000 senior physicists from 90 countries — 4,000 of them from 65 developing countries. It has truly created something of a revolution so far as the studies of physics are concerned, so far as the developing world is concerned. Over the years it has tended more and more to emphasise technology transfer in physics. In this, we have particularly been helped by a Solid State Committee headed by Professor J. Ziman of Bristol and Professor S. Lundqvist of Chalmers in Gothenburg. Two weeks ago, we inaugurated the first ever extended three-month Course on the Physics of Oceans and Atmosphere attended by some 60 senior physicists, meteorologists and ocean-scientists from some 30 developing countries. The Centre, however, still remains a singularity — the one isolated Centre within the United Nations family of its kind in the entire spectrum of advanced scientific knowledge.

After 1963, the disillusionment with the existing international order came fast. You know the history of this decade as well as I do. President Kennedy with whom, rightly or wrongly, the liberal aspirations — also about world development — got linked was assassinated.

Around 1968 was the beginning of the student revolt and the realisation that the environment was being wrecked. I felt then and still feel — and this is why I am speaking to you today — that the developing world lost a great moment, lost a great potential alliance, a great potential source of strength when the protesting energy of world youth concentrated on the one issue of environment and did not espouse at the same time the more embracing cause of world development.

In between these years came the repeated failure of UNCTAD conferences convened to propose a redress for relatively ever-falling commodity prices. It is good to be reminded today that the price of petroleum fell decisively between 1950 and 1970 — down to one dollar a barrel, stimulating a growth in energy use of between 6% to 11%. The reception of UNCTAD's proposals — its fervent appeals for some stability and indexation of commodity prices — were received with derisory scorn, typified even today by the influential London *Economist* writing in its issue of 30 August this year on the eve of the United Nations Conference, "The notion that the price of each commodity can be tied, not to the demand for it, but to the average rise in the price of manufactured goods, is a proposal to try to repeal, by some conference fiat, the Laws of Supply and Demand. The industrial countries should simply refuse any concessions to this proposal." And this in a year which saw the index of manufactures' prices go up to 140, while commodity price index hovered around 114. Thus, in this one year alone, the poor have subsidised the welfare economies of the rich to the extent of 26% of their earnings.

In 1972 came the great Conference on Environment in Stockholm. It was significant not just for pinpointing that the environment was being wrecked and that some countries were contributing more than their fair share towards wrecking it. Even more important, it thrust into prominence the interdependence of the human community in solving the issues raised.

In 1972 came also the Club of Rome report on Outer Limits to Growth — with the thesis that world resources are finite and simply could not sustain infinite growth of industrialised economies. It is not well known that the poor countries had received a sharp reminder of this — as early as mid-1972 — in the form of a precipitous doubling of the price of wheat. This had happened because the failure of the crops in USSR made them buy 30 million tons of grain, nearly exhausting the world grain reserves. This was one of the contributory causes of the threefold increases in oil prices; followed by yet another doubling of grain prices. Add to this the waning of the resource transfers of foreign aid programmes — the one collective commitment of the western countries — and you can understand the origins of the short-term crisis — the financial bankruptcy of poor humanity — with which I started my lecture.

To complete the story of foreign aid, the 17 richest nations allocated .3% of their GNP to overseas development last year, compared with .52% in 1960. While Sweden generously earmarked .72%, UK and USA provided .30% and .25% respectively. The World Bank estimates that by 1980 the average of the 17 countries will be .28% and of US .18%. Contrast this with the US contribution at the beginning of the Marshall Plan of 2.79% of GNP. Ministries in rich countries usually dismiss as unrealistic the United Nations target allocating .7% of their GNP to aid. Yet this target could probably be reached in the second half of this decade if they merely devoted 2% of the increased wealth — the $1,000 per capita growth I spoke of before — which is expected to accrue to the industrialised nations in the next few years. At the United Nations Conference just concluded, EEC ministers did announce their readiness to try to meet the .7% target in 1980, though unfortunately, UK and USA expressed reservations.

Realising these stark facts, and realising that the developed world was unlikely to produce a Messiah — or even a Keynes — who would preach social justice between nation and nation, the developing countries decided in 1974 to use the forum of the United Nations for calling for a New International Economic Order.

The New International Economic Order

What is the International Economic Order? The Rio Declaration, perhaps somewhat more radical than the United Nations Resolution, starts with the preamble: "Developed countries — by and large — have shown remarkable reluctance to initiate and support change. Having derived much of their wealth from cheap resources and raw materials of developing countries, they still refuse to give access to their markets to the Third World. They refuse to recognise the inevitability of modifying their life styles, and scale and patterns of consumption, the maintenance of which requires a disproportionate share of world resources. They have used the power provided by science and technology to pursue policies shaped by selfish interests over the world's oceans, and they are squandering a vast fraction of mankind's resources, in scientific manpower as well as materials, in stockpiling of weapons of mass destruction." The document then goes on to say that "The struggle of the Third World is for *Economic Liberation*, greater equality of opportunity and securing of right to sit around the bargaining tables as equals with a redistribution of *future* growth *opportunities*. In the last analysis, we must look on the demand for the New International Order as a part of a historical process, as a movement, to be achieved over time."

The United Nations Resolution on the New Order is perhaps somewhat more muted; it starts with a call for a commitment from mankind for the banishment of poverty and prevailing disparities; it calls for a just and equitable relationship between prices of raw materials and manufactured goods; it calls for access to the achievements of modern science and technology; it calls for an end to wasteful consumption — particularly in respect of food and expenditure on armaments.

In order to see how the ideals expressed in the United Nations Resolutions are carried out into realities it is perhaps worthwhile to consider food and military expenditures in somewhat greater detail.

Food

In November 1974, the United Nations convened a Conference in Rome on Food. This conference adopted the following declaration: "Within a decade no child will go to bed hungry and no family will fear for its next day's bread and no human being's future will be stunted by malnutrition." To achieve this target, a World Food Council was set up, with the minimum target of distributing 10 million tons of grain a year as food aid, and achieving average 3.6% increases per year in food production by poor countries, through an international provision of agricultural inputs.

On 29 June 1975, the *London Times* reported "The World Food Council ended its inaugural meeting here in Rome at 2.00 a.m. yesterday. It was saved from being an obvious farce and a failure only by some quick facesaving footwork by western diplomats. France, Germany and Italy have so far refused to endorse an increase in EEC's food aid from 1.3 million tons to 1.6 million tons. This was bitterly attacked in Rome, not the least by UK which threatened to increase its own bilateral aid if the insensitivity of its partners continued. The commitment to 10 million tons — even though well short of food aid levels of 1960 — has still not been reached."

Is there a real absolute overall shortage of food in the world, which makes the contribution of this 10 million tons impossible — and with this the inevitability of starvation in poor countries? The answer is NO.

It should be emphasised again and again that the grain is physically available. It is simply being consumed by well-fed people. Since 1965, the richer nations have added 350 pounds per head to their annual diets, largely in the form of beef and poultry. This was stimulated by a special pricing policy at a time when US surpluses of food gains were running in excess of world demand by some 60 million tons a year, in spite of a curtailment in area cropped, by one-half. This is very nearly the equivalent of an Indian's total diet for a whole year. Few will maintain that the industrialised countries were undernourished in 1965. A cut in consumption, for example the suggested equivalent of one hamburger a week, could provide all the grain needed to support a population as large as one-third of the Indian subcontinent.

Let us next consider the question of armaments and arms reduction. In 1973 the world military expenditure came to 245 billion dollars. This sum is 163 times greater than that spent on international co-operation for peace and development through the United Nations system; (this sum stands at approximately 1.5 billion, excluding the World Bank). The superpowers spent 50% of these 245 billion dollars, while another 30% was spent by military alliances. The share of the Third World also, unfortunately, increased between 1955 to 1975 from 6% to 17% — and we are not entirely blameless. The world military expenditures are now greater than the GNP of all of Africa and all of South Asia. During the two decades, 1960's and 1970's the total military expenditure was $4,000 billion, which is greater than all goods and services produced by all mankind in one year.

When we consider the situation with materials and men, the situation regarding expenditures appears even grimmer. Close to 7% of all raw materials in the richer countries are consumed by the armament industry. This includes oil, iron, tin, zinc, copper and bauxite. It is estimated that about 50 million people are employed for military purposes in armed forces and defence activities. Close to half a million scientists and engineers, almost half of the world's scientific and technological manpower, is devoted to military research and development, costing between $20 to $25 billion. These sums represent 40% of all public and private research and development expenditure mankind appropriates. Contrast this with the half million dollars we have been able, after five years of continuous effort, to collect for the International Foundation for Science whose first General Assembly is taking place in Stockholm today. The situation is clear; it is not the poor countries who jeopardise the global balances, it is the rich and their rivalries and their desire to hold monopoly military power.

To summarise, the demand for a New International Order is a demand of a basic minimum standard of living and of economic security for all citizens; a deliberate policy of development and re-distribution to achieve this. Just as on a national scale the achievement of the social and economic goals is left not entirely to individual effort and initiative but is prompted actively by the combined efforts of the entire community, so also on an international level the aspirations of the nations of the world in the social and economic sphere should be made easier to achieve by a concentrated effort of the world community as a whole — by the Family of Man, acting as a whole.

What the developing countries really want on a psychological plane is to regain their sense of dignity and self-respect which they enjoyed for long centuries and which they lost only during the brief period of western domination; a domination based essentially on an industrial and technological revolution which is hardly two centuries old. The fact that country after country in all parts of the world has successively and successfully mastered technology is also not overlooked by those who are still left behind. What the developing countries are asking for is not unlimited migration, to the open, uncultivated, under-utilised areas of this globe; they have never asked for transfer of income, wealth and resources at any exorbitant level. It is rather a meaningful sharing of technology and equitable trade they are really after.

Perhaps the time has come for supplementing national transfer with some international sources of revenues — the international commons, taxed for the benefit of the poorest strata of poor countries. This would be a first step towards the establishment of an international taxation system and an international treasury aimed at providing automatic transfers of resources for development assistance. I remember this suggestion being voiced by Linus Pauling at the 1969 Nobel Symposium on the Place of Value in a World of Facts, held in Stockholm in 1969, and the somewhat cool reception the idea received. It appeared too radical at that time. But perhaps its time has come; perhaps one may start with an international commons provided by resources of world oceans, the one resource not yet finally carved out among nation states.

Table II

Grain Surpluses and Deficits for 1973 (in million tons of grains)	
North America	+ 91
Latin America	− 3
Asia	−43
Africa	− 5
East Europe	−27
West Europe	−19

Oceans

The 138-nation Law of the Sea Conference held in Caracas, Venezuela in 1974 has negotiated at its last session in Geneva a single informal negotiating text which fortunately is still subject to amendment. The comprehensive treaty envisaged for 1975 has been called the most vital document the United Nations will produce, since 1945. The treaty envisages extension of territorial waters from 3 to 12 nautical miles, an exclusive economic zone under coastal state jurisdiction extending up to 200 miles, plus 200 metres depth whichever is farther. This, if finally approved, would be an unmitigated disaster, even though some developing countries will benefit from this.   The sea bed contains perhaps 1,500 billion barrels of petroleum; at present some 15% of the world's oil and gas comes from the oceans, but they contain perhaps the major portion of future oil potential.   Some 18 billion dollars of high-protein fish are caught annually and then there is the possibility of easy dredging from the North Pacific deep ocean-floor of some 400 million tons of copper, manganese, nickel and cobalt nodules every year.   Compare this 400 million tons with 10 million of these minerals consumed annually now. The exciting thing about these nodules at the ocean bed is that they obligingly renew themselves all the time — either because they are organic materials like coral or because some obscure process of ionisation is at work at the sea bed.

The effect of the proposed treaty will be to place 62% of the sea bed oil under the jurisdiction of 10 of the most fortunate coastal states — most of which have per capita incomes already exceeding $1,000 — while 51 countries with little or no continental shelf will get only 1%.   I am no legal expert, but to any internationalist it is clear that what is truly needed is the replacement of the outmoded concept of national sovereignty by a concept of "functional sovereignty" which permits the interweaving of national and international jurisdiction within the same territorial space. At present, the only agreement which has been reached envisages that there will be an International Sea Bed Resources Authority which will provide environmental protection over deep sea mining and that it might be allocated revenues collected directly from production of deep sea minerals.   However, regarding the more immediate resource, oil, there is still a discussion going on, if royalty revenues from sea bed oil may be pooled through an international fund, to be used primarily for developing countries.   Canada has suggested collection of 1% of sea bed oil revenue. The US Government suggests a small percentage of revenue from oil *beyond* 200 miles limit.   But there is no forceful voice, yet, calling for a significant reapportioning of these new windfalls such that a meaningful international commons is paid — for global development.

This trend of thinking must be reversed.   Substantial revenues from sea bed oil could go to the international community.   Twenty percent of these could provide a sum for developing countries of up to as much as 6 – 12 billion dollars a year.   The International Sea Bed Resources Authority could become a model for world institutions, dealing with arms control, disarmament or global resource management.   Geneva 1975 may be the last and the only opportunity of ensuring that the concept of "common heritage of mankind" becomes and remains — not just an empty concept.

I should perhaps conclude by telling you what actually happened at the United Nations Conference. What is it that has been achieved? Dr. Henry Kissinger, alive to the danger of a confrontation more virulent, more destructive than any cold war, urged the recognition that if there was no action on the demands from the poor, "Over the remainder of this century, . . . the division of the planet between north and south could become as grim as the darkest days of cold war. We would enter an age of festering resentment, of a resort to economic warfare, a hardening of new blocks, the undermining of co-operation, the erosion of international institutions — and failed development."

Dr. Kissinger and the US have promised a multiplicity of institutions to meet the needs of co-operative world development. Two of these are:

(i) A "Development Security Facility" to stabilise prices of commodities against crude cycles of export earnings though "indexing" was decisively ruled out;

(ii) Measures to improve access to capital technology and managerial skills — and in particular an International Energy Institute, an International Centre for Exchange of Technological Information and an International Industrialization Institute.

At long last the physicists we have trained at Trieste will find a rightful role in the development of their countries, though I do hope these new Centres do not suffer the frustrations which 11 years of running a United Nations institute have taught me to fear. In dealing with the United Nations one finds to one's frustration that the promises which one department of a national government makes go unheeded by the other departments of the same Government; that each department of each donor country wishes to investigate *ab initio* what the United Nations Centre is achieving. So far as Trieste is concerned, this year there have been five commissions reporting on the Centre; there will be two more before the year ends. And this happens every year. The point is that the United Nations funds are extremely limited, the organisation is an orphan and the energy needed to keep any initiative alive through the United Nations is often out of all proportion to the results achieved.

To come back to the Conference, regretfully there was no new commitment of resource transfers — these new institutions will presumably divide the old cake differently; their very multiplicity will unfortunately not make any realisation of the ideals of world development any easier.

I come back to you — my audience — you are our only hope to realise the ideals I spoke of into realities in a meaningful manner. For make no mistake, for world development much more sacrifice will be called for and will have to be made. But I am a believer in man's moral state and I shall conclude with the words of a mystic who expressed the international ideal of Family of Man in the 17th century, John Donne: "No man is an island, entire of itself; every man is a piece of the continent, a part of the main; if a clod be washed away by the sea, Europe is the less, as well as if a promontory were, as well as if a manor of thy friends or of thine own were; any man's death diminishes me, because I am involved in mankind; and therefore never send to know for whom the bell tolls; it tolls for thee."

International Commons: Sharing of International Resources

Address by Professor Abdus Salam at the Royal Moroccan Academy Meeting, 26 April 1983.

1. In 1945 Europe was devastated. Soon after, the United States took a remarkable initiative with the launching of the Marshall Plan to finance European recovery. Some 32 billion dollars were generously provided, amounting, in the beginning, to a contribution of around 2.79% of the Gross National Product of the USA. A magnificent act of magnanimity, it was pure altruism, because the USA knew that by building up Europe, it was contributing to the future prosperity for the entire Western world, including enhanced prosperity of the United States itself, through trade and commerce. It is unfashionable nowadays to speak in these terms, but one may have called this act Keynesianism at its best, inspired by the earlier success of the New Deal in the United States itself. One of the results of this all too rare act of economic wisdom was that during the next decades — the sixties and the seventies — after Western Europe was back on its feet, the prosperity of all countries — including the donor country of the USA — increased to levels unmatched in world history before.

The Marshall Plan led to similar ideas for US and European aid to be extended to the developing countries. Here, of course, the needs were greater. Perhaps the sheer magnitude of the development tasks meant that the donors felt shy of doing as much as they had done for Western Europe. The aid packages were more meagre. And there was one other limitation. Those were also the days of the hottest phases of the "cold war". The aid packages extended to the developing countries were not purely economic aid. The cold war had imposed a selectivity; the most generous economic help went hand in hand with military aid. The donors also wanted the sums alloted to be spent in helping Western interest including the Western exports.

As I mentioned earlier, the small quanta of aid funds were inadequate with the needs. Instead of 2.79% of the GNP of the Marshall Funds, this time the funds (contributed by OECD countries) never went beyond .5%, falling to around .34% by the 1970's. Even though the Pearson Commission set up in 1969, made its recommendation of aid quantum being fixed at .7% of the GNP of donor countries — a recommendation later endorsed by the Brandt Commission — these figures have never been met except by a very few of the donors. Thus the US share has fallen to less than .2% with a fall also in the shares of the UK, France, Federal Republic of Germany, Japan and others. And, furthermore, the Eastern bloc have never joined the aid consortia; their aid (.14% of GDP) — much smaller — is disbursed bilaterally. The OPEC countries started in the early 1970's with 1.18% of their GDP, went up to nearly 3% in 1975 and then declined (with a total of 7.7 billions) to 1.4% of GDP in 1981.

I am not so much concerned in this note with the precise aid percentages. I am much more concerned with the way in which the conceptual basis of such transfer of resources is presented. It is my belief, and I am sure members of this Academy share this belief with me, that unless an idea has a sound, generally accepted conceptual currency behind it, it does not win adherence.

2. Some of the considerations on which a theory of transfer of resources to the developing world may be based may run as follows.

(*i*) Economic self-interest. As emphasised earlier, in the case of US help to Europe, Keynesianism theories which inspired the New Deal, may have been at the back of the Marshall Plan. Involved in this is firstly the idea that in order that societies should be economically well-off, one needs a large base of economic activity. Secondly, the securing of this large base needs, in its turn, the prosperity of all sections of the society, leaving no pockets of poverty, within the society. Thus, prosperity for all, an inter-dependence and a perceived mutuality of interests of all sections of the society, is the key idea with the Marshall Plan extending the scope of the society covered, from USA alone, to embrace also the continent of Western Europe. The Plan based itself on the view that the prosperity of the US would increase if Europe became prosperous, and able to exchange goods and services with the US.

What we are speaking about today is the carrying of these ideas further to include the developing countries. In the words of Willie Brandt: "The mutuality of interests can be spelled out clearly in the areas of energy, commodities and trade, food and agriculture, monetary solutions, inflation control . . . and ground space communications. The depletion of renewable and nonrenewable resources, throughout the planet, the ecological and environmental problems, the exploitation of the oceans, not to forget the unbridled arms race, which both drains resources and threatens mankind — all of these also create problems which affect peace and will grow more serious in the absence of a global vision" . . . "Whoever wants a bigger slice of an international economic cake cannot seriously want it to become smaller . . . Most industrialised countries, even during the biggest boom in human history, have not tried hard enough to get near the minimum aid target to which most of them had solemnly agreed. That record is not only disappointing but also reminds us that, had the target been met, several developing countries would not be importing more goods and services thus mitigating economic difficulties of the North."

(*ii*) To highlight the economic interdependence, particularly in the context of producing job opportunities in the developed countries, Brandt continues: "Perhaps one can illustrate part of the problem from the development of some of the present industrialised countries in the nineteenth and early twentieth centuries. A long and assiduous learning process was necessary until it was generally accepted that higher wages for workers increased purchasing power sufficiently to move the economy as a whole. Industrialised countries now need to be interested in the expansion of markets in the developing world. This will decisively affect job opportunities in the 1980's and 1990's and the prospect of employ-ment."

This sentiment was echoed also by J. Tinbergen *et al.* in their paper on "A new world employment plan". According to them, "The second element of a new world employment policy consists of an increase in international income transfers to Third World countries in order to increase employment in these countries. The resulting increase in welfare and purchasing power in these countries will lead to higher imports from the industrialised countries. This then will be an important stimulus to higher employment in the industrialised countries too."

The same statement comes from Masaki Nakajima: "In the past, of course, we had the Keynesian approach to demand development. That, unfortunately, was geared to the development of a single economy. Today, in order to solve a massive problem like worldwide unemployment, I think we have to expand the Keynesian approach so it can be applied on a global scale. One possibly effective area would be toward a solution of North-South problems.''

(*iii*) One may look upon aid as a compensation for the decline in commodity prices. Year after year we have seen that the weakness — economic, as well as political — of the developing countries has meant that the commodity prices have not kept up with the increase in prices of industrial goods. As Michael Manley, the Ex-Prime Minister of Jamaica, once explained: "In the 1950s, ten tons of sugar brought a Jamaican farmer a Ford tractor. In the 1970s the same tractor costs 25 tons of sugar. Why? Is it that the Jamaican peasant is subsidising by a factor of 100% the social security and welfare of Ford plant workers? And not only have the commodity prices not kept up with industrial prices, they have also seen such ups and downs that for a developing country there is no possibility of any rational planning of its economic future. These "vagaries" of price cycles are attributed to the vagaries of stock exchanges. Speaking plainly, is this not a type of organised brigandage which the rich societies have permitted their stock market speculators to indulge in?

As is well-known, this economic weakness of the developing countries has led them to the brink of bankruptcy. The facts in respect of the economic situation of the developing world are stark. The non oil-producing countries have suffered a deterioration in their export earnings of some 100 billion dollars annually between 1980 and 1983. At least 50 billion dollars of this are attributable to lower commodity prices. The pleas of the developing countries to have some consideration given to the commodity prices have consistently fallen on deaf ears. On the testimony of the Ex-Chancellor Helmut Schmidt of Germany who wrote in *The Economist* of 26 February 1983, the German Federal Republic put forward a proposal for stabilising developing countries' exports of raw materials for international discussion as early as 1978. Unfortunately, according to him, this proposal was not taken up at the Agenda of any international conferences. It is time, he says, to raise the proposal again. In the meanwhile the developing countries may be forgiven if they consider aid as part compensation for this decline of commodity prices.

(*iv*) One may look upon aid as part of the compensation for the 19th century exploitation of the riches of the developing countries — a transfer of resources from the ex-Colonies, ex-Empires which enriched particularly some of the European countries and gave them economic prosperity.

(*v*) One may point to the disparity of distribution of world resources and the instability it creates. There is, at present, a tremendous disparity (Table I) between the rich and the poor in the ultimate criteria of prosperity — the reserves of arable land, of forest, coal and iron. It is important to realise that the full scale occupation of empty regions of the globe — Siberia, Canada, Australia — took place in the 19th century and that it is of relatively recent origin. Some among those who plough the exhausted soils of Asia and Africa may not for long be able to avert their hungry gaze from the virgin soils of some fortunate and empty corners of this globe. It is hard for them to comprehend that there can exist parts of the world where 15 and 20% of agricultural land has to be "banked" and the farmers paid not to cultivate it in order that world prices of grain can be kept up. It is hard for them to believe that open spaces still exist in Canada, Australia, Siberia, and elsewhere, and that material rewards must be paid to those willing to pioneer their colonisation. There is one lesson from history we must not forget: a world as polarised as ours is unstable; it cannot endure this way forever.

It was perhaps in recognition of this disparity and the instability which it breeds that Lyndon Johnson expressed himself thus: "Many of our most urgent problems do not spring from the cold war or even from the ambitions of adversaries. They are the ominous obstacles to man's effort to build a great world society — a place where every man can find a life free from hunger and disease. Those who live in the emerging community of nations will ignore the problems of their neighbours at the risk of their own prosperity . . . . There is no simple solution to these problems. In the past there would have been no solution at all. Today, the constantly unfolding conquests of science give man the power over his world and nature which brings the prospect of success within the purview of hope." Lyndon Johnson had the courage, pursuing this line of thinking, to allocate the funds which he saved from the defence budget of the United States to his programmes against poverty in that country. One wishes there were more men like him who would declare that a similar consequence would follow global disarmament and cuts in military expenditure will mean more funds for global development.

Table I The Disparity in Natural Resources (per capita)

	Asia	North America	USSR	Europe	Oceania	World
Agricultural area (hectares)	.54	2.63	2.8	.55	30	1.4
Accessible forest area	.20	2.07	5.4	.33	1.6	.96
Coal reserves (tons)	6.3	2000	90	960	812	365
Oil reserves (tons)	.8	27.8	16.9	4		12.5
Iron ore reserves (tons)	16.4	389.6	502	59.8	25.0	102

(Estimate made by the United Nations in 1950 and quoted in *World Population and Production*, by W. S. Waytinsky and E. S. Waytinsky.)

3. These were some of the arguments for the transfer of resources. But what is needed is some sort of automaticity in these transfers and I wish to speak about this now.

In 1969, the Nobel Laureate, Linus Pauling, speaking at the Nobel Symposium held in Stockholm, on the Place of Value in a World of Facts, made a plea for international taxation whereby the world distribution of wealth among the nations of the world may be adjusted by levying a tax on the nations with a higher GNP and providing the funds to those in the developing countries. Pauling spoke of the transfer of resources of the order of 200 billion dollars per year, about 8% of the world's then total income, which he thought was the right figure for an international income tax. I remember listening to Pauling and thinking to myself; this is a totally Utopian proposal. None took it very seriously at the meeting. Pauling thought it was possible to formulate a fundamental principle of morality, independent of revelation, superstition, dogma and creed and acceptable by all human beings in a scientific, rational way by analyzing the facts presented to us by the evidence of our sense. He said a major fact of our lives was that there is so much suffering in the world, much of it unnecessary and avoidable. To minimise this suffering we must provide every person not only with adequate food and shelter but also with education.

To produce the funds required, Pauling identified militarism as one of the major causes of human suffering. Militarism then cost the world over 250 billion dollars per year. It costs three times that much today. This amount of wealth wasted on military conflicts each year is greater than the total annual personal income per year of two-thirds of mankind. An elimination of these conflicts would enable these funds to be spent on minimising the suffering from deprivation of the majority of mankind. Pauling suggested that scientists and scholars should begin to formulate a practical schedule of progress toward the goal of such transfers. He said that it was only the intellectuals and scientists of the world who could analyse this problem in a sufficiently thorough way; they should take political actions as individuals, as science advisers, as educators and by applying pressure on governments and voters.

After 1969 when Pauling spoke, there was a formulation of what is called the New International Economic Order which was adopted in 1974 by the 6th Special Session of the UN General Assembly. Unfortunately, just after these proclamations were made, came the increase in oil prices and rise of the monetary of economics. Few people today remember the work which was done on the international economic order. But, in any case, to my knowledge, not much thinking went into emphasising Pauling's ideas of world taxation. I have now come to believe that Pauling's idea was one of the most important ideas which came out of the last decade. It seems to me a great pity that it was not given a proper economic formulation by the economists of the world either from the rich or the poor nations, and that the concept of world international taxation has not become common currency to replace aid commonly thought of as charity and which in the ultimate analogies depends on the vagaries of national governments.

One of the few who have discussed this issue in recent times is again Willie Brandt who said in 1980 in the introduction to his Commission's Report: "It is our conviction that we will have to face more seriously the need for a transfer of funds . . . with a certain degree of automaticity and predictability disconnected from uncertainties of national budgets and their underlying constraints. What is at stake are various possible forms of international levies."

"Why should it be unrealistic," he asks, "to entertain the idea of imposing a suitable form of taxation on a sliding scale according to countries' ability? There could be even a small levy on international trade, or a heavier tax on arms exports. Additional revenues could be raised on the international commons, such as seabed minerals."

4. Brandt advances the idea of "international commons" as a prelude to full-fledged taxation. That certain resources of the Seas should be declared as belonging to mankind as a whole is an idea which the developing countries have espoused since 1968. A convention to regularise this has now been embodied in a draft for "Law of the Sea" which the UN Conference has recommended all UN states to adopt. This was at Montego Bay, in the last days of December, after nine years of patient negotiations marked by a willingness to compromise as a necessary part of the search for a larger solution. One hundred and nineteen nations have so far found it possible to overcome their individual reservations and marginal disappointments and sign the instruments which make the Law of the Sea a new international fact of life, giving substance to a "Sea Bed Authority" which will be sited in Jamaica.

The US Government, however, has decided to stand out and vote against adoption of the recommendation. This decision was taken by the Republican Administration in 1981, repudiating the skillful negotiations by the Carter Administration to arrive at a compromise formula for the draft convention. After the US had decided to back out on the Carter Administration's pledges, Britain too has decided to stand out. Since these two nations represent a substantial slice of the world's economic power and technological capabilities, their decision to stand out in the interest of their multinationals can be a very serious wrecking manoeuvre indeed.

In this context the remarks of Mrs. Jean J. Kirkpatrick, the US Ambassador to the UN, on 3 March 1983, are significant. Writing in a journal on Regulation published by the *American Enterprise Institute,* she complains: "The UN regulatory initiatives extend quite literally from the depths of the oceans to the heavens, from the Law of the Sea Convention to an Agreement Covering the Activities of State on the Moon and other Celestial Bodies." According to her, the US balked at signing the Law of the Sea Convention which required that mining companies and other undersea ventures be licensed by a new international authority, pay what would amount to royalties to it and be bound by its decision on production and the like. In Mrs. Kirkpatrick's view, the big push within the UN stems from a sort of class-welfare, poor nations versus rich with regulation as a weapon for the redistribution of wealth. According to her, this type of thinking guides many of the participants in a UN political process. "There is a good deal of vote-trading, arm-twisting, demagoguery, playing to the galleries and the result is that proposed agreements which are supposed to benefit all nations often turn out to be, above all, instruments for global redistribution of wealth and a new global paternalism. In a world body of 157 nations, the US and the capitalist West are an outnumbered automatic minority. The UN agencies then are the scene of a struggle that we seem doomed to lose. The international bureacracy functions as the new class to which power is to be transferred. Global socialism is expected and, from the point of view of many, is the desired result." As her remarks clearly show, there is an urgent task for us, particularly the intellectuals from developing countries, to invest the idea of International Commons with a theoretical basis so that it comes to be accepted by the populations of the developed countries.

5. One way to make these ideas more acceptable may be to declare that these commons will be used only for global tasks. And among these urgent global tasks is the application of science and technology to global problems. If, for example, these International Commons were used for building up research and development capabilities, now sadly neglected, in the sphere of energy and in the sphere of environment, there may be less opposition to these than there has been.

To take the case of environmental tasks, everyone speaks of the degradation of the biosphere. People speak of the disappearing rain forests and the imminent disappearance of large numbers of animal and plant species. The Report "Year 2000" commissioned by President Carter states that "in the next 17 years, 1/4 million plant and animal species will disappear because the developing countries will be forced to cut their forest wealth in order to make up for scarce fuel and in order to grow more food." One may in this context ask if it is not the concern of the environmental groups also in the developed countries to help to preserve this global heritage? Should they not come to the rescue of the developing countries? Should this type of global assistance not be a first charge on the International Commons?

As a scientist, I would like the International Commons to be used for research on global scientific problems. This was one of the suggestions made at the Vienna Conference on Science and Technology in 1979. The global problems suggested were research on diseases of the developing countries, research on greening of deserts, research on weather modification particularly for developing countries, research on alternative energy, research on productivity of marginal soils, research on earthquake predictions and the like. Recently, in December 1982, there was a meeting of chemists — Chemrawn — held in the Philippines on the subject of Chemistry and Development. We at Trieste are going to hold a Conference on Physics and Development in 1984 in conjunction with IUPAP. I hope many of the ideas expressed at these forums for further research do not fade away for lack of resources.

Perhaps I should mention Masaki Nakajima's "Dream for Mankind". In Table II is his list of global super-infrastructure projects which may constitute elements of a "Global New Deal"; the implementation by the richer nations of the superprojects would lead to stimulation of constructive demand in manufacturing industries, as well as technological incentives, in lieu of arms production. Hopefully, this would be accompanied by increase in GNP and employment opportunity both in developed and developing countries. According to Nakajima, "Now is the time for mankind to exert a bold, new and brave, long-range vision, a vision which transcends narrow short term national interest. . . . As the prophet-King Solomon said in the Bible: 'Where there is no vision mankind perishes.'"

It is the grand vision behind these projects which is so commendatory and it is this global vision which alone will solve our problems of the future.

To summarise, I would like to see the ideas of world taxation brought into the sphere of fundamental economic thinking. As a prelude to this, I would like to see the ideas of International Commons and the sharing of global resources (for example, the riches of the seas, and of the Antartica) taken up vigorously on a theoretical, as well as at a political level — for example, at the Summit Conferences.

Personally, I would like to see these Commons expended on scientific research on global problems in the first instance. I would like to see much greater emphasis given to the global programmes like International Geophysical or Biospherical Programmes which, at the moment, organisations like ICSU undertake on a meagre budget with contributions from the not too affluent UN bodies like UNESCO. As an example of joint scientific projects, there are the programmes of collaborative research for European nations like the fusion research programme and laboratory at Culham in the UK, or the European Molecular Biology programme and laboratory at Heidelberg. On the other hand, there are very few global laboratories for global problems research. These should be the first charge on the International Commons.

I have been speaking about a theoretical basis for international taxation based on well-reasoned appeals to economic self-interest and to man's rationality. However, as the great religions of the world teach us, in the end the most potent human actions stem from man's ethical sense. I am a firm believer in man's moral and spiritual state and I shall conclude with the words of a mystic who thus expressed the international idea of Family of Man in the 17th century — John Donne — "No man is an island, entire of itself; every man is a piece of the continent, a part of the main; if a clod be washed away by the sea, Europe is the less, as well as if a promontory were, as well as if a manor of thy friends or thine own were; any man's death diminishes me, because I am involved in mankind; and therefore never send me to know for whom the bell tolls; it tolls for thee."

Table II

Name	Nations (Areas) Involved	Outline of the Proposal
1. Greening of deserts	North African nations and Arab States	Greening of the desert in the Sinai and the Arabian penisula.
2. Collection station for solar heat		Erect a large-scale installation for the collection of solar energy in a remote part of the World. Total investment in land, pipelines, and accessory equipment would reach $20 to $50 trillion. Its total annual output would be equivalent to 200 billion barrels of oil.
3. Electric power generation using sea currents		There are 12 promising areas along undeveloped ocean shores extending from the equator to the temperate zones. Maximum generating potential of one area, 35 million kW. Total for 12 areas about 200 million kW.
4. Himalayan hydroelectric project	India China Bangladesh	Damming of the Sanpo River on the upper reaches of the Brahmaputra in the frontier area between China and the Indian province of Assam to make it flow into India through a tunnel across the Himalayas. Potential generating capacity 50 million kW in maximum 37 million kW in average. Annual generating capacity 240 billion to 330 billion kWh.
5. African central lake	Central African nations	Control the flow of the Congo River by building a dam to create a vast lake in Congo and Chad regions of Central Africa to improve natural conditions in the area.

Nuclear Security, Disarmament and Development

Speech delivered by Professor Abdus Salam at the Groupe de Bellerive Conference on Nuclear War, Nuclear Proliferation and their Consequences at Geneva, 27–29 June 1985.

1. The world's stock of nuclear weapons, which was three in 1945, has been growing ever since and is 50,000[a] in 1985. *Nearly two trillion dollars of the public funds have been spent over the years to improve their destructive power, and the means of delivering them.* One indicator of the awful power of these weapons is that the explosive yield of the nuclear weapons stockpiled today by the US, USSR, UK, France, and China is equivalent to *one million Hiroshima bombs.* Less than 1,000 of these 50,000 weapons could destroy USA and USSR. A thousand more in an all-out nuclear exchange could destroy the world as a habitable planet, ending life for the living and the prospects of life for those not yet born, sparing no nation, no region of the world.

Hannes Alfven has suggested that the word "annihilators" should be used for nuclear weapons, to bring home to mankind their real nature. The awful point about nuclear annihilators is that their destructiveness has only superficially sunk in. We continue to think of a nuclear war in terms of the historical experience of mankind with wars in the past. "Thus, though it is recognised that the only value of nuclear weapons is deterrence, how many weapons are necessary for deterrence has never been made clear by either superpower. Can the objective of deterrence be expressed in absolute terms? Does it represent a minimum of ten city-destroying weapons or 1,000? How many lives, what proportion of the enemy's industrial capacity, have to be assured of extinction?

[a]These numbers include "strategic", "intermediate" and "tactical" weapons. The average yield is 1/3 megatons of TNT per weapon.

The uncertainty of what the opponent may do next encourages the worst-case assumptions and makes the competition *open-ended.*"

"The military strategists have gone on from the doctrine of 'deterrence' to 'damage limitation'. *'Damage limitation' meant that nuclear weapons must be given the capability to destroy other nuclear weapons before they could sow death and destruction. Thus one could justify unremitting efforts at modernisation: goals of pin-point accuracy, an MX with the power to blow up missiles in their silos, the ultimate Star Wars defence through satellite systems.*"[b]

2. I should not be misunderstood; I am not criticising one superpower versus another. I will mostly use the US figures in this article only because they are the more readily available. The truth is that both superpowers, equally, as well as all members of the nuclear club, stand indicted before the bar of humanity. A most pertinent question in this regard has been asked by the UN Secretary General in his speech to the General Assembly on 12 December 1984, voicing the thoughts of all of humanity:

"As I look across this hall, I see the delegation of 159 member nations. Almost all the world's peoples are represented here. And all of them — all of us — live under the nuclear threat. As Secretary General of this organisation, with no allegiance except to the common interest, I feel the question may justifiably be put to the leading nuclear-weapon powers: by what right do they decide the fate of all humanity? From Scandinavia to Latin America, from Europe and Africa to the Far East, the destiny of every man and every woman is affected by their actions ... The responsibility assumed by the great powers is now no longer to their populations alone: it is to every country and every people, to all of us.

"No ideological confrontation can be allowed to jeopardise the future of humanity. Nothing less is at stake: today's decisions affect not only the present, they also put at risk succeeding generations. Like supreme arbiters, with our disputes of the moment we threaten to cut off the future and extinguish the lives of the innocent millions as yet unborn. There can be no greater arrogance. At the same time, the lives of all who live before us may be rendered meaningless. For we have the power

[b]I have used extensive quotations from Ruth Leger Sivard's excellent book, *World Military and Social Expenditures*, 1983.

to dissolve in a conflict of hours or minutes the entire work of civilisation, with all the cultural heritage of humankind . . .

"At a time of uncertainty for the young and despair for the poor and the hungry, we have truly mortgaged our future to the arms race — both nuclear and conventional . . . The arms trade impoverishes the receiver and debases the supplier. Here there is a striking resemblance to the drugs trade. Yet we continue on the same course even when faced with the silent genocide of famine that today stalks millions of our fellow men and women. The international community has to focus and act on the link between disarmament and development. We should take concrete and far-sighted steps towards the conversion of arms industries from military to civilian production. And we should begin to redress some of the enormous imbalance between research on arms and research on arms limitation and reduction . . . ".

3. In the context of security, it is pertinent to reflect from the point of view of developing countries.

Security for us in the developing world means not only security from a nuclear winter which may be unleashed on mankind by accident or through the design of "homicidal maniacs", but also security from conventional wars waged on our soils. Since 1945, there have been 105 wars (with deaths of 1,000 or more per year), with or without superpower involvement. These have been fought in 66 countries — *all of them in the Third World.* Twelve of these are being waged today, in more than one third of which the richer countries are implicated. "On the average, each has lasted three and a half years. They have caused 16 million deaths, the majority of them in Asia. Cambodia lost two million, over one quarter of its population; Vietnam 2.5 million or 6% of its population; Nicaragua, by the end of 1983, had lost 1.5% of its population with 35,000 deaths and El Salvador 45,000 deaths with 1% death toll. Most of these deaths took place among civilians with incalculable material and social costs; for example, in Iran, where damage to the civilian economy may be over one hundred billion dollars; or in Afghanistan", — next door to my country — where daily carnage continues, with superpower involvement. "With already four out of five adults illiterate, 17,000 schools have been destroyed."[c] Thus, loss of security for us in the developing world is something which is more immediate, something we constantly live with.

[c]From Ruth Leger Sivard's *World Military and Social Expenditure*, 1983.

4. Regarding the military expenditure, conventional and nuclear, reduced to numbers, "the world's arms race and its effect on human life easily loses touch with reality". The current global war-budget is around $700 billion per year, out of a total global GNP of 12,000 billion. Of this expenditure, 550 billion is attributable to the developed and 150 billion to the developing countries.

Twenty-five million men under arms; one billion — one quarter of humanity — live under military-controlled government; and more than nine million *civilians* have been killed in "conventional" wars since Hiroshima.

How much are the superpowers spending on war today? "At $855 per capita in 1982, military expenditures in the US are to be compared with $75 in comparable prices just before World War II. The military effort has risen faster than GNP, with US spending 6.5 percent of GNP as compared with one percent pre-war." The implications this has in respect of diminishing social expenditures need not be spelled out. Similar figures, even more stark in proportion, apply to the USSR.

"Between 1960 and 1983, the world as a whole spent some eight trillion (out of a total GNP of around 154 trillion) on military expenditures. *While military expenditures of developed countries (including the centrally-planned) rose by more than $400 billion, the foreign economic aid rose by no more than $25 billion in 1982;* the superpower military expenditures were 17 times larger than their extensions of aid to countries in need."

And it is not just the superpowers which have spent lavishly on war. With the rich countries' drive to sell arms, the developing world has been equally profligate. "Among the 25 countries which since 1981 have had to negotiate to reschedule their debt, six had spent more than $1 billion each for arms imports, in the five years preceding. Between them the 25 piled up a bill of $11 billion for arms in that period. Among 20 countries with the largest foreign debt, *arms imports* between 1976 and 1980 were equivalent to 20 per cent of the increase in debt."

5. What does this military expenditure mean in terms of unmet global human needs? Again I quote from Ruth Leger Sivard:

Poverty Two billion people live on incomes below $500 per year.[d] At least one person in five is trapped in absolute poverty, a state of destitution so complete that it is silent genocide.

Jobs In the Third World, one in three who wants to work cannot find a regular job. In all countries it is the young people who are hardest hit by unemployment, in the US half of black teenagers are jobless.

Food 450,000,000 persons — one tenth of mankind — suffer from dire hunger and malnutrition. There are an estimated 15 million deaths yearly from malnutrition and infection, conditions which are preventable and which society has both the knowledge and means to prevent. Every minute 30 children die for want of food and inexpensive vaccines and every minute the world's military budget absorbs $1.3 million of the public funds.

Education 120,000,000 young children of school age have no school they can go to. Educational neglect in fact begins at the earliest ages. One-third of the children between the ages of 6 and 11 are not in school. Over 250 million children in the world have not received even a basic education. To get a comparative estimate, the cost of a single new nuclear submarine equals the annual education budget for *160 million school-age* children in 23 developing countries.

6. I shall not go on labouring these points, but concentrate on just a set of proposals which had been made from time to time at the Forum of the United Nations by the Governments of France, Mexico, Senegal, USSR, and others, as well as by the Brandt Commission, to redress this balance. The proposals concern the creation of an International Disarmament Fund whose proceeds would be used for development tasks. It is hoped that such a fund may discourage war spending. In any case, it would ease warring humanity's guilt-feelings and its conscience. The most detailed proposal is that from France presented by its President, Mr. Giscard d'Estaing in his address to the UN General Assembly in 1978. This proposal, elaborated in a memorandum by the French Government (United Nations Document A/S−10/AC.1/28), envisioned the fund as a new United Nations specialised agency which would constitute a practical manifestation of

---

[d]Of these 1/2 billion live on incomes below $100 per year.

the relationship acknowledged by the world community to exist between disarmament and development. Contributors to the fund would be those states which were both most heavily armed and most developed; beneficiaries of the fund would be those States which were least heavily armed and least developed.

In principle, the fund would be based on the disarmament dividend approach, that is, on resources released by disarmament measures. However, the French proposal also provided for a transitional phase of the fund with an initial one-time endowment of $1 billion, until resources derived from disarmament savings could become its long-term basis. In its transitional phase, contributions to the fund would be assessed on the basis of a State's level of armament, measured by the possession of certain types of weapon systems the existence of which, according to the proposal, could be objectively determined. The fund would make grants or loans to developing countries, utilising as much as possible existing international agencies for the adiministration of its loans and grants.

Various criteria could be used in the transitional stage to identify "the richest and most heavily armed" countries. Assuming that the five permanent members of the Security Council would be automatically included, the following criteria have been suggested (1977 statistics in 1977 US dollars):

As a criterion of wealth: a per capital GNP of more than $1,000;
As criteria of armaments: a level of military expenditure in excess of 2 per cent of GNP; a volume of military expenditure in excess of $US1.5 billion.

Of the sum foreseen as the contribution during the transition stage, amounting to $1,000 million, 50 per cent should be based on the States' nuclear armaments and 50 per cent on conventional armaments. The criterion used to determine the relative contributions for the nuclear sector of the United States of America and the Union of Soviet Socialist Republics could be their numbers of vehicles, based on the SALT Agreement. The joint participation of these two countries would amount to 80 per cent of the whole sum. China, the United Kingdom and France would jointly contribute 20 per cent.

7. One may disagree with the details but the intention of setting up such a fund is clear and commendable. For today's discussion, I propose that

we should spend our time discussing the modalities of setting up such a fund and its uses. I believe the two questions are linked. My remarks are mainly addressed to the last issue.

Why have such proposals for disarmament funds so far fallen[e] by the wayside? Is it that such resolutions at the United Nations forum among delegates to UN bodies seldom bear fruit? If this is so, I believe that for effectiveness, we must engage the young men and women — particularly in the developed countries — in the crusade of getting their government to act. The public outcry, focused principally against nuclear weapons, is today firmly backed. How soon it will begin to affect government priorities is not yet clear but it has already become a healthy counter-weight to *official policy that has lost touch with reality.* It is imperative that this movement would embrace the constructive issue of development linked to nuclear and conventional disarmament.

But to effect this, we must remember that the young people today are not moved by the *totality* of global development tasks; nor always by dangers of unchecked population increase, nor by polarisation of poverty and riches nor by world illiteracy. I believe that what immediately moves the young today are

(*i*) the environmental problems;

(*ii*) death from hunger and

(*iii*) the desirability of eradication of dread diseases like leprosy and trachoma.

My feeling is that if we ask for funds for eradication of illiteracy or generally for bridging the gap between the rich and the poor or for solving the energy problems of the world — however worthy all these causes may be — it is unlikely that we shall succeed in getting enough public support. I would therefore like to argue that — at least at the outset — we should set out the goals of Disarmament Fund to deal with

(*i*) problems of global environment;

(*ii*) death from hunger and

(*iii*) eradicable human disease.

8. Consider global environment: The biosphere has been likened in its thinness to "dew upon an apple". Its survival intact is mankind's survival.

---

[e]I suppose the setting up of such a fund unilaterally by their own nations never strikes the statesmen-proponents.

Take one aspect of its health, connected with the preservation of rainforests situated in the Third World. According to the Report presented to President Carter giving global projections for the year 2000, significant losses of world forests — particularly those in the tropics — are predicted for the next 20 years, principally as demand for fuelwood and food, both from the poor and the rich, increases. The world's forests are now disappearing at the rate of 18-20 million hectares a year (an area half the size of California), with most of the loss occurring in the humid tropical forests of Africa, Asia and South America. The projections indicate that by the year 2000 some 40 percent of the remaining forest cover in less developed countries will be gone.

I wish to emphasise that an important factor in this disappearance is man's greed. According to a recent study made by Catherine Caufield: "In the face of big business, the environment would seem to stand little chance of survival. The beef-producing industry provides a perfect example. Because Latin American beef is half the price of America's own home-produced grass-fed variety, more and more of the land is being turned into pasture — much of it at the expense of the rainforests."[f]

The question should be asked: Is the saving of this global heritage to be left to the poor impoverished countries of the South only? Should this not be a charge on a global fund — possibly linked with disarmament?

Why is the disappearance of the tropical forests, proceeding now at 2% of the forest per annum, disastrous for mankind as a whole? One of the many ecological reasons is the anticipated annihilation of a large fraction of species and organisms which inhabit these forests.

"Approximately 1.5 million kinds of organisms have been named and classified, but these include only about half a million from the tropics. Many tropical organisms are very narrow in their geographical ranges and are highly specific in their ecological and related requirements. Thus, tropical organisms are unusually vulnerable to extinction through disturbance of their habitats." Since more than half of the species of tropical organisms are confined to lowland forests, in most areas, these forests will be gone within the next 20−30 years; and with them most of the three million of these organisms."

"With the loss of organisms, we give up not only the opportunity to study them, but also the chance to utilise them to better the human

[f]*The Rainforest* by Catherine Caufield, published by Heinemann, London, 1985.

condition, both in the tropics and elsewhere. The economic importance of wild species, a tiny proportion of which we actually use, has been well documented. Suffice to say that the entire basis of our civilisation today rests on a few hundred species out of the millions that might have been selected, and we have just begun to explore the properties of most of the remaining ones. Unfortunately this process of extinction cannot be reversed."

But how does a Global Development Fund come into this? In a recent issue of the *Bulletin of Atomic Scientists*, P. H. Ravan has argued that if the West cannot find the means to eliminate real poverty in the ecologically devastated areas, the people living there will topple any Government; be it friendly or unfriendly! Thus it is no coincidence that El Salvador is ecologically the most devastated of all countries of Central America and yet the authors of the Kissinger Report pay no attention to the ecological problems which force peasants to shift to and destroy permanently, through cutting of forest cover, the productivity of their marginal lands. One may in this context ask if it is not the concern of the environmental groups in the developed countries to help to preserve this global heritage? Should they not come to the rescue of the developing countries? Should this type of global assistance not be a first charge on the international communities and on a Global Development Fund?

9. A second area for which such a fund may be used is freedom from hunger, with the Ethiopian tragedy still on public conscience. Favourable climate, water, good arable land and chemical inputs are the four factors essential for enhancing food production. According to the Carter Report the global area of arable land will increase only four percent by the year 2000, so that most of the increased output of food will have to come from higher yields. During the same period expected increases in world population are nearly in the 30-40% range — from 4.5 billions to 6 billions. Unfortunately, to feed this enhanced population, and to avoid recurrence of other Ethiopias, most of the elements which contribute to higher yields of food crops — fertiliser, pesticides, energy for irrigation, and fuel for machinery — depend on the scarce resources of oil and gas.

To make the problem more difficult, regional water shortages will become more severe. In the 1980 – 2000 period population growth alone will cause requirements for water to double in nearly half the world. Still greater increases would be needed to improve standards of living. In many less developed countries, water supplies will become increasingly erratic by the year 2000 as a result of extensive deforestation. Development of new water supplies will become more costly virtually everywhere.

"Unless action is taken, serious deterioration of agricultural soils will occur worldwide, due to erosion, loss of organic matter, desertification, salinisation, alkalinisation, and waterlogging. Already, an area of cropland and grassland approximately the size of the state of Maine is becoming barren wasteland each year," and the spread of desert-like conditions is likely to accelerate. Offices of India's Planning Commission reported recently, "We in India are on the verge of an enormous ecological disaster, with our water reserves drying up. What is happening in Africa is going to happen in India within a few decades".

10. So far as the chemical inputs for enhanced agriculture are concerned, in December 1982 in the Philippines, under the auspices of the International Union of Pure and Applied Chemistry, 600 topranking chemists of the world met and drew up a plan of action where chemistry could be utilised to raise world food productivity through chemical inputs in 15 years — the goal being to increase world food productivity by 50% by the year 2000. A number of world institutes for teaching, training and research for chemists from the Third World were to be created. If they are not, it will mainly be because of shortage of funds.

11. The World Resources Institute of Washington, D. C. has made a list of some of the truly serious ecological problems which are deserving of wide international attention:

(*i*) Loss of crop and grazing land due to desertification, erosion, conversion of land to non-farm uses, and other factors. The United Nations reports that, globally, farm and grazing land are being reduced to zero productivity at the rate of about 20 million hectares a year.

(*ii*) Depletion of the world's tropical forests, which is leading to loss of forest resources, serious watershed damage (erosion, flooding, and siltation), and other adverse consequences. Deforestation is projected to claim a further 100 million hectares of tropical forests by the end of this century.

(*iii*) Mass extinction of species, principally from the global loss of wildlife habitat, and the associated loss of genetic resources. One estimate is that more than 1,000 plant and animal species become extinct each year, a rate that is expected to increase.

(*iv*) Rapid population growth, burgeoning Third World cities and ecological refugees. World population will most likely double by the early decades of the next century, and almost half the inhabitants of developing countries will live in cities — many of unmanageable proportions.

(*v*) Mismanagement and shortages of fresh water resources. Water borne diseases are responsible for perhaps 80% of all illness in the world today.

(*vi*) Overfishing, habitat destruction, and pollution in the marine environment. Twenty-five of the world's most valuable fisheries are seriously depleted today due to overfishing.

(*vii*) Threats to human health from mismanagement of pesticides and hazardous substances and from pathogens in human wastes and aquatic vectors. An estimated 1.5 to 2.0 million persons in developing countries suffer acute pesticide poisoning annually and pesticide-related deaths are estimated at 10,000 a year.

(*viii*) Climate change due to the increase in "greenhouse gases" in the atmosphere. The steady build-up of carbon dioxide and other gases in the atmosphere, due principally to fossil fuel burning, is predicted to create a "greenhouse effect" of rising temperatures and local climate change — the question increasingly is not "if?" but "how much?" For a variety of reasons, poor countries are likely to suffer disproportionately from the consequence of climate change.

(*ix*) Acid rain and, more generally, the effects of a complex mix of acids, ozone and other air pollutants on fisheries, forests, and crops.

(*x*) Mismanagement of energy fuels and pressures on energy resources, including shortages of fuelwood, the poor man's oil. Although the energy crisis is in temporary remission in the developed countries, the high costs of oil imports and fuelwood shortages continue to plague much of the developing world.

12. And this brings me finally to one of the most crucial charges on a Disarmament Fund — relevant scientific research in the areas of global environment, food and dread disease.

"At the cutting edge of the military competition between the major powers is a mobilisation of research resources without parallel in history. The results of this research provide an irresistible momentum to the arms race. The postwar take-off in weapons research was even more spectacular than the rise in military expenditures in general. In the US, government-financed military R & D jumped from $1.7 billion in fiscal year 1947 to $22.1 billion in fiscal 1983 (both in 1980 prices). The 13-fold increase in research expenditures was four times as fast as the already very rapid growth in US military spending over the same period."

The stark fact is that one half of mankind's research effort is on military R & D. The scientific talents of these men and women together with the resources devoted to military R & D might have been spent on research on ecology, climate, food and disease.

In this context, consider climate studies. I do not know whether one can really hope to change the climate; but surely it is a scandal that there is no scientific study of the climate of the Sahelian area over a long-term period. The universities in these areas have Departments of Meteorology but these Departments are weak, ill-organised and without any funds. They could be made stronger. Can one hope for a Global (Disarmament Development) Fund to organise the building up of indigenous scientific communities for carrying out such studies in the countries concerned? *The truth is that mankind is already engaged in the "The Third World War" — the war against our heritage of resources — against life on Planet Earth. And we are winning it ".*

13. Nine hundred years ago, a great physician of Islam, Al Asuli, living in Bokhara, wrote a medical pharmacopaea which he divided into two parts: "Diseases of the Rich" and "Diseases of the Poor". If Al Asuli were alive and writing today about the afflictions wrought upon itself by mankind, I am sure he would divide this pharmacopaea into the same two parts. One part of his book would speak of the affliction of Nuclear Annhilation inflicted on humanity by its richer half. The second part of his book

Table I
Summary of Global Expenditures (1960-1983)

	Industrialised countries (28 countries; 1/3 of mankind)	Developing countries (114 countries, (including OPEC countries 2/3 of mankind))
	(Trillion = million million = thousand billion = 10^{12}) (Expressed in 1982 dollars)	
GNP	$124 trillion	$30 trillion
Military expenditures (including nuclear)	$7 trillion	$1.2 trillion (Middle East $0.6 trillion)
Nuclear expenditure[1] (Superpowers only)	$2.0 trillion	
Development aid	$0.3 trillion	$0.06 trillion (donated by Middle East countries)
Public Scientific Research and Development[2] (of this attributable to military R&D)	$2.5 trillion $1.3 trillion	$0.1 trillion

[1] Estimated SDI expenditures (up to 2000 AD), which need a special explanation, are in excess of 1.5 trillions. With such outlays and with an intensive technological effort, a *partial* SDI programme may be feasible. The major objection to it is that to make up for the kill ratio lost, all that the opposing side has to do is to increase its offensive capacity by making more nuclear weapons (i.e. a Strategic Offensive Initiative (SOI)) at a fraction of the cost of SDI.

[2] At the 1979 UN held Vienna Conference on Science and Technology, the poor countries presented a case for 2 billion dollars for Science and Technology aid (particularly for applicable sciences) to be spent in addition to the $2 billion a year from their own resources. They received an offer of assistance of 70 million dollars, which dwindled to 40 millions a year by 1981 and has dwindled to zero in 1986. Unfortunately, the scientific community of the poorer countries has received scant organised help from their colleagues and peers in the richer countries to redress this situation.

would speak of the affliction which poor humanity suffers from —
underdevelopment, undernourishment and famine. He would add that
both these diseases spring from a common cause — excess of science and
technology for the case of the rich, and a lack of science and technology
for the case of the poor. He might also add that the persistence of the
second affliction of mankind — underdevelopment — was the harder
to understand, considering that the remedies for it are readily available
in that the world has enough resources, technical, scientific, and material,
to eradicate poverty, disease and early death for the whole of mankind,
if it wishes to do so. It has only to eschew deployment of these resources
towards aggravating the first affliction.

Appendix

I reproduce the vision of Ruth Leger Sivard — World Military and Social
Expenditures 1985 — on how she would spend $100 billion for world
development.

"Bringing Star Wars Down to Earth

A world scarred by poverty and violence calls for a brave new vision
and a sharp turnabout in political direction. Unlikely as change on this
scale may seem at present, dreaming is sometimes useful in expanding
a vision of what might be compared with what is. Imagine if, instead of
chasing the chimeras of star wars and evanescent military gaps, the political
leadership of the two greatest powers agreed to focus on the real gaps,
the universal needs of humanity on planet Earth . . .

The dream scenario goes as follows:

Each superpower independently and cleverly decides that the only
sane objective is to lower the opponent's military threat, rather than
racing endlessly to keep ahead of him. Each grasps the opportunity in
negotiations to make substantial cuts in the opponent's nuclear forces
and to put an end to all nuclear and anti-satellite testing. Budgetary
savings for the two countries are conservatively estimated at two trillion
dollars over the next 10 years.

In the euphoria of success, the superpowers decide to invite the whole
world to share in half of the savings. (The other half is to be used to
reduce the budget deficit in the US and to improve the lot of consumers

in the USSR.) Together the US and USSR announce that **$100 billion a year** will be made available for an unprecedented human development program, to which other nations are asked to contribute in proportion to their military expenditures. Global response is immediate and wildly enthusiastic. A World Peace Corps is formed.

Impressed by reports of the remarkable mobilisation of public support in Colombia and Sri Lanka to carry out successful country-wide drives for child immunisation, the superpowers agree that every programme financed must include provision for spirited participation by the public on the same scale. They also agree on the immediate objective: a direct attack on poverty, to bring help to the very poorest first. Planning is left to an international board of dedicated specialists. Meanwhile, the traditional international development agencies will continue their efforts on structural adjustments and long-term development programmes.

As the governing board of wise men and women meet to determine priorities, the mails are flooded with world-wide expressions of support and a growing list of unmet needs with initial annual price tags. A random sampling shows the scope of poverty and neglect awaiting attention.

Growing long-term employment, now common in industrialised as well as poorer countries, has serious psychological effects on society. Skills and motivation decline. Visible projects, clinics, housing, schools, generate employment and community pride. Highly publicised development programmes in some slum and squatter areas could stimulate public action in others.

Community projects for job creation. . . . . **$18 billion**

Hunger is the enemy faced daily by hundreds of millions of people throughout the world. Children are the major victims. Those who survive malnutrition in early life are mentally and physically handicapped. As a supplement to emergency food aid, nutrients such iron, Vitamin A, and iodine, should be provided to fortify staple foods.

Nutrients to supplement staple foods . . . . **$2 billion**

In rural areas of the Third World, the majority of the population does not yet have access to health services even of the most basic kind. A cash programme for training of bare-foot doctors and medical auxiliaries, and provision of small health posts is essential for improving health care outreach.

Community-based health service . . . . **$10 billion**

In the coming decade the increase in the world's school-age population will center in the Third World, where only half the children of school-age are now attending school. Even to maintain present enrolment ratios, 100 million new school places will be needed by 1995. Training and construction programmes should target areas of greatest deficiencies.

New schools and teacher training . . . . **$12 billion**

The earth's forests are shrinking at an alarming rate, the result of rapid population increases in regions dependent on wood for fuel, and urban and industrial spread. Deforestation represents a fuel crisis in developing countries, and also leads to soil erosion and desertification. A widespread tree-planting programme is needed.

A major increase in tree planting . . . . **$3 billion**

Young people constitute 40 percent of the unemployed in many areas. By the end of the century the youth population is expected to exceed 700 million in the Third World where unemployment is acute and chronic. Training and apprenticeship programmes emphasising new skills must be expanded.

Training programmes for young people . . . . **$ 5 billion**

Subsistence agriculture, the major employer of the labour force in the Third World, is also the key to eliminating food shortages. Planning and assistance must include women, who are the major food growers in many areas, and include sufficient acreage to support a family. Technical aid, irrigation, pest control, fertilisers, are essential for increasing yields.

Development of small farm holdings . . . . **$15 billion**

Unsafe water and inadequate sanitation account for three fourths of illnesses in poor countries. Over half the population lacks an adequate supply of safe water; even larger numbers are without sanitation facilities. Water taps and sewage systems could sharply reduce the incidence of diseases like malaria, and add 10 years to low life expectancies. .

Safe water and sanitation for all . . . . **$20 billion**

Industrialisation has neglected labour-intensive, light-capital manufacturing which could increase employment and improve living conditions especially among poorest families. There is a universal need for such low-cost technology as simple stoves, small biomass plants, hand-powered grinders for grain.

Labour-intensive, low-cost technology . . . . **$4 billion**

Repeated child-bearing, short birth intervals, and pregnancy at an early or late age, all pose high risks to the health of women and their children. Programmes for family planning, food supplements in pregnancy, prenatal and perinatal care are limited especially in South Asia and Africa where mortality is highest and population growth most rapid.

Health-care network for women . . . . **$6 billion**

Street people are increasing in numbers in cities throughout the world. Estimates for Latin America alone indicate that there may be as many as 40 million street children, many abandoned, most suffering from malnutrition. Community-support programmes are needed for housing, health care, and education.

Community care for street people . . . . **$2 billion**

The lag in basic educational opportunities for women is evident in an increasing gap between male and female literacy. The problem requires specifically targeted programmes for deprived groups, particularly in rural areas. Young school graduates should be trained to help in these programmes.

Literacy drive emphasising skills for women . . . . **$3 billion**"

II.1
SCIENCE AND TECHNOLOGY
IN THE DEVELOPING COUNTRIES

Notes on Science, Technology and Science Education in the Development of the South

Prepared for the 4th Meeting of the South Commission, 10 – 12 December 1988, Kuwait.

Abstract

This globe of ours is inhabited by two distinct types of humans. According to the UNDP count of 1983, one quarter of mankind, some 1.1 billion people are developed. They inhabit 2/5ths of land area of the earth and control 80% of the world's natural resources, while 3.6 billion developing humans — "Les Misérables" — the "mustazeffin" — live on the remaining 3/5ths of the globe. What distinguishes one type of human from the other is the ambition, the power, the élan which basically stems from *their differing mastery and utilisation of present day Science and Technology*. It is a *political* decision on the part of those (principally from the South) who decide on the destiny of developing humanity if they will take steps to let Les Misérables (the "mustazeffin") *create, master* and *utilise modern Science and Technology*. These notes are devoted to this topic.

"In the conditions of modern life, the rule is absolute: the race which does not value trained intelligence is doomed ... Today we maintain ourselves, tomorrow science will have moved over yet one more step and there will be no appeal from the judgement which will be pronounced ... on the uneducated".

Alfred North Whitehead

THE TRIESTE DECLARATION
On Science and Technology as an
Instrument of Development in the South*

"Recognising the fundamental importance of Science in socio-economic and cultural development and technological progress, and keeping in view the recommendations of the South Commission pertaining to the crucial role of Science in the Third World, as mankind approaches the 21st century, the members of the Third World Network of Scientific Organisations present at the meeting held in Trieste from 4—6 October 1988, resolve to work towards giving Science and Technology a position of highest priority in their own countries and to strengthen their collaboration with other countries of the South as well as of the North".

1.   Science and Technology, A Shared Heritage of Mankind

The first thing to realise about the Science and Technology gap between the South and the North is that it is of relatively recent origin. In respect of sciences, George Sarton, in his monumental History of Science, chose to divide his story of achievement into Ages, each Age lasting half a century. With each half-century he associated one central figure. Thus 450—400 BC Sarton calls the Age of Plato; this is followed by the half-century of Aristotle, of Euclid, of Archimedes and so on. From 600 AD to 650 AD is the Chinese half-century of Hsüan Tsang (and of the Indian mathematician, Brahmagupta). From 650 to 700 AD is the age of I-Ching, followed by the Ages of Jabir, Khwarizmi, Razi, Masudi, Wafa, Biruni (and Avicenna), and then Omar Khayam — Chinese, Hindus, Arabs, Persians, Turks and Afghans — an unbroken Third World succession for 500 years. After the year 1100 the first Western names began to appear; Gerard of Cremona, Roger Bacon — but the honours are still shared equally with the Third World men of science like Ibn-Rushd (Averroes), Tusi and Sultan Ulugh Beg.

---

*There were 15 Ministers of Science and Technology, 12 Presidents of Academies and 17 Chairpersons of National Research Councils, representing 36 Third World countries, present at the inaugural meeting of the Third World Network of Scientific Organisations in Trieste.

The same story repeats itself in Technology, for China and the Middle East, at least until around 1450 when the Turks captured Constantinople because of their mastery of superior cannonade. No Sarton has yet chronicled the history of medical and technological creativity in Africa — for example of early iron-smelting in Central Africa 2500 years ago. (Scientific American, June 1988 issue). Nor of the pre-Spanish Mayas and Aztecs — with their independent invention of the zero and of the calendars, of the moon and Venus, as well as of their diverse pharmacological discoveries (including quinine). But one may be sure, it is a story of fair achievement in Technology and Science. From around 1450, however, the Third World begins to lose out (except for the occasional flash of individual brilliant scientific work), principally because of lack of tolerant attitudes to the diversity and creation of Science.

And that brings us to the present century when the cycle begun by Michael the Scot who went, around the year 1220 A.D., from his native glens in Scotland South to Toledo and then to Sicily in order to acquire knowledge of the works of Razi, Avicenna (and even of Aristotle) turns full cycle — and it is we in the developing world who must turn Northward for Science. Science and Technology are cyclical. They are a shared heritage of all mankind. East and West, South and North have all equally participated in their creation in the past as, we hope, they will in the future — the joint endeavour in Sciences becoming one of the unifying forces among the diverse nations on this globe.

2.   The Widening Gap in Science and Technology

Today, the Third World is only slowly waking up to the realisation that in the final analysis, *creation, mastery* and *utilisation of modern Science and Technology* is basically what distinguishes the South from the North. On Science and Technology depend the standards of living of a nation. The widening gap in Economics and influence between the nations of the South and the North is essentially the Science and Technology gap.[a]

---

[a]One aspect of the South's deprivation is the fact that Science (even as contrasted with Technology) has been treated as a marginal activity for the South and by the South. The emphasis has been on the "transfer of technology". (The words "Science" or "Science Transfer", for example, do not occur in the Brandt Commission Report). Very few within the developing world appear to realise that science must be broad-based in order to be effective for applications and that the science of today is the technology of tomorrow.

Nothing else — neither differing cultural values, nor differing perceptions of religious thought, nor differing systems of economies or of governance — can explain why the North (to the exclusion of the South) can master this globe of ours and beyond.[b] Why does this gap exist and why is it growing so fast? Why is the size of Science and Technology sub-critical and their utilisation in the South so meagre?

We shall endeavour to answer these questions in this report. But before we do this, let us start with two remarks. One concerns the long-term nature of Science and Technology as applied to development. We are not likely to see the benefits for a long time. The year 2000 would be a good year to aim at *if we start today.*

Our second remark concerns the widespread feeling that the acquiring of Science and Science-based High Technology is hard. We would like to say emphatically that this is not the case. In eloquent phrases, C.P. Snow, in his famous lecture on "The Two Cultures", made the point that Science and Technology are the branches of human experience "that people can learn with predictable results ... For a long time, the West misjudged this very badly. After all, a good many Englishmen have been skilled in mechanical crafts for half a-dozen generations. Somehow, we, in the North, have made ourselves believe that the whole of technology was a more or less incommunicable art."

In Snow's words: "... There is no evidence that any country or race is better than any other in scientific teachability: there is a good deal of evidence that all are much alike. Tradition and technical background seem to count for surprisingly little.

"There is no getting away from it. It is ... possible to carry out the scientific revolution in India, Africa, South-East Asia, Latin America, the Middle East, within fifty years. There is no excuse for Western man not to know this".

---

[b]The role of superior technology in the rise and fall of nations is a relatively neglected subject. Thus, for example, when we think of the British conquest of India, the part played by the superior fire power of Clive's British-made arms is forgotten. It is equally forgotten that the British "trained" themselves to manufacture such arms. Nor is the role played by the navigational skills (developed at the secret Centre in Sagres in Portugal through the personal interest of Prince Henry the Navigator) — which permitted European ships to sail straight into the oceans, rather than hug the coast lines — ever taken into account. Why men like Prince Henry the Navigator arise from time to time among some peoples is, of course, one of Nature's mysteries.

3. The Four Areas of Science and Technology

We shall, in this paper, be concerned with the *Research and Developmental (R&D) aspects of Science and Technology and, most crucially, with their utilisation of society's benefit.*

Civilian Science and Technology may perhaps be divided into four categories of (*i*) Basic Sciences; (*ii*) Sciences in Application; (*iii*) Conventional "Low" Technology; and (*iv*) Science-based "High" Technology.

Let us consider each of these areas in turn.

(*i*)   *Basic (curiosity-oriented [c]) Sciences*

There are five sub-disciplines comprised among these:

(1) Physics (including Geophysics and Astrophysics); (2) Chemistry; (3) Mathematics; (4) Biology; plus (5) Basic Medical Sciences.

Research and training for Basic Sciences is conducted in the Universities or in the Research Centres specifically created for this purpose in the North. As a rule, these are funded by National Science Foundations or by Academies of Sciences (which are also responsible for international contacts of scientists).[d]

So far as developing countries are concerned, by and large we have tended to neglect this area of Science assuming, for some reason that

---

[c]The curiosity-orientation of Basic Sciences had been beautifully expressed by David Hilbert, the great Goettingen mathematician (who lived in the early part of the 20th century). According to Dr. Wolfgang Wild, the former Minister of Science of the State of Bavaria, Hilbert had the following inscribed for his tombstone: "WE MUST KNOW. WE SHALL KNOW".

The present has been called the greatest century of Basic Sciences because there have been absolutely fundamental advances made (like "quantum theory, relativity, anti-particles, curved space-time") in the first part of the century and Standard Models elaborated in Particle Physics, in Astrophysics (Big Bang Model), in Earth Sciences (Plate Tectonics) and in Biology (Double Helix Model) during the second half. The tasks of future generations of researchers from the Third World to improve on these will, we are afraid, be that much harder.

---

[d]To obviate the problems of Research unconnected with teaching, the best modality has been invented in the US where *all* scientific research, whether for Basic or Applied purposes, is conducted either in the universities directly, or in institutes which may be federally financed but are almost always linked with the universities. For example, the major laboratories of the US Department of Energy — the Brookhaven National, the Argonne National and the Los Alamos Laboratories — are operated on behalf of the Department of Energy, with funding provided by it, by *consortia of universities.*

we could live off the scientific results obtained by others. This has been an unmitigated disaster in that it has also deprived us of men and women who would know about the basics of their disciplines, who could act as references to whom one could turn, to discuss the inevitable scientific problems which arise when applications of Science are made.[e]

(ii)   Sciences in Application

One may list five areas of Sciences in Application: These are (1) Agriculture (including Livestock, Fisheries and Forests); (2) Medicine and Health; (3) Energy (including Nuclear, Fusion, Solar and Non-conventional); (4) Environment and Pollution; (5) Earth Sciences (including Irrigation and Soils, Meteorology and Oceanography, Minerals, as well as Seismology).

As a general rule, Research and Development in Applied Sciences are carried out in the North under the auspices of Research Councils or by private industry.[f] *This includes Research, Development (Adaptation and Modification) and Application of Scientific Methodology to developmental problems.*[g] *The Research effort, in order to be effective,*

---

[e]We are not recommending here the setting up of the likes of the 200 inch telescope at Mount Palomar which was a gift from the Rockfeller Foundation, nor the setting up of the great (but costly) laboratories like CERN in Geneva for Particle Physics set up by a consortium of European nations. However, we definitely do believe that a profound knowledge of the basics is absolutely vital for applications and that research is *sine qua non* for assuring such profound knowledge.

It is crucial that our young men and women should receive training for research, in the highest traditions possible internationally. This may be done at United Nations-run International Centres for Sciences (see later), where also the costlier items of equipment which developing countries may not be able to afford individually (like super-computers) could be installed for joint work.

[f]Inspite of the large technological content of some of these areas, it is important to realise that these are not areas of technology, but of science. Different societies have differing problems and sub-areas which naturally need emphasising more than others.

[g]We must once again emphasise the important of harnessing Applied Sciences (and the technologist for economic growth and for the betterment of the human condition — be it in improving agricultural output, medical advances or efficient transportation and housing. (This is done under the name of "Scientific Humanism" in the North although as a rule such "Scientific Humanism" is not concerned with the plight of the South).

must be supplemented with first-class extension services.[h]

What is emphasised more in any given country depends on a nation's priorities and cannot be spelt out here.

An essential part of the research process is the free availability of scientific literature.

It is important to realise that the distinction between Basic and Applied Sciences is not absolute. There inevitably are gradations.

(iii)   Conventional "Low" Technology

The five sub-areas of this are:

(1) Bulk Chemicals; (2) Iron and Steel and Fabrication with Other Metals; (3) Design and Fabrication in (indigenous) Industries (like Cotton and Leather manufactures or Automative industries etc.); (4) Petroleum Technologies; (5) Power Generation.

Here no new scientific principles remain to be discovered. However, Developmental work relating to design, and of adaptation and modification, may be important. This is the traditional area of craftsmanship and skills — the science employed is of yesteryears. Thoroughness (in all aspects in the manufacture and after services), beauty of design, the quality of workmanship and cost, and manufacturing-competitiveness are all-important. These are just the areas where developing countries are *not* deficient.

This is also the classical area of *"negotiated Technology Transfer"*[i] and the area on which centrally-planned economies of the second world as well as the developing countries have placed their strongest emphasis. *Any country which wishes to industrialise will have to develop one or more of the technologies listed above (as, for example, USSR and Japan initially had to).*[j]

---

[h] Such extension and "one-window services" are extremely important ingredients of the type of political input which we have spoken about in the Abstract of this paper. Scientists themselves must play a role in ensuring that extension services do exists and their research does not go waste. (This is the additional cross which must be carried by the applied scientist in a developing country).

[i] So beloved of our diplomats!

[j] In this paper, we have used the generic word Technology, whenever we wish to refer to the whole area of Technology — whether "Low" or "High". ("Low" Technology is not used in any pejorative sense). Like Basic versus Applied Sciences, the distinction

[continued overleaf]

(*iv*) Finally, there are five areas of *Science-based "High" Technology* which, in the conditions of today, may comprise:

(1) New Materials (including composite materials and High Temperature Superconductors); (2) Communication Sciences which consist of two types of sub-disciplines:

(2a) *Microelectronics* (including Development of Software; Micro-processors, Computer-aided Design and eventually, Fabrication of Microchips); and (2b) *Photonics* (including Lasers and Fibre Optics); (3) Space Sciences; (4) Pharmaceuticals and Fine Chemicals (5) and finally, for the 21st century, Biotechnology,[k] and gene-splicing, *so full of promise of a true revolution in the methods of Agriculture, Energy and Medicine.*[1]

between the two areas of Technology, "Low" versus "High", is often blurred. (By "High" Technology, we mean specifically the "Science-based Technology" of today). *One should say it clearly and emphatically that "Low" Technology is like "Basic Sciences" — it must be developed by any nation wishing to industrialise* — particularly the "design" and fabrication part of it. Thus it is important of recognise that a development of "Conventional Low" Technologies alone could lead to economic well-being especially in the short-term if Technology Transfer has already been negotiated. A nation may need engineering expertise and a skilled and disciplined workforce alone in the first instance — i.e. only Development and no Research. Such an attitude towards Research will, of course, eventually prove short-sighted — particularly in the areas of Sciences in Application, and for modern "High" Technology (which the developed countries will not part with).

[k] "Biotechnology thrives on new knowledge generated by molecular biology, genetics and microbiology, but these disciplines are weak, often nonexistent, in the under-developed world. Biotechnology springs from universities and other research institutions — centres that generate the basic knowledge needed to solve practical problems posed by society. But the universities of the underdeveloped world are not research centres. ... And the few creative research groups operate in a social vacuum; their results might be useful abroad, but are not locally. ... Biotechnology needs dynamic interactions among the relevant industries. These interactions, however, are weak in countries in which science is perceived as an ornament, not as a necessity. ... Biotechnology requires many highly skilled professionals, but ... underdeveloped nations lack sufficient people well-trained in the pertinent disciplines. ... Economic scarcity and political discrimination induce professionals and graduate students to emigrate or abandon science altogether". (Guest editorial from the journal *Biotechnology*, September 1986).

[1] "During the past year or so the bio-revolution has begun to spin off significant new developments in areas of agriculture that are far apart. These include the following: 1) A gene splicing breakthrough that *could shortly revolutionise the economics of dairy farming* with the first bovine somatotropin (BST), a genetic growth hormone

[continued overleaf]

"High" Technology differs from Classical "Low" Technologies in that high expertise in the relevant Basic Sciences (like Physics or Chemistry, or Biology, or Mathematics) is crucial. The materials used are minimal in their bulk and size. Very few of the developing countries, with the exception of the "Confucian belt" countries — like China or Singapore — or Brazil are conscious of the need for or have made progress in "High" Technology, the general feeling being that this whole area is beyond them. It is this feeling of lack of interest and faith in their own scientists that one must wish to fight against since the future undoubtedly lies here. This is on account of the enormous value-added potential of the industries based on "High" Technology and the possibilities of exporting the products of "High" Technology. There can be little "Technology Transfer" from the North in this area (unless this is of yesteryears's Technology) because no one will now want to sell — one has to learn to reinvent from published literature.[m]

that offers increases in milk yields of 15 to 20 per cent without raising feed costs; 2) *Calves can now be 'harvested'* from cows at a greatly increased rhythm thanks to embryo duplication techniques that enable a single cow to produce twin calves five times a year; 3) *Industrial tissue-culture techniques* may soon eliminate the need to grow whole plants ... Biotechnology specialists, notably the UK company *Plant Science,* are already producing digitalis, opium, ginseng and pyrethrum by culturing root cells in a fermentation vessel.

"Big chemical companies like Monsanto and Sandoz have bet ... on their strategies of switching emphasis away from industrial chemicals into biotechnology. Their sights are firmly set on an industry that is forecast to grow from its present turnover of around $25 billion a year to an annual $100 billion by the year 2000." — Giles Merritt (November 1987).

[m] This is what the Japanese apparently did. " ... we at Sony took the basic transistor and redesigned and rebuilt it for a purpose of our own that the originators hadn't envisioned. We made a completely new kind of transistor, and in our development work, our researcher, Leo Esaki, demonstrated the electron tunneling effect, which led to the development of the tunnel diode for which he was awarded a Nobel Prize seventeen years later, after he had joined IBM. ... The highly educated work force of Japan continues to prove its value in the field of creative endeavor. In the recovery from the war, the low cost of this educated labor was an advantage for Japan's growing *low-technology industry.* Now that the industrial demand is for *high technology,* Japan is fortunate to have a highly educated work force suited to the new challenge". — Akio Morita, *Made in Japan.* (I would like to recommend this book by the man who founded "Sony" of Japan, to anyone interested in Science and High Technology).

If one still persists in thinking of Japan as a country which has lived off borrowed technology, creating little basic knowledge, it is worth pointing out that the situation is fast changing. Thus, the finest (creative) Encyclopaedic Dictionary of Mathematics today is the Japanese, translated into English by the Massachusetts Institute of Technology Press (1977).

Of the four aspects of Sciences and Technology which have been mentioned above, the first to be developed so far as our countries are concerned, is Conventional "Low" Technology. The next may be Science in Application. (This is assuming that expertise in Basic Sciences is available). The last to develop, as a general rule, is Science-based "High" Technology.[n]

4. Why has Science and Technology Lagged Behind in the South?

There are three reasons why Science and Technology in the Third World countries have suffered.

(*a*) (*i*) *Lack of Meaningful Commitment towards Science, either Basic or Applied.* By and large, there has been scant realisation that Science can be applied to development as, for example, there was in Japan at the time of the Meiji Restoration around 1870 when the Emperor took five oaths. One of the oaths set out a national policy towards science — "Knowledge will be sought and acquired from any source with all means at our disposal, for the greatness and security of Japan".[o] The consequences of this lack of commitment have been little expenditures on Science, whether Basic or Applied (discussed in Sec. 5), weak universities, few research centres for Applied Sciences, sub-critical and isolated communities of scientists (with scant provision for infrastructure and for scientific literature) and weakness in scientific (and technological) education (discussed further in Sec. 7).

(*ii*) *No Commitment to Self-Reliance in Technology.* In technology, by and large, few of our Governments have made it a national goal to

---

[n]There are, of course, entrepreneurial considerations for the development and export of products of one technology versus another. These, though extremely important, (for example, the crucial choice before a country between *import substitution* versus *emphasis on exports,* as discussed by Dr. Hyung Sup Choi, former Minister of Science and Technology of South Korea) will not be elaborated here. (His paper — reproduced in Appendix II — was specially written for the South Commission). We would like to suggest that administratively our countries should try to develop the three Science-based areas (Basic Sciences, Sciences in Application and Science-based High Technology) together because of the similarities of approach. Thus, it may repay us if the crucial industrial areas of Classical and Conventional "Low" Technology are the province of a separate administration within our countries.

[o]Japan's "electron" into the rich "white man's club" — the OECD — did indeed follow upon mastery and utilisation of modern Science and Technology.

strive for self-reliance. The situation may be better for "Low" Technologies but is pitiful where High Technology is concerned for we have paid little heed to the scientific base of high technology, i.e. to the truism that science transfer must always accompany high technology transfer, if such transfer is to take.

(*b*)   *Inadequate Institutional and Legal Framework.* In respect of political actions needed, there is the necessity to have institutional and legal enactments. For example, Dr. Hyung Sup Choi spearheaded, on the institutional side, the creation of the Korea Institute of Science and Technology (KIST), the Korea Advanced Institute of Science (KAIST), the Korea Technology Finance Corporation (KTFC) and others, while, on the legal side, there was the enactment of several important laws for the development of science and technology. These included: the Law for the Promotion of Technology Development of 1972,[p] which provided fiscal and financial incentives to private industries for technology development; and the Engineering Services Promotion Law of 1973 to promote local engineering firms by *assuring markets on one hand and performance standards on the other.*

(*c*)   *The Manner in which the Enterprise of Science has been Run.*

Science depends for its advances on towering individuals. An active enterprise of Science must be run by scientists themselves and not by bureaucrats or by those scientsists who may have been active once, but have since ossified.[q] Science flourishes on criticism and toleration

---

[p] Apparently, among these, the most important was the Law for the Promotion of Technology Development of 1972. This Law was passed to encourage the private sector to adapt and improve imported science and technology, and to develop domestic science and technology through the R&D activities of government subsidised laboratories. The government took a follow-up step in 1977 by extending this law encouraging tax and financial incentives to a wider range of industries, while making R&D activities mandatory for strategic industries. Presumably all these laws were enforced. It is this type of care and concern for the *utilisation* of Applied Sciences and Technology without which *no amount of expenditure on Science and Technology is likely to be meaningful.*

[q] It must be realised in this context that young men and women are attracted to a career in Sciences, mostly on account of their innate curiosity and their desire to discover the basic Laws of Nature. (This must not be discouraged in our countries if we want to keep producing good scientists, eventually to work on Applied problems and of Science-based High Technology).

of opposing views. This has not been jealously safeguarded within our societies.

5.   The Sub-Critical Size of Science and Technology in the South[r]

(*a*)   One of the revealing indices of the size of Third World Science and Technology is the funding which the South provides for Research, Development and Utilisation of Science and Technology. To appreciate this, one has only to look at Tables I, II and III which give the Defence, Education, Health and Science Expenditures as percentages of GNP, both in the South and the North.

The point about the Tables is the following: Both the industrialised and the developing countries spend 5.6% of their respective GNP's on defence. The educational expenditures are also similar — 5.1% for the industrialised versus 3.7% for the developing countries. For health, it is 4.8% for the industrialised versus 1.4% for the developing countries — admittedly, a difference, but not as striking as for Science Technology. The figures for this latter differ from each other by nearly an order of magnitude.

The industrialised countries spend 2—2.5% of their GNP's on Research, Development and Modification, Adaptation plus the Utilisation of Science and Technology: versus less than 0.3% (on UNESCO's estimates, Table III) for most developing countries. (There are some few exceptions — the most notable ones being Argentina, Brazil, Cuba, India, Mexico and South Korea, which spend more than 0.5% on Science and Technology.[s] Even though one may argue that spending on Science

---

[r]According to an empirical law discovered by the late Professor Jolla Price of Yale University, with few exceptions, a country's output of scientific research is directly proportional to its spending on Science and is correlated with its GNP.

[s]We were heartened to hear from our colleagues on the South Commission that from 1989, during the next five years, Venezuela's expenditures on Science and Technology may go up to 2% of GNP from its present 0.4%, with state action for the utilisation of scientific and technological research. We were told that the Philippines' expenditures may likewise go up from 0.3% to 1.5% and Brazil's, from its present 0.6% to 2% of GNP by 1990. Cuba, we were told, is already spending 0.9% of its GNP on Science and Technology. Iran is expected to raise its spending on Science and Technology from its present 0.5% to 1% of GNP with immediate effect and to 2% later. South Korea is also expected to raise its spending on Science and Technology to 2% of GNP with immediate effect and to 3% by 1992.

TABLE I

Defence, Education and Health Expenditures in US dollars (1984) (as % of GNP)

	Population (× 1,000)	GNP (million US$)	GNP Capita (US$)	Defence (%)	Education (%)	Health (%)
Industrialised countries	1,125,033	11,019,363	9,795	5.6	5.1	4.8
Developing countries	3,651,353	2,697,982	739	5.6	3.7	1.4
Africa[1]	517,588	356,774	651	4.4	3.8	1.1
Middle East[2]	100,901	314,518	3,117	18.7	6.2	2.6
South Asia	992,628	266,330	268	3.5	2.8	0.8
Far East[3]	1,513,771	726,496	480	5.9	3.2	0.9
Latin America and Caribbean	394,718	752,688	1,907	1.6	3.7	1.3

Based on *World Military and Social Expenditures*, by Ruth Leger Sivard, World Priorities, Inc., Washington D.C., 1987.

[1] Less South Africa
[2] Less Israel
[3] Less Japan

TABLE II

Industrialised Countries' Expenditure on Science and Technology

Country	Population	GNP per capita (US$) 1984	Public Expenditures in Education (% in GNP)	Scientists/ Engineers in R&D (per million inhabitants)	Expenditure on R&D[4] (% on GNP)
France	55.17 (1985)	9,540	5.8 (1983)	1,363 (1980)	2.25 (1980)
Federal Republic of Germany	61.02 (1985)	10,940	4.6 (1984)	2,178 (1984)	2.54 (1984)
Japan	120.75 (1985)	11,300	5.6 (1984)	4,436 (1984)	2.65 (1984)
Netherlands	14.48 (1985)	9,290	6.9 (1984)	2,170 (1984)	1.97 (1984)
U.K.	56.49 (1984)	8,460	5.2 (1984)	1,545 (1980)	2.3 (1984)
U.S.A.	1,293.30 (1985)	16,690			2.69 (1984)

[4] Based on UNESCO statistics (1987). (These figures may include: application, diffusion and commercialisation and venture capital for technology provided by some governments as well as export credits to high technology companies).

TABLE III

Estimated Expenditure for Research and Development in 1980
as Percentage of G.N.P.

(Selected Countries)

ASIA		LATIN AMERICA AND CARIBBEAN		AFRICA	
India	0.9	Brazil	0.6	Algeria	0.3
Pakistan	0.2	Argentina	0.5	Nigeria	0.3
Bangladesh	0.2	Peru	0.2	Egypt	0.2
Sri Lanka	0.2	Chile	0.4		
Indonesia	0.3	Mexico	0.6		
Philippines	0.2	Cuba	0.7		
Singapore	0.5	Venezuela	0.4		
Republic of Korea	1.1				
Iran	0.5[5]				
Iraq	0.1[5]				

(From UNESCO Statistical Yearbook 1987)
[5] 1975 figures

and Technology is only a *necessary* condition for the developmental
aspects of Science and Technology and not a *sufficient* one (on account
of other motivational (for example, cultural) factors which may be
just as important), it remains a fact that the industrialised countries
are expanding (in GNP terms) on the average seven to nine times more
every year on Science and Technology than the Third World. *We in
the Third World are just not serious about Science and Technology.*
The profession of Science and Science-based Technology is hardly a
respectable — hardly a valid profession in the South.[t]

(*b*)   A second index of the sub-critical size of Science and Technology
is the numbers of those engaged actively in this activity in the Third

[t]In the British Colonial Empire, Britain did not leave behind the concept of a Scientific
Civil Service which, incidentally, had been part of the United Kingdom's own
administrative structure.

World. The UNESCO figures once again paint a different picture for the South and the North. In the North, an order of 2000 or more inhabitants per million are — engaged in Research and Development — while those similarly engaged in the South seldom exceed more than a few hundred. Once again, there is a factor of at least one or two orders of magnitude between the respective numbers. The Chinese figures in this context are revealing. According to Professor Lu Jiaxi, former President of the Chinese Academy of Sciences — speaking on Chinese Science at the Second General Conference of the Third World Academy of Sciences in Beijing in September 1987 — the Chinese had fewer than 500 researchers in 1949 altogether — less than one per million of population. The situation in most developing countries today is similar to that in China in 1949. (There are now 300,000 researchers in China and the country is approaching international norms, with a factor of 600 increase in 40 years.)

Regarding the numbers engaged in Science and Technology promotion, Developmental adaptation and modification, plus extension and utilisation, the situation is the same.

The socialist regimes by and large have taken to Science and "Low" Technology as a Religion, and with the same fervour. (They are now slowly waking up to the possibilities in civilian "High" Technology.)

6.   Steps Needed to Make Science and Technology Strong in Developing Countries

(*a*)   *The Five Classes of Communities in a Developing Society which must Cooperate.* There are five classes of communities in a developing society which could be involved in the building up and utilisation of the enterprise of Science and Technology in our countries. First, there are our rulers who determine the priorities. Second, there are the planners and the economists who advise them. Third, there may be the entrepreneurs with their management skills and risk capital. Fourth are the educators — plus the religious leaders[u] in some of the developing

---

[u] The religious leaders in some of our countries can, of course, play a leading role in popularising Science and Technology for the masses. The Scriptures uniformly emphasise the value of science in recognising Allah's design and of technology in mastery of nature. If only the divines could be persuaded to introduce modern Sciences and Technology into the curricula of their own seminaries!

countries who interact directly with the public. Fifth, come the scientists and the technologists.

Different societies have differing experiences in regard to the primary of one or more of these classes. For example, the Brazilian experience has been one of closest collaboration between the rulers — the military men in the past — and the economists and the scientists and the ecomomists. For India, Nehru's influence — no doubt conditioned by his background as a Science student at Cambridge — was paramount in laying down the traditions for scientific and technological research. The Chinese experience has been similar, where statesmen like Chou-en-lai and the scientist have collaborated actively. The same happened in the USSR (where, starting with Peter the Great) Lenin and others were responsible for the building up and the utilisation of Science and Technology.

Then there is the case of Japan where the ambitions of the Meiji statesmen (and now MITI) coincided with the patriotic feelings of the scientists and technologists themselves.[v] In all these examples, the fortunate scientist and technologist worked closely, in execution and advice, with rulers, who set the priorities for the country's development.

---

[v] This paper is about civilian Science and Technology but there is a strong correlation between strong local Science and Technology and Defence. We were told that one of the drives behind the Japanese acquiring Science and Technology in the Meiji era was to make their country strong for defence. They had before them the sad examples of India and China succumbing to the foreigner because of the inferior scientific and technological base from which they operated.

A similar sentiment has been expressed in a recent (August 1988) issue of The Herald of Pakistan: "The Gulf war has demonstrated one important fact: that religious fervour alone is no match for hi-tech weaponry. All religious 'fundamentalism' is today powerless if it is unable to arm itself with the instruments and weapons that only modern science and technology can provide". As if this had not already been shown by the experience of the Sudanese at the battle of Omdurman in the year 1898!

In this context, the role of research versus technological development must not be forgotten. As early as 1799 — against the opposition of the Ulema and surprisingly, even of a section of the military establishment — Sultan Selim III introduced the subjects of algebra, trigonometry, mechanics, ballistics and metallurgy into Turkey. He imported French and Swedish teachers for teaching these disciplines. His purpose was to rival European advances in gun-founding. Since there was no corresponding emphasis on research in these subjects, and particularly, in materials research, Turkey could not keep up with the newer advances being made elsewhere. The result was predictable: Turkey did not succeed. Then, as now, technology, unsupported by science, will not flourish.

This is the type of political action needed for the entire South, without which nothing will happen.

(*b*)   *Generous Patronage and Minimal Expenditures on Science and Technology*. No Science and Technology — Research, Development and their meaningful Extension and Utilisation — is possible without a nation spending an inescapable minimum of funds on it. In the industrialised countries, as a general rule, some 2 – 2.5% of GNP is made available — by the State as well as by private industry — for the four broad areas mentioned earlier. So far as *absolute* expenditures are concerned (barring funds spent on low technologies), the funds spent on *Basic Sciences Research in the North amount to some 4 – 10% of a nations's educational budget — taken as a unit — while roughly the same amount is spent on Applied Science Research*[W] *and twice as much on Research and Development related to "High" Technology.*[X] *If the developing countries adopt, as a desirable minimum, the lower figure of (4 + 4 = 8%) of their educational expenditures to be spent on Sciences (both Basic and Applied)* (including training for research as well as for international contacts) this should give us the colossal figure of 7.25 billion dollars from the South's own domestic resources, according to the estimates made by the Third World Academy of Sciences (TWAS). (See Appendix IV). For Science-based technology, one may

---

[W] These moderate expenditures are necessary to secure critical sizes of Science and Technology communities, and for providing them with tools to do their work. It comes as a surprise — certainly to the author — when such outlays are frowned upon, particularly by the economists, as wasteful luxury, even after it is demonstrated that these would increase the GNP manifold — *if only (in the sphere of Applied Sciences) by bringing about agricultural plenty — another Green Revolution — and better health.*

[X] I am told that in Egypt, three million dollars was spent in setting up a factory for the manufacture of thermionic valves. The factory was built in the same year that transistors were perfected and began to invade the world markets. The recommendation to set up the thermionic valve factory was, of course, naturally made by foreign consultants. It was, however, accepted by Egyptian officials who were not particularly perceptive of the way science was advancing and who presumably never consulted the competent physicists in their own country.

consider a further figure of 8% of the *educational expenditure,*[y] bringing the desirable minimum total for Science and High Technology to around 14.5 billion dollars for the South as a whole. (No Science-based development will accrue and no enhancement of GNP, unless we make these basic outlays.[z] Once again it is to be emphasised that this is NOT a recommendation to diminish the education budget but to find the additional spending on Science and Technology through general belt-tightening).[aa]

(c) *The Modalities of Growth of Sciences and High Technology, including Training and International Contacts.* If one were charged with running Science in a typical developing country of modest size, one would allocate in the first place (extra) funds roughly equal to 4%

[y] It is important to emphasis again and again that we are recommending these expenditures to be made, additional to the present spending on education.

We had the choice of linking the fields of Science and Science-based High Technology, at the earlier stages *when they are not yet money spinners,* with education or with defence. We decided to link this area with Education expenditures treated as a unit — something which everyone is likely to comprehend. Assuming that education budgets are (roughly on the average) 4% of GNP, we have thus recommended (on the average) $4 \times (0.08 + 0.08)\% = 0.64\%$ of the GNP to be spent on *Science and High Technology* in the Third World. (The amounts which each country will require, on this formula, to spend on Science and High Technology are shown in Appendix I). This (approximate) 0.64% of GNP is lower than the blanket expenditure of 1% of GNP on Science and Technology which UNESCO has recommended (this presumably also includes expenditures for development, modification and adaptation plus extension and utilisation (0.36%), involved in the Low Technology area). The 0.64% of GNP may be saved from defence expenditures in the first instance although *such monies will be repaid many times over to the nation in the course of ten years or so through the inevitable increase in the overall GNP,* which will follow spending on Science and Technology. (Agricultural plenty — another Green Revolution — in Africa, if attained through the applied and the extension scientist's efforts may, for example, repay all expenditures on Science).

[z] The scientists must try their best to assure that adequate returns on the funds spent on Science and Technology are paid back to the nation. This is because one looks upon such spending as a sacred trust, — in particular for the 36 countries of the "real South" (in Gerald Segal's definition), each with a population of one million or over and a GNP less than $400 per capita and which together account for nearly half of the world's population.

[aa] As an administrative measure, it may further be recommended that the Science and Technology budgets should be shown as a separate item of the developing country budgets and not amalgamated with that for education, as is commonly done now.

of a country's education budget to build up Basic Sciences in the universities, to maintain international contacts, and *for the appropriate training of cadres of scientists and technicians so as to assure a critical size. At the same time, one would commission and blue-print a comprehensive plan for Applied Sciences, allocating and spending (in the third or the fourth year) extra funds roughly equal to another 4% of the education budget for these.* (What one spends on, depends on a nation's priorities and could be in one or more of the following areas: agriculture, health, livestock, energy, materials and minerals, environment, soil sciences, oceans, communications. This is assuming that trained manpower has by now been assembled.)

Finally, one would spend on training of personnel and on Research and Development in the area of Science-based high technology, sums which are equivalent to 8% of the education budget, *as the quickest way to produce wealth,* making up the desirable minimal total of 16% of the education budget *after* a lapse of five years or so. (We are assuming, of course, that in parallel, there would be the crucially important emphasis on Conventional Technology to look after "Low" as contrasted to "High" Technology (with a national strategy for short term "low" technology utilisation plus its adaptation and modification to local conditions.))

(*d*)   *Utilisation of and Reciprocal Commitment and Responsibility of Scientists.*

(*i*) *Feeling of inferiority regarding indigenous science and technology.* Technological dependence of the Third World is the mental subordination that arises from a strong sense of inferiority towards Science and Technology. This feeling, which is particularly serious among decision-makers, tends to inhibit scientific and technological initiatives in the South. This is an important barrier to be overcome, so far as autonomous *and self-reliant development is concerned.* The economists in our countries must learn that their community should show respect to and employ the scientist and the technologist within the country before thinking of hiring scientists and technologists from abroad. Wherever this has happened, like in Brazil, where indigenous economists have worked hand in hand with the indigenous scientists and technologists, the country has taken phenomenal steps in growth.

It is important that scientists and technologists from the South should band together and exercise pressure on their governments for their due recognition. This is certainly the case for scientific communities in the North.

I would like to suggest that the time has come when our courts of state should once again be adorned with scientists and technologists. I am reminded of the story of King Arthur of legendary fame; at his Court there was a Court Magician; his name was Merlin. Merlin was responsible for using magic for forging steel for swords and for providing magical medicinal potions. The scientists are the Merlins of today. They can perform feats of magic undreamt of by Merlins of yesteryears. They can, indeed, transform society. But in our Third World countries, these Merlins have no place in the affairs of State. Should they not be invited to perform their professional role?

Some will say — perhaps rightly — that the Merlins in developing countries are gormless amateurs — they hardly know their applied craft. They choose to live in their own ivory towers, and our Southern societies are thereby forced to import the real Merlins from the North. This may be true, but why is this so? Could this emasculation have come about through the fact that our own Merlins are so few in numbers, and even these few have never been invited to make a contribution to development in their own countries. Not even by their colleagues — the professional economists — who in this metaphor are the High Priests of development. Only experience can teach the Merlin-Scientist the craft of developmental problem-solving, even if he knows his science. This vicious cycle of lack of mutual trust must be broken, hopefully, before the year 2000.

(ii) Reciprocal responsibility of scientists and technologists. A parallel sense of responsibility must also be instilled by and into the scientists and the technologists in developing countries. Scientists and Technologists are at present, an insignificant proportion of our populations. They constitute a particular social subculture and as such occupy a particular niche of every society. The relevance of this niche is a function of its explicit articulation and integration with the national development process. The top-ranking scientist and technologist must, in

particular, feel that he is part of a team which is engaged in an exciting venture. Such an articulation depends upon the conscious involvement of the scientific and the technological community in the tasks of socioeconomic development, as well as upon the image of science and technology (and of scientists and technologists) in the minds of the non-scientific population — in particular of decision-makers (politicians, entrepreneurs, managers). This two-way interaction depends as much on the attitude of scientists and technologists towards development, as upon the reciprocal attitude of the administrators towards the scientists.[ab]

(*e*)   *The Universality of Science and the Brain Drain.* One of the troublesome points and one which often arouses unwholesome jealousy for the scientific profession is the international character of Science and Technology and the perceived possibility of migration of scientists.[ac]

(*i*) We believe that one of the best anti-brain drain devices is the one which has been pioneered by the IAEA-UNESCO-run International Centre for Theoretical Physics at Trieste — the Associateship Scheme — whereby distinguished scientists working and living in developing countries are guaranteed to spend from six weeks to three months at the Centre at times of their own

---

[ab]*Creation of a Favourable Science and Technology Climate.* "Science and Technology development gains momentum when a suitable environment for its popularization is created. The creation and promotion of such an environment is a prerequisite for science and technology development, particularly in a country where social and economic patterns and customs are bound by tradition. ... Korea has launched a movement for the popularization of science and technology as an integral part of its long-range science and technology development plan. The movement aims to motivate a universal desire for scientific innovation in every one in all aspects of their lives. It has been led by the Ministry of Science and Technology, the Korea Science Promotion Foundation, and the Saemaul Technical Service Corps in cooperation with concerned government agencies, industry, academic circles, and the mass-communication media. ... This movement is in no way conceived as the special province of scientists and engineers, although this group can provide key support and resources in view of its pertinent talent and knowledge" — Hyung Sup Choi.

[ac]It is important to state it clearly and emphatically, that scientists (unlike, for example, medical men) do not leave their countries for trivial monetary reasons in the first place. They almost always do so because of isolation, because of lack of similar individuals to talk to and work with and on account of paucity of scientific literature. (This, in turn, compounds the problem of attaining a critical size of the scientific enterprise).

choosing, three times during a period of six years. Their fares and living expenses in Trieste are met by the Centre. No salaries are paid. Some 350 physicists working in the Third World are presently Associates of the Centre. After 18,000 visits made during the last 24 years by *research physicists* of the Third World, there has not been a single case of brain-drain from among the Associates and others who have come to work at this prestigious Centre. There is the need for similar schemes for research workers from other disciplines besides Physics.[ad]

(*ii*) There are other schemes like TOTKEN, devised by UNDP, which recognise the existence of expatriates and *that they have a role to play in the development of their own countries.* This scheme makes it possible for expatriate scientists and technologists to visit their countries on a regular basis.

(*iii*) It is worth pointing out that the developed countries have played a role in intensifying the brain drain problem. The National Academy of Sciences of the USA, for example, has estimated that only one half of foreign nationals that obtain their Ph.D. in physics in the USA return to their home country. This is because the US Science Community has recognised that it cannot meet the demands for physicists in the USA from its native-born Ph.D's. There has been much effort spent and numerous reports written stressing the desirability of getting as many as possible of the foreign nationals — and not only[ae] from the developing countries — to stay.

(*iv*) In this context, we would humbly like to suggest that (even if there is no governmental action on the part of the North's Aid Agencies — and there should be) the expatriates' own feelings of indebtedness to their societies for having educated them may at least partly be assuaged by suggesting to them to contribute to a (privately-run) Foundation for Science which may be set up

---

[ad] Even for Third World economists. For example, the United Nations University Economics Institute (WIDER) in Finland may consider building up a long-term Associateship-Scheme of its own.

[ae] "If we capitalise the value of those who have left the British Isles for America since the war, we have very much more than paid back the whole of the Marshall Aid". — Lord Bowden.

by each developing country. Such Foundations for Science may receive donations, in cash or kind, from expatriate scientists at least equivalent to the educational benefits they may have obtained from their own country — thus enriching their countries' scientific and educational system. (This should be done with a finesse and not in a ham-handed manner.)

(*f*)   *The Role of Private Foundations for Science Technology.* The role of private foundations for Science and Technology (for all but the centrally-planned economies of the Second World) cannot be over-stressed. There are 22,000 such foundations in the United States alone. Provision must be made for these foundations to receive generous government help in terms of taxation relief. (The role of private foundations is particularly important for the countries of the Middle East where once personal foundations for research and education were legion. Today one may see private palaces being errected all over but hardly a Palace for Science.)

(*g*)   *Recognition that Creativity in Scientific Research is not Easy and that it has its own Mores and Modalities.* (See Appendix VI). Finally, it should be borne in mind that scientific research cannot always deliver solutions to all the problems all the time, particularly to the time scale set by the administrators. Witness, for example, the situation in the North, regarding SDI, or cancer research, or AIDS research — this, in spite of the expenditures of billions of dollars in rich countries and the provision of hundreds of researchers. This must be kept in mind when making (what may be unreasonable) demands on the small and pro-visionless scientific communities and holding them to vindictive ransom (as has happened in some developing countries).

7.   Science and Technology Education Policy

In the Table (Table IV) appended to this section, we give the World Bank figures for educational enrolment for developing countries. (These figures do not distinguish between Science and non-Science studies.) We strongly believe that scientific studies should start as early as possible together with the teaching of the three R's and the instilling of a feeling for experimentation. (Of the three R's, the easiest to teach in my opinion is arithmetic — addition, subtraction and multiplication (but not division). Everyone must learn to count. Next comes reading, particularly readings

TABLE IV

Percentage of Age Group Enrolled in Education

		Primary (6 to 11 years of age)		Secondary (12 to 19 years of age)		Tertiary (20 to 24 years of age)	
		1965	1985	1965	1985	1965	1985
	Low income economies (GNP/Cap less than 400 US$)	74 w*	99 w	22 w	34 w	2 w	. . .
	China and India	83 w	110 w	25 w	37 w	2 w	. . .
	Other low-income	44 w	67 w	9 w	22 w	1 w	5 w
1	Ethiopia	11	36	2	12	0	1
2	Bhutan	7	25	0	4	. . .	0
3	Burkina Faso	12	32	1	5	0	1
4	Nepal	20	79	5	25	1	5
5	Bangladesh	49	60	13	18	1	5
6	Malawi	44	62	2	4	0	1
7	Zaire	70	98	5	57	0	2
8	Mali	24	23	4	7	0	1
9	Burma	71	102	15	24	1	. . .
10	Mozambique	37	84	3	7	0	0
11	Madagascar	65	121	8	36	1	5
12	Uganda	67	. . .	4	. . .	0	1
13	Burundi	26	53	1	4	0	1
14	Tanzania	32	72	2	3	0	0
15	Togo	55	95	5	21	0	2
16	Niger	11	28	1	6	. . .	1
17	Benin	34	65	3	20	0	2
18	Somalia	10	25	2	17	0	. . .
19	Central African Rep.	56	73	2	13	. . .	1
20	India	74	92	27	35	5	. . .
21	Rwanda	53	64	2	2	0	0
22	China P R.	89	124	24	39	0	2
23	Kenya	54	94	4	20	0	1
24	Zambia	53	103	7	19	. . .	2
25	Sierra Leone	29	. . .	5	. . .	0	. . .
26	Sudan	29	49	4	19	1	2
27	Haiti	50	78	5	18	0	1
28	Pakistan	40	47	12	17	2	5
29	Lesotho	94	115	4	22	0	2
30	Ghana	69	66	13	39	1	2
31	Sri Lanka	93	103	35	63	2	5
32	Mauritania	13	. . .	1	. . .	. . .	. . .
33	Senegal	40	55	7	13	1	2
34	Afghanistan	16	. . .	2	. . .	0	. . .
35	Chad	34	38	1	6	. . .	0
36	Guinea	31	30	5	12	0	2
37	Kampuchea, Dem.	77	. . .	9	. . .	1	. . .
38	Lao PDR	40	91	2	19	0	1
39	Viet Nam	. . .	100	. . .	43	. . .	. . .

TABLE IV (Cont'd.)

	Primary (6 to 11 years of age)		Secondary (12 to 19 years of age)		Tertiary (20 to 24 years of age)	
	1965	1985	1965	1985	1965	1985
Middle-income economies	85 w	104 w	22 w	49 w	5 w	14 w
Lower middle-income (GNP/Cap between 400 and 1600 US$)	75 w	104 w	16 w	42 w	4 w	13 w
40 Liberia	41	...	5	...	1	...
41 Yemen PDR	23	66	11	19	...	...
42 Indonesia	72	118	12	39	1	7
43 Yemen Arab Rep.	9	67	0	10	...	...
44 Philippines	113	106	41	65	19	38
45 Morocco	57	81	11	31	1	9
46 Bolivia	73	91	18	37	5	20
47 Zimbabwe	110	131	6	43	0	3
48 Nigeria	32	92	5	29	0	3
49 Dominican Rep.	87	124	12	50	2	...
50 Papua New Guinea	44	64	4	14	...	2
51 Cote d' Ivoire	60	78	6	20	0	3
52 Honduras	80	102	10	36	1	10
53 Egypt Arab Rep.	75	85	26	62	7	23
54 Nicaragua	69	101	14	39	2	10
55 Thailand	78	97	14	30	2	20
56 El Salvador	82	70	17	24	2	14
57 Botswana	65	104	2	29	...	1
58 Jamaica	109	106	51	58	3	...
59 Cameroon	94	107	5	23	0	2
60 Guatemala	50	76	8	17	2	8
61 Congo, People's Rep.	114	...	10	...	1	...
62 Paraguay	102	101	13	31	4	10
63 Peru	99	122	25	65	8	24
64 Turkey	101	116	16	42	4	9
65 Tunisia	91	118	16	39	2	6
66 Ecuador	91	114	17	55	3	33
67 Mauritius	101	106	26	51	3	1
68 Colombia	84	117	17	50	3	13
69 Chile	124	109	34	69	6	16
70 Costa Rica	106	101	24	41	6	23
71 Jordan	95	99	38	79	2	37
72 Syrian Arab Rep.	78	108	28	61	8	17
73 Lebanon	106	...	26	...	14	...
Upper middle-income (GNP/Cap between 1600 and 4000 US$)	97 w	105 w	29 w	57 w	7 w	16 w
74 Brazil	108	104	16	35	2	11
75 Malaysia	90	99	28	53	2	6
76 South Africa	90	...	15	...	4	...
77 Mexico	92	115	17	55	4	16

TABLE IV (Cont'd.)

		Primary (6 to 11 years of age)		Secondary (12 to 19 years of age)		Tertiary (20 to 24 years of age)	
		1965	1985	1965	1985	1965	1985
78	Uruguay	106	110	44	70	8	32
79	Hungary	101	98	. . .	72	13	15
80	Poland	104	101	58	78	18	17
81	Portugal	84	112	42	47	5	13
82	Yugoslavia	106	96	65	82	13	20
83	Panama	102	105	34	59	7	26
84	Argentina	101	108	28	70	14	36
85	Korea Rep. of	101	96	35	94	6	32
86	Algeria	68	94	7	51	1	6
87	Venezuela	94	108	27	45	7	26
88	Gabon	134	123	11	25	. . .	4
89	Greece	110	106	49	86	10	21
90	Oman	. . .	89	. . .	32	. . .	1
91	Trinidad & Tobago	93	95	36	76	2	4
92	Israel	65	99	48	76	20	34
93	Hong Kong	103	105	29	69	5	13
94	Singapore	105	115	45	71	10	12
95	Iran, Islamic Rep.	63	112	18	46	2	5
96	Iraq	74	100	28	55	4	10
97	Romania	101	98	39	75	10	11
	Developing Countries	78 w	101 w	22 w	39 w	3 w	8 w
	Oil Exporters	69 w	107 w	14 w	44 w	2 w	10 w
	Exporters of manufactures	86 w	109 w	27 w	40 w	3 w	. . .
	Highly indebted countries	88 w	104 w	21 w	47 w	5 w	16 w
	Sub-Saharan Africa	41 w	75 w	4 w	23 w	0 w	2 w
	High income oil exporters	43 w	86 w	10 w	56 w	1 w	11 w
98	Saudi Arabia	24	69	4	42	1	11
99	Kuwait	116	101	52	83	. . .	16
100	United Arab Emirates	. . .	99	. . .	58	0	8
101	Libya	78	127	14	87	1	11
	Industrial market economies (GNP/Cap higher than 4000 US$)	107 w	102 w	63 w	93 w	21 w	39 w
102	Spain	115	104	38	91	6	27
103	Ireland	108	100	51	96	12	22
104	New Zealand	106	106	75	85	15	35
105	Italy	112	98	47	75	11	26
106	United Kingdom	92	101	66	89	12	22
107	Belgium	109	95	75	96	15	31
108	Austria	106	99	52	79	9	27
109	Netherlands	104	95	61	102	17	31

TABLE IV (Cont'd.)

		Primary (6 to 11 years of age) 1965 1985	Secondary (12 to 19 years of age) 1965 1985	Tertiary (20 to 24 years of age) 1965 1985
110	France	134 114	56 96	18 30
111	Australia	99 106	62 95	16 28
112	Germany, Rep. of	. . . 96	. . . 74	9 30
113	Finland	92 104	76 102	11 33
114	Denmark	98 98	83 103	14 29
115	Japan	100 102	82 96	13 30
116	Sweden	95 98	62 83	13 38
117	Canada	105 105	56 103	26 55
118	Norway	97 97	64 97	11 31
119	United States	. . . 101	. . . 99	40 57
120	Switzerland	87 . . .	37 . . .	8 22
	Nonreporting nonmembers	102 w 105 w	66 w 92 w	27 w 21 w
121	Albania	92 97	33 69	8 7
122	Angola	39 93	5 13	0 1
123	Bulgaria	103 102	54 100	17 18
124	Cuba	121 105	23 85	3 21
125	Czechoslovakia	99 97	29 39	14 16
126	German Dem. Rep.	109 101	60 79	19 31
127	Korea, Dem. Rep.	. . . . . .	. . . . . .	. . . . . .
128	Mongolia	98 105	66 88	8 26
129	USSR	103 106	72 99	. . . 21

Weighted Percentage of Age Group Enrolled in Education,
for males and females, by group of countries

	Primary (6 to 11 years of age)				Secondary (12 to 19 years of age)				Tertiary (20 to 24 years of age)	
	Males		Females		Males		Females		Totals	
	1965	1985	1965	1985	1965	1985	1965	1985	1965	1985
Low income economies (GNP less than 400 US$)	. . .	110	. . .	88	. . .	41	. . .	26	2	. . .
China and India	. . .	121	. . .	98	. . .	45	. . .	29	2	. . .
Other low-income	58	75	31	56	13	28	4	16	1	5
Middle-income economies	92	109	79	101	26	57	19	51	5	14
Lower middle-income (GNP between 400 and 1600 US$)	84	111	66	100	21	50	12	41	4	13

TABLE IV (Cont'd.)

	Primary (6 to 11 years of age)				Secondary (12 to 19 years of age)				Tertiary (20 to 24 years of age)	
	Males		Females		Males		Females		Totals	
	1965	1985	1965	1985	1965	1985	1965	1985	1965	1985
Upper middle-income (GNP between 1600 and 4000 US$)	100	108	93	102	31	66	26	63	7	16
Developing Countries	84	110	62	92	28	45	14	33	3	8
Oil Exporters	78	113	59	101	20	53	9	42	2	10
Exporters of manufactures	...	119	...	98	...	48	...	33	3	...
Highly indebted countries	91	108	84	99	23	57	20	57	5	16
Sub-Saharan Africa	52	85	31	67	6	26	2	14	0	2
High income oil exporters	60	82	25	69	15	55	5	41	1	11
Industrial market economies (GNP higher than 4000 US$)	107	101	106	101	65	91	61	92	21	39
Other countries with an upper middle income	103	...	102	...	60	...	72	...	27	21

Note The above data refer to a variety of years, generally not more than two years distant from those specified, and are mostly from UNESCO. However, disaggregated figures for males and females sometimes refer to a year earlier than that for overall totals.

The data on *primary* school enrolments are estimates of children of all ages enrolled in primary school. Figures are expressed as the ratio of pupils to the population of school-age children. While many countries consider primary school age to be 6 to 11 years, others do not. The differences in country practices in the ages and duration of schooling are reflected in the ratios given. For some countries with universal primary education, the gross enrolment ratios may exceed 100 percent, because some pupils are younger or older than the country's standard primary school age. The data on secondary school enrolments are calculated in the same manner, but again the definition of *secondary* school age differs among countries. It is most commonly considered to be 12 to 17 years. Late entry of more mature students as well as repetition and the phenomenon of *bunching* in final grades can influence these ratios.

The *tertiary* enrolment ratio is calculated by dividing the number of pupils enrolled in all post-secondary schools and universities by the population, age 20 to 24. Pupils attending vocational schools, adult education programmes two-year community colleges, and distance education centres (primarily correspondence courses) are included. The distribution of pupils across these different types of institutions varies among countries. The *youth* population, that is 20 to 24 years, is used as the denominator since it represents an average tertiary level cohort. While in higher income countries, youths aged 18 to 19 may be enrolled in a tertiary institution (and are included in the numerator), in developing and in many industrialized countries, many people older than 25 years are also enrolled in such an institution. These data and definitions come from UNESCO.

The *summary measures* in this table are country enrolment rates weighted by each country's share in the aggregate population.

From: "World Development Report 1988", World Bank 1988.
Published for the World Bank, Oxford University Press; 1988.

from Science, the provision of which should be the responsibility of our scientific communities as part of their duties to help to create a favourable climate about Science among the general public.)

As can be seen from the Table, there are wide variations between the different countries, as well as between the industrialised and the developing countries. The starkest variations, however, are in the average numbers we educate between the ages of $12-19$ (secondary education level) and the ages between $20-24$ (tertiary education). The low income developing country averages ("the real South" with a GNP/capita less than \$400) are particularly small compared with those of the developed countries (22% to 37% versus 93% for secondary education and 5% versus 39% for the tertiary education). This means that at the earning stage, a student in most of our countries is ill-equipped for the modern world. It is important not only to cure this imbalance, but also to change the thrust of our secondary education. The following observations are relevant in this context.

(*i*)   *Higher Secondary Education*

After a period of compulsory lower secondary education (which may finish at the age of 15 or so) most modern societies provide for two parallel educational systems. Using the U.K. terminology of the 1970's, these two systems may be called (1) the system of professional education comprising technical, vocational, agricultural and commercial courses, and (2) the system of higher education comprising courses which lead on to the university level, in the sciences, engineering, medicine and the arts.

(*ii*)   A major structural failing of the Third World educational system has been that, in general, no credible professional system has developed. It is true that a half-hearted system of polytechnic institutions and vocational schools has been built up in recent years in a number of Third World countries, but this system has had scant prestige attached to it. (As a general rule, such systems have been run by the Ministries of Labour and Employment, rather than the Ministries of Education.)

(*iii*)   To see how inadequate such a system has been, one may recall that in industrialised countries the proportion of those enrolled for the two streams is of the order of 50 : 50. In the Third World, however, the proportion of the professional versus the university level enrolment is normally of the order of 10 : 90. This preponderance of the technologically illiterate is the major cause of unemployment and of the Third World's technological backwardness.

(*iv*)   One of the main educational tasks before the Third World is to change this ratio of 10:90 to 50:50.[af] (In the conditions of today, the "professional system" should be accorded equal status with the better known "liberal" educational system and should include courses on modern materials, as well as courses on microelectronics.)

(*v*)   Our first task will be to bring a measure of prestige to the professional system of education. One will need to give serious consideration to the institution of National Certificates — or preferably, decide to identify these with the prevailing awards. What we have in mind is this. Parallel with the present system of education in arts and sciences, we should create a second — the professional system of education. Each award — the matriculation, or the Bachelor's Degree — may be obtained either after the present "liberal" courses in arts or sciences as now, or after technical, agricultural, or commercial courses from a polytechnic, an agricultural, or a commercial school. So far as job opportunities in administrative services are concerned, all Bachelor's Degrees (general, technical, or commercial) — likewise all matriculates of whatever variety — would count as equivalent. Only thus will the exclusive hold on the public mind of the present prestigious "liberal" system of education be broken.

(*vi*)   *University Level*

The proportion of those following Science and Engineering versus those following the arts at the "liberal" university level is of the order of 50:50 in most of the industrialized countries. This is certainly not the case for most developing countries. One must aim at a 50:50 ratio for the developing countries also. This will need equipping the institutions of higher learning adequately.

[af] In the report of the United Kingdom University Grants Commission (issued in 1984) the figure of 52.48 is cited for populations of scientists and technologists versus art students. And at the secondary level, whereas in China, or Japan, all science subjects are compulsory — in the USSR, even the future musicians or footballers or seamstresses must study physics, chemistry, mathematics and biology till they are sixteen — there is no such compulsion in most of the South countries' educational systems.

(vii) Specialisation

One proposal which may be considered in this context is that of specialisation. Could, for example, a consortium of Universities in the US and UK be helped by their Governments and encouraged to take care of University Science in all those developing countries which desire this? Could the Netherlands and Belgium look after the building up of libraries and laboratories? Could Germany and Japan look after technical education at all levels? Could Scandinavia look after the scientific aspects of ecology? Could Switzerland and Austria (with their well-known pharmaceutical expertise) look after medical education? Could Italy with its experience of setting up International Centres in Physics and Biotechnology, look after the creation of similar institutions in all disciplines of science in concert with developing countries? Could the US, Canada, Australia and New Zealand look after education for agriculture and education for prospecting? Could one envisage the USSR taking care of primary, secondary and vocational education Third-World-wide? Could France, Canada and Spain carry out all these actions for the French and Spanish speaking developing countries if desired by them? This is merely an illustration of what a possible division of the relevant tasks could be. Eventually, of course, these suggestions would have to be tailored and modified when detailed projects are elaborated.

(viii) What I have in mind is something patterned along the lines of the success which India achieved in the decade of the sixties when it created four Indian Institutes of (Science and) Technology. The one in Kanpur was created by a US consortium of universities which helped to raise and furnish it, besides supplying the higher cadres of teaching staff for a number of years. The one in Delhi was helped by a consortium of British universities, the one in Bombay by the USSR and the one in Madras by the Federal Republic of Germany. Each nation helped to build up the institute under Indian auspices, contributed staff and left behind a tradition in teaching and research which has continued even after the original contracts have expired. There was a healthy rivalry between the donor nations vying with each other; this guaranteed the excellence and standards of quality. What we envisage in the proposal above is something like this except that it is to be carried out on a much wider canvas. One would hope that by the year 2000, if the

plans are drawn up now, many of the objectives we have mentioned will have been achieved. In this, one must not forget that (even though not affluent enough to contribute materially) countries like Argentina, China, India, Brazil, Egypt, Mexico, Nigeria, South Korea, Cuba, Yugoslavia and many others could make highly valued intellectual inputs to these specialised efforts.

8.   International Modalities for the Growth and Utilisation of Science and Technology

(*a*)   *Recognition that the Growth of Sciences and Technology is basically a Problem for the South and a Long-Term One.* In the end, the growth of science and its utilisation by the South is a Southern problem, though outside help — particularly if it is organised — can make a crucial difference. The modalities for growth and utilisation of science and technology entail two types of actions: those needed to be adopted in and by the South and those that need to be carried out in concert with the North.

It has been the *general* experience of the communities in the South that whereas scientific communities in the North are quite generous in accepting trainees for research and in helping to build up corresponding communities in developing countries, the same cannot be said for *technologists.* This is understandable in view of the fact that technology provides the visible raison for better standards of living of the industrialised countries. One cannot blame the North for not wanting to part with such an advantage lightly. Accepting this, it is important that the aid asked for by the Third World is more in terms of "science transfer" *if "technology transfer" cannot be negotiated.* (One must keep reminding oneself that the Science of today is the Technology of tomorrow.)

(*b*)   *Concerted Action between the South and the North.* We list here some of the modalities through which the North can help the South to build up its scientific base.

(*i*) *10% of the aid funds to be earmarked for science and technology.*[ag] The linking of the aid for Science and Technology with the total bilateral aid from the North is an important political modality. It should come to be established (principally through the type of requests made by the South, as well as the favourable reception to such requests by the North), that 10 per cent of the aid given by every developed country is to be spent to enhance Science and Technology in the Third World.

(*ii*) *Birthright of southern scientific communities: free access to scientific literature.* It should be a considered as part of a birthright of scientific communities in a developing country that the country should have at least one complete Central Science Library containing all Science journals and all scientific books. Arrangements (by the Aid Organisations or the World Bank) should be made with publishers in the North that such books and journals are made available at a fraction of their present price — at least one copy for each country — and sent to a designated Central Library in at least fifty of the developing countries, which can make use of this literature right away.

(*iii*) *United Nations agencies and international centres for science.* In multilateral cooperation, the United Nations agencies, including the United Nations University, should have a prominent role in building up scientific infrastructure in their areas of competence. Developing countries need international research institutions on the applied side like the Wheat and Rice Research Institutes in Mexico and the Philippines and the International Centre for Insect Physiology and Ecology in Kenya. There is also the experience on the basics side, of UNESCO and IAEA in relation to the International Centre for Theoretical Physics, Trieste (with visits of 4000 physicists last year — 2500 of them from developing countries and 1500 from the industrialised countries), or, of UNIDO for the International Centre for Bio-technology and Genetic Engineering, at Trieste and Delhi, on the

---

[ag] This would amount to $3.5 billion. $3.5 billion (in foreign currencies) for Southern Science and Technology would make up, together with the domestic $14.5 billion, a total of $18 billion. This would constitute some 10% of the world's spending on Science and High Technology — and even though small — this could, without doubt, transform the South.

applied side. These Centres are run by the scientists for the scientists. The South should, at the least, utilise those trained at these Centres and urge other United Nations Organisations to set up international centres in disciplines relevant to their competence.

In this context, there is the proposal to create in Trieste an International Centre for Science which shall have five components 1) the existing International Centre for Theoretical Physics; 2) the existing Interantional Centre for Genetic Engineering and Biotechnology; 3) a new International Centre for High Technology and New Materials; 4) a new International Centre for Chemistry, Pure and Applied; and finally, 5) a new International Centre for Earth Sciences (for research in and for imparting knowledge of the recent advances in geology, prospecting, soils, as well as for the environmental aspects of Earth Sciences, including man-made global change). The International Centre for Science will be created with United Nations sponsorship by the Italian Government. The Third World Academy of Sciences — a non-Governmental organisation — convened in October 1988 the first meeting of the Third World Network of Scientific Organisations. The meeting was attended by Ministers of Science and Technology and Higher Education, Presidents of Academies and Chairmen of National Research Councils from 36 Third World countries. The discussions focussed on the Third World joint action on global scientific problems like Fusion Research, the Human Genome problem, the Greenhouse Effect and Waste Disposal in the South. This initiative is the scientific analogue of the Founding of the Group of 77. At present 92 scientific organisations from 60 developing countries have joined this Network.

(*iv*) *Multinational corporations.* We wish there was some mechanism for persuading multinational corporations (which use the South as a manufacturing base for their products of High Technology), to conduct some of their Research in the developing countries themselves.[ah]

[ah] Apparently, the Brazilians succeeded in persuading the multinational corporations in Brazil to do precisely this but the rest of the developing world still has to learn the secret behind this.

(*c*) *South-South Collaboration.*[ai] South-South collaboration in Science and Technology is important for Science and Technological Education and for Higher Training, for Science in Application and for building up of Technology. This is because of similarity of the problems and the experiences.[aj]

Some of these ideas are elaborated below.

(*i*) The Third World Academy of Sciences, which has as its Fellows (or Associate Fellows) 139 of the Third World's most prestigious scientists from 50 countries (9 of them Nobel Laureates born in the Third World), has among its projects one for South-South collaboration in Sciences. More than two hundred Fellowships have so far been made available by the Scientific Institutions in Argentina, Brazil, Chile, China, Colombia, Ghana, India, Iran, Kenya, Madagascar, Mexico, Vietnam and Zaire for hospitating such visits. The Third World Academy of Sciences pays for the travel of Third World scientists to a Third World country. Clearly, the efforts of the Third World Academy of Sciences — a non-governmental organisation — need to be enhanced a hundred fold.

(*ii*) South-South collaboration of scientists takes place automatically at International Centres, for example, at the International Centre for Theoretical Physics, where, during the last 24 years, there have been 18,000 visits of developing country physicists who have met and have had the opportunity to collaborate with each other. (Even for the larger countries like India and China, scientists from different parts of the country hardly meet except in the international locations.) What is urgently needed is provision of funds for carrying on this collaboration after the physicists have left the Centre.

[ai] South-South Collaboration is important but one must not forget that Science and Technology are being created mostly in the North or in Japan. Thus, it would be counterproductive to speak of South-South collaboration in this area to the exclusion of South-North collaboration.

[aj] Whereas at present one knows all about the scientific establishments and the organisations which exist in the developed countries, one has very little knowledge of what facilities are available for scientific training and research in the developing world. This needs to be rectified as a top priority.

(*iii*) Another possibility of South-South collaboration arises through the possible joint programmes for Higher Technological Education. For example, the Indian Institutes of (Science and) Technology would, we are sure, be even more responsive to the needs of the whole South, if they were mandated to do so.

9. Desirable Regional Arrangements

(*i*)   Appendix I gives lists of developed and developing countries arranged according to *population,* and indicating GNP per capita and defence, health, education and the present science spending as per-centages of GNP. This Appendix also gives actual science spending (in millions of dollars) as well as the desirable minimum — 16% of a country's educational budget — in accordance with the recommenda-tions of the Third World Academy of Sciences.

(*ii*)   Leaving aside Luxembourg, Iceland and Malta, with populations of less than 1 million inhabitants, the minimum population for a Developed Market Economy country is 3.2 millions (for New Zealand). Of the developing countries, 10 have less than one million inhabitants; 16 have populations which range between 1 and nearly 3 millions, while 32 developing countries have populations ranging between 3 millions and 10 millions. (Above 10 millions there are 44 developing countries — this includes the Asian socialist countries.)

(*iii*)   We feel that it would be desirable if the richer of the developing countries like Brunei, Qatar, Kuwait, United Arab Emirates and others, which fall into the first category, would set up foundations for Science and Technology to help other Third World countries. They would also be excellent sites for United Nations-run International Centres for training and research of the type we have mentioned in Section 8, (*b*) (*iii*), acting as modern "Athenses" of the developing world (assuming that these countries are willing to help generously in the setting up of such centres).

(*iv*)   Clearly the countries in the first category, (that is, with populations between 1 and 3 millions), are too small to set up independent institutions for Science and Technology. They will need linking up with neighbouring countries with similar problems. This cannot be spelt out in any detail here because the possibilities depend on the local preferences.

(*v*)   In the second category of countries with populations of between 3 and 10 millions, it is interesting to note that Switzerland (with a population base of 6.4 millions) is a world leader in pharmaceuticals, specialised engineering and high technology. If there is the will, there is every possibility for Third World countries belonging to this second category to emulate the Swiss example in time.

10. Global Science

The South must play an appropriate role in the area of Global Science, like the Global Change Programmes of ICSU, the UNESCO Programmes relating to the Biosphere, the management of the Antarctica, or the more highly technological programmes of the IAEA in respect of Fusion and, (for some of the more advanced developing countries) Space Sciences. It should be pointed out that both the North and the South are currently engaged in wrecking the environment. By playing its appropriate[ak] role in the North-led programmes for rectifying this, the South can make a distinctive contribution of great significance to the solution of this global problématique.

11. Envoi

In this note, we have pleaded that *political* action is needed in order to build up and utilise the so-far neglected community of its own scientists and technologists, by the developing world. Their numbers need to be multiplied so that they constitute a critical mass; they have to be given proper recognition, provided with scientific literature, contacts and provisions for their work and guaranteed tolerance for their beliefs *by those who run our countries.*[al] They may be ill-prepared at the present for

---

[ak] With financial commitment of its own!

[al] In a recent College on Biophysics held at the International Centre for Theoretical Physics, a Brazilian scientist (who won the Centre's Heisenberg Prize) listed the following items for success with his research work in Brazil: 1) Imagination; 2) Hard Work; 3) Provision of Equipment; 4) Contact with Scientists of Developed Countries; 5) Interdisciplinarity (particularly necessary for his subject, Biophysics).

taking on this role but with careful nurturing and proper trust, they surely have the capability of transforming the South.[am]

In this context, judging from the actual versus desirable levels of educational versus science expenditures and other evidence, it is clear that while Latin American as well as the Confusian-Belt and other Asian countries are indeed taking steps to enhance their scientific and technological communities, *the African, the Arab, and the Islamic countries* by and large have a long leeway to make up.

PARTICULARLY IF THE TRYST WITH THE YEAR 2000 IS TO BE KEPT.

[am] As their colleagues have done in the North! There is no question but that our present world is a creation of modern science in application. We tend to forget that it was the Science of Physics in application, which brought about the modern Communications Revolution and gave a real meaning to the concept of One World and its mutual interdependence. We tend to forget that it was the Science of Medicine which brought about the Penicillin Revolution, leading to the present level of world population. We tend to forget that it were the Sciences of Chemistry and Genetics in application which brought about the Fertilisers and the Green Revolutions, to feed part of this population. And we tend to forget that it is to these same sciences — the wealth-producing sciences of physics and geophysics, and the survival sciences of medicine, molecular biology, cell culture and chemistry — to which the Third World must turn for resolution of some of its current problems.

Appendix I

Comparative Resources and Expenditure versus Proposed Funding for Science
(16% of the Education Budget) by country.

Developed Market Economy Countries

Country	Population (X 1,000)	GNP/Capita (US$)	Defence % of GNP	Health % of GNP	Educ. % of GNP	Science Expenditure (Millions US$)	16% of Educ. Budget (Millions US$)
1 Iceland	241	10,720	n.a.	6.80	3.50	20.6	19
2 Luxembourg	366	13,380	0.80	0.80	5.65	n.a.	46
3 New Zealand	3,246	7,310	1.90	4.90	4.40	214	184
4 Ireland	3,560	4,840	1.80	7.05	6.70	155	194
5 Norway	4,144	13,890	2.90	6.40	6.80	921	643
6 Israel	4,296	4,920	27.10	3.50	8.40	528	320
7 Finland	4,919	10,870	1.50	5.30	5.50	802	484
8 Denmark	5,101	11,240	2.40	5.80	6.50	688	614
9 Switzerland	6,421	16,380	2.20	5.60	5.00	2,313	828
10 Austria	7,545	9,150	1.20	4.60	5.80	828	648
11 Sweden	8,330	11,890	3.10	9.10	8.00	2,575	1,329
12 Belgium	9,853	8,450	3.10	5.90	5.90	1,166	810
13 Greece	9,937	3,550	7.20	3.60	2.40	71	161

Appendix I (Cont'd.)

Country	Population (× 1,000)	GNP/Capita (US$)	Defence % of GNP	Health % of GNP	Educ. % of GNP	Science Expenditure (Millions US$)	16% of Educ. Budget (Millions US$)
14 Portugal	10,198	1,970	3.50	3.00	4.80	80	182
15 Netherlands	14,486	9,180	3.20	6.70	7.00	2,660	1,504
16 Australia	15,789	10,840	3.20	5.20	6.00	1,880	1,712
17 Canada	25,414	13,670	2.30	6.40	7.40	4,863	3,954
18 South Africa	32,432	2,010	4.00	0.50	2.70	n.a.	360
19 Spain	38,730	4,360	2.40	4.60	2.50	844	736
20 France	55,133	9,550	4.10	6.70	5.30	9,477	4,721
21 United Kingdom	56,539	8,390	5.40	5.40	5.10	9,962	4,041
22 Italy	56,945	6,520	2.70	5.90	5.60	4,084	3,554
23 Germany Fed. Rep.	61,065	10,940	3.30	8.10	4.60	16,701	4,952
24 Japan	120,579	11,330	1.00	4.60	5.10	35,520	10,168
25 United States	238,780	16,400	6.40	4.30	5.00	101,818	29,203

For Science and Technology Ireland, Greece and Portugal are spending less than 16% of their Education Expenditures.

Appendix I (Cont'd.)

Socialist Countries

Country	Population (X 1,000)	GNP/Capita (US$)	Defence % of GNP	Health % of GNP	Educ. % of GNP	Science Expenditure (Millions US$)	16% of Educ. Budget (Millions US$)
1 Albania	2,943	n.a.	4.40	2.60	3.00	n.a.	23
2 Mongolia	1,909	n.a.	10.50	1.40	5.00	n.a.	15
3 Bulgaria	8,980	n.a.	4.00	4.00	6.20	1,429	448
4 Hungary	10,660	n.a.	2.20	2.80	5.00	1,413	500
5 Czechoslovakia	15,497	n.a.	4.00	5.20	5.20	3,533	809
6 German Dem. Rep.	16,716	n.a.	4.90	2.90	4.50	5,099	964
7 Korea Dem. Rep.	20,357	n.a.	10.20	0.90	3.20	n.a.	115
8 Romania	22,866	n.a.	1.40	2.00	2.00	145	260
9 Poland	37,288	2,120	2.50	3.90	3.90	949	1,054
10 Vietnam	61,640	n.a.	n.a.	n.a.	n.a.	n.a.	n.a.
11 USSR	277,563	n.a.	11.50	3.20	4.70	85,054	14,688
12 China P.R.	1,041,094	310	7.00	1.40	2.80	n.a.	1,395

Appendix I (Cont'd.)

Developing Countries — Populations up to nearly 3 million

Country	Population (× 1,000)	GNP/Capita (US$)	Defence % of GNP	Health % of GNP	Educ. % of GNP	Science % of GNP	Expenditure (Millions US$)	16% of Educ. Budget (Millions US$)
1 Brunei Darussalam	294	17,580	7.90	0.70	2.00	n.a.	n.a.	12
2 Qatar	320	15,980	5.80	n.a.	4.90	n.a.	n.a.	44
3 Bahrain	423	9,560	3.60	2.20	3.30	n.a.	n.a.	25
4 Cyprus	660	3,790	2.40	1.90	3.80	0.1	2.5	15
5 Fiji	702	1,700	1.20	2.70	6.60	n.a.	n.a.	13
6 Gambia	737	230	2.10	3.00	4.40	n.a.	n.a.	1
7 Swaziland	758	650	1.50	1.80	5.80	n.a.	n.a.	6
8 Guyana	806	570	4.80	3.80	7.40	0.2	1	5
9 Guinea Bissau	886	170	n.a.	n.a.	n.a.	n.a.	n.a.	n.a.
10 Gabon	997	3,340	2.10	1.3	4.20	n.a.	n.a.	24
11 Mauritius	1,036	1,070	0.30	2.20	4.20	0.2	2.2	7
12 Botswana	1,070	840	3.30	2.30	8.40	0.2 (1975)	1.8	12
13 Oman	1,181	7,080	27.70	2.30	3.70	n.a.	n.a.	44

Appendix I (Cont'd.)

Country	Population (× 1,000)	GNP/Capita (US$)	Defence % of GNP	Health % of GNP	Educ. % of GNP	Science % of GNP	Science Expenditure (Millions US$)	16% of Educ. Budget (Millions US$)
14 Trinidad & Tobago	1,187	6,010	1.00	1.80	6.00	0.8	57	72
15 United Arab Em.	1,381	19,120	7.40	1.00	1.80	n.a.	n.a.	73
16 Lesotho	1,515	480	2.90	1.30	3.30	n.a.	n.a.	4
17 Mauritania	1,693	410	6.60	1.31	7.33	n.a.	n.a.	8
18 Kuwait	1,736	14,270	5.30	2.40	4.20	0.9	223	182
19 Congo	1,872	1,020	2.60	1.40	5.40	n.a.	n.a.	18
20 Yemen PDR	2,086	540	17.00	n.a.	7.00	n.a.	n.a.	12
21 Panama	2,180	2,020	2.10	6.30	5.30	0.2 (1975)	8.8	35
22 Liberia	2,196	470	2.60	1.80	4.50	n.a.	n.a.	7
23 Jamaica	2,227	940	1.40	3.50	6.40	0.1 (1975)	2.1	31
24 Singapore	2,557	7,420	5.70	1.60	5.30	0.5	94.8	158
25 Central African Rep.	2,583	270	2.00	1.10	5.40	0.2	1.5	6
26 Costa Rica	2,593	1,290	0.00	1.44	6.00	0.1	3.3	33

Appendix I (Cont'd.)

Developing Countries — Populations from 3 to 10 million

	Country	Population (× 1,000)	GNP/Capita (US$)	Defence % of GNP	Health % of GNP	Educ. % of GNP	Science % of GNP	Science Expenditure (Millions US$)	16% of Educ. Budget (Millions US$)
27	Uruguay	3,004	1,660	2.90	0.90	2.40	0.2 (1975)	10	30
28	Togo	3,038	250	2.50	2.20	6.20	1.4 (1975)	10.6	7
29	Nicaragua	3,263	850	12.40	4.60	6.00	0.3	8.3	31
30	Lebanon	3,301	1,833	7.30	1.20	5.80	n.a.	n.a.	55
31	Paraguay	3,388	940	1.20	0.60	1.60	0.2 (1975)	6.3	10
32	Papua New Guinea	3,499	710	1.70	3.20	6.90	n.a.	n.a.	27
33	Jordan	3,512	1,560	14.10	1.70	7.80	2 (1980)	54.8	49
34	Libyan A. Jamahiriya	3,600	7,500	12.90	1.30	3.70	0.2 (1982)	45	169
35	Sierra Leone	3,745	370	0.70	1.10	2.60	n.a.	n.a.	7
36	Benin	4,043	270	2.30	1.40	5.00	n.a.	n.a.	8
37	Honduras	4,396	730	5.30	1.70	4.00	n.a.	n.a.	136
38	Burundi	4,696	240	3.50	0.80	3.40	0.4	4.5	6
39	Chad	4,982	n.a.	10.40	0.70	1.80	0.3 (1975)	1.8	1
40	Somalia	5,384	270	10.00	0.60	1.60	n.a.	n.a.	6
41	Haiti	5,451	350	1.60	1.00	1.20	n.a.	n.a.	3

Appendix I (Cont'd.)

Country	Population (× 1,000)	GNP/Capita (US$)	Defence % of GNP	Health % of GNP	Educ. % of GNP	Science % of GNP	Science Expenditure (Millions US$)	16% of Educ. Budget (Millions US$)
42 El Salvador	5,564	710	5.10	1.50	3.00	0.9	35.5	19
43 Rwanda	6,026	290	1.50	0.60	3.10	0.1	4.9	8
44 Guinea	6,049	320	3.10	1.20	3.30	n.a.	n.a.	10
45 Dominican Rep.	6,261	810	1.20	1.40	2.00	n.a.	n.a.	20
46 Bolivia	6,383	470	2.60	1.50	4.90	n.a.	n.a.	20
47 Niger	6,391	200	0.80	1.00	2.90	0.1 (1975)	1.5	7
48 Senegal	6,558	370	2.80	1.30	4.90	1 (1975)	24	19
49 Zambia	6,640	400	4.10	2.10	5.40	0.5 (1975)	13.3	29
50 Malawi	7,044	170	1.70	2.30	2.60	0.2 (1975)	2.4	5
51 Tunisia	7,143	1,220	5.60	2.60	5.80	n.a.	n.a.	83
52 Mali	7,511	140	4.90	1.50	3.30	n.a.	n.a.	6
53 Burkina Faso	7,885	140	2.70	0.80	2.70	0.5 (1975)	5.5	4
54 Yemen Arab Rep.	7,955	520	17.60	1.70	7.00	0.3 (1975)	12.4	46
55 Guatemala	7,966	1,240	2.90	0.90	1.80	0.5	49.4	26
56 Zimbabwe	8,406	650	6.20	2.20	7.50	n.a.	n.a.	79
57 Angola	8,756	n.a.	14.20	1.20	5.20	n.a.	n.a.	58
58 Ecuador	9,367	1,160	1.60	1.10	4.10	0.4 (1975)	43.50	87

Appendix I (Cont'd.)

Developing Countries — Populations from 10 to 20 million

Country	Population (× 1,000)	GNP/Capita (US$)	Defence % of GNP	Health % of GNP	Educ. % of GNP	Science % of GNP	Science Expenditure (Millions US$)	16% of Educ. Budget (Millions US$)
59 Côte d'Ivoire	10,072	620	1.20	1.30	5.00	0.3 (1975)	18.7	52
60 Cuba	10,097	1,852	7.10	3.40	6.30	0.7	130.9	193
61 Cameroon	10,191	810	2.00	1.00	3.50	0.8 (1988)*	49.5	48
62 Madagascar	10,312	250	2.40	2.20	3.80	0.20 (1980)	5.1	17
63 Syrian Arab Rep.	10,483	1,630	16.60	0.40	6.10	n.a.	n.a.	161
64 Saudi Arabia	11,521	8,860	21.70	3.10	7.80	n.a.	n.a.	1299
65 Chile	11,990	1,440	4.20	2.70	4.80	0.4 (1980)	68.50	172
66 Ghana	12,710	390	0.60	0.80	1.50	0.9 (1975)	44.6	15
67 Uganda	15,474	n.a.	1.10	0.20	1.30	n.a.	n.a.	13
68 Malaysia	15,611	2,050	5.60	1.20	6.10	n.a.	n.a.	309
69 Iraq	15,654	1,861	50.00	0.80	3.40	0.1 (1975)	27	152
70 Sri Lanka	16,143	370	1.70	1.30	2.80	0.2	11.9	29
71 Nepal	16,527	160	1.30	0.80	2.70	n.a.	n.a.	11
72 Venezuela	17,323	3,110	1.60	2.00	6.20	0.4	215	629
73 Mozambique	17,791	n.a.	4.80	0.80	1.90	n.a.	n.a.	17
74 Peru	18,653	960	6.90	1.10	2.90	0.2	35.8	95

* recent data.

Appendix I (Cont'd.)

Developing Countries — Populations from 20 to 50 million

Country	Population (× 1,000)	GNP/Capita (US$)	Defence % of GNP	Health % of GNP	Educ. % of GNP	Science % of GNP	Science Expenditure (Millions US$)	16% of Educ. Budget (Millions US$)
75 Kenya	20,375	290	4.10	2.10	6.00	0.8 (1975)	47.2	61
76 Algeria	21,865	2,530	1.80	1.40	4.70	0.3 (1975)	165.9	401
77 Morocco	21,924	610	5.60	1.00	7.20	n.a.	n.a.	156
78 Sudan	21,931	330	3.30	0.20	4.70	0.2 (1980)	26.7	66
79 United Rep. Tanzania	22,242	270	3.30	1.40	3.30	n.a.	n.a.	37
80 Yugoslavia	23,100	2,070	3.70	4.10	3.50	n.a.	382	381
81 Colombia	28,418	1,320	1.40	0.80	3.10	0.1	37.5	230
82 Argentina	30,531	2,130	2.50	1.40	4.15	0.5 (1980)	325	483
83 Zaire	30,557	170	1.20	0.40	3.50	n.a.	n.a.	48
84 Burma	36,831	190	3.60	1.00	2.00	n.a.	n.a.	21
85 Korea Rep.	40,646	2,180	5.40	0.30	4.80	1.1	886	654
86 Ethiopia	42,271	110	9.30	1.40	3.00	n.a.	n.a.	22
87 Iran Islamic Rep.	45,160	1,778	13.30	1.60	7.50	0.5 (1975)	401.4	966
88 Egypt Arab Rep.	47,108	710	8.50	1.20	4.30	0.2	67	236
89 Turkey	49,406	1,130	4.80	0.60	3.30	0.6 (1980)	335	n.a.

Appendix I (Cont'd.)

Developing Countries — Populations from 50 to 100 million

Country	Population (× 1,000)	GNP/Capita (US$)	Defence % of GNP	Health % of GNP	Educ. % of GNP	Science % of GNP	Science Expenditure (Millions US$)	16% of Educ. Budget (Millions US$)
90 Thailand	50,950	830	4.00	1.10	4.10	0.3	126.80	268
91 Philippines	54,725	600	1.80	0.70	1.80	0.2	65.67	91
92 Mexico	78,820	2,080	0.70	0.40	2.60	0.6	983.6	985
93 Pakistan	94,933	380	6.00	0.40	1.80	0.2 (1980)	72.1	109
94 Nigeria	99,669	760	1.80	0.60	2.00	0.3 (1980)	227.2	241

Appendix I (Cont'd.)

Developing Countries — Populations more than 100 million

Country	Population (× 1,000)	GNP/Captia (US$)	Defence % of GNP	Health % of GNP	Educ. % of GNP	Science % of GNP	Science Expenditure (Millions US$)	16% of Educ. Budget (Millions US$)
95 Bangladesh	100,592	150	1.70	0.40	1.80	0.2 (1975)	30.2	43
96 Brazil	135,539	1,640	0.80	1.60	4.00	0.6	1,333.7	1,242
97 Indonesia	162,212	530	3.90	0.60	3.40	0.3	257.9	549
98 India	765,147	250	3.20	0.90	3.10	0.9	1,721	1,004

In the population range of 20 millions upwards, it is to be noted that Argentina, the Republic of Korea, Turkey, Mexico, Brazil and India are the only countries where the actual science expenditures are higher than 16% of their education budgets — the desirable minimum for developing countries recommended by TWAS (Third World Academy of Sciences).

Population and GNP/Capita figures are from "World Bank Atlas 1987" and refer to 1985; Defence, Health and Education figures are from "World Military & Social Expenditures 1987–88" and refer to 1984; Science (% of GNP and Expenditure) are from "UNESCO Statistical Digest 1987" and generally refer to 1984 or 1985 — the tables show the year referred to when no figures for 1984 or 1985 are available.

Appendix II

Science and Technology Policies in the Industrialisation of a Developing Country — Korean Approaches

by Hyung Sup Choi

(Paper for the Background Documentation at the Task Force Meeting on the Role of Science and Technology in the Development of the South, South Commission, May 31, 1988, Geneva, Switzerland.)

Introduction

Since World War II, many countries have made great efforts to industrialise their economies, reflecting the aspirations of their people. Although some failed, there were a few cases where special conditions prevailed for success or were made to prevail. Each success has been unique, and no all-purpose formula or guidelines seem to exist for others to emulate. Nevertheless, inquiry into the developmental process from the Korean point of veiw may be worthwhile.

The prospects for entering into industrialisation today by developing countries are not so evident; nor can it be taken for granted that late-comers have the advantage. On the contrary, strenuous efforts are required to build up capacity for embarking upon industrialisation. Elimination of the obstacles to industrialisation is a long-term affair, which ideally should precede the industrialisation process. If one were to try to accomplish the process all at once, the task would be formidable, or even impossible. Thus, a step-by-step approach should be taken to build the capacity for industrialisation. In the process, one could capitalise on the advantages of being a "late-comer".

In any approach to industrialisation in a developing country, great care must be taken in selecting the field to be developed and in deciding the extent of industrialisation. These decisions must be based on a clear understanding of the country's potential and the constraints to which it is subjected. A country richly endowed with natural resources needed for industrialisation may use a different approach from a country rich in human resources but possessing few natural resources.

To properly chart the path to industrialisation requires consideration of numerous socio-political, cultural, and economic factors. The case of Korea will be discussed here as an example.

Korea opted for development of light, labour-intensive industries by absorbing the labour force from the primary sector. However, the demand for industrial products in the primary sector was too slight, so it was necessary to look outward for capital, markets, and technology. Korea, therefore, did not choose to develop industry first by pursuing import-substitution and then export-promotion policies; instead, the two were undertaken almost simultaneously, particularly when the first long range economic development plan went into effect. The apparent success of this bold approach can be attributed to several factors: 1) amenability to training and the absorptive capacity of the labour force in dealing with relatively sophisticated technologies, 2) close trade relations with the U.S.A. and Japan, both big markets, 3) full exploitation of the technical advantage of being a late-comer in industrialisation, and 4) capacity to adapt to the international economic environment, which was actively supported by the government through creation of a favourable investment climate for foreigners[1].

The most conspicuous constraints for the rapid industrialisation scheme were the deficiencies in the social overhead sectors. Because the infrastructure for industrial development was very poor, the government placed a great emphasis on quick and decisive action to build roads, ports, communications and other facilities essential to development including expanding facilities for technical education. About 50 percent of the total foreign capital induced was spent for these facilities, on top of over 70 percent of the total public loan funds from overseas.

The First Five Year Economic Development Plan (1962–1966) called for selective industrialisation on one hand and the establishment of social infrastructures on the other, allowing the national economy to find a proper berth. Such industries as power, cement, fertiliser, and coal mining were among the targets selected by the government. Light industries, such as textile, plywood and consumer goods evolved largely from initiatives by private entrepreneurs who saw domestic captive markets for such products.

The Second Five Year Economic Development Plan (1967–1971) pushed forward the continued expansion of basic chemicals, petrochemicals, and iron and steel industries. Growth momentum was established through these sectors so that their dynamism could be felt within all Korean industries. These industries are highly capital intensive by nature and need a huge infrastructure which has to be supported by the government. They would not necessarily develop sufficient linkage effects directly, but they are essential to the foundation upon which the high linkage industries can be built.

One pressing problem in developing these leading industries was whether or not they could be operated at, or at least near, full capacity. The leverage was found to be extremely small, because the cost of capital for these industries, most of which originated abroad, was much higher than that required in advanced countries. This hard fact of life had much to do with the design of every industrial project.

The Third Five Year Economic Development Plan (1972–1976) more or less followed the same direction of industrialisation. Greater economies of scale were needed along with development of agriculture and social services, to capitalise on the previous experience of advanced countries and of Korea itself and thus to maximise the advantages of the late-comers. This orientation necessitated the introduction of newer and higher level technologies on an order of magnitude never experienced. It was an irreversible decision and an answer to the issue of survival or extinction in an ever-stiffening international competition.

The technological development of a country usually starts with the importation of advanced foreign technology and proceeds, through development of domestic variants of this imported technology, to the final goal of technological self-reliance. However, little can be expected from imported technology in the absence of a capability to modify and improve it for domestic application. Therefore, to achieve viable results from the technology transfer, a corresponding effort must be made to assimilate and adapt imported technologies.

With these thoughts in mind, I will try to relate the role of technology in Korea's experience. I shall be giving more attention to the role of technology at the national level, not because I believe in a highly centralised system, but because of the impact it can have at the national level. The national government has a crucial role to play during various stages of industrialisation and the choice of technologies to achieve those goals.

Marco-Micro Linkages for Science and Technology Development

In developing countries, technology has come to be viewed as one of the most important means of achieving national progress. The Korean government has planned intensive policies and strategies for the development of science and technology with many innovative supporting measures. Particular attention will be directed toward the use of high technology, for this was the path chosen by Korea to industrialise and to evolve an outward oriented economy[2]. While less sophisticated technology can surely serve the needs to some aspects of national development, Korea decided that the high technology path could afford it the most options in reaching development goals.

In this plan, Korea adopted a three-pronged approach, emphasising a capability build-up, particularly manpower development at various levels, accelerated introduction of foreign advanced technologies, and stimulation of domestic R&D activities. To this end, the implementation plan was formulated in consideration of both institutional and legal factors.

Institutional Set-Up

The institutional framework of this approach was somewhat daring. It included the establishment of: 1) the Ministry of Science and Technology (MOST) in 1967 as the central policy making, planning, coordinating and promotional body in the government; 2) the Korea Institute of Science and Technology by a special law (KIST Assistant Act) in 1966 as an autonomous multidisciplinary industrial research institute; and 3) the Korea Advanced Institute of Science (KAIS) in 1971, in addition to existing universities and colleges, to be a mission-oriented post-graduate school. Finally, many vocational training institutes and high schools were established to meet the rapidly rising, almost explosive, demand for skilled workers and technicians.

In recognition of the the need for an institution to bridge academia and industry, the traditional and the contemporary, and the domestic and the foreign, the concept of an intermediary agent was introduced. Accordingly, Korea first established an independent, multidisciplinary industrial research institute (KIST). The second step was the reinforcement of an information clearing house (KORSTIC) for industrial research. The third was the establishment of quality control and instrument calibration service centres as short-term measures in selected fields of industry. As a long term measure, the Korea Standard Research Institute (KSRI) was organised and reinforced to support industry.

The Korea Institute of Science and Technology (KIST) was created to bolster the industrial sector, particularly in those areas where the national economic development plan emphasised elimination of bottlenecks hindering further growth. This institute was established, through special legislation, as a contract research organization to make researchers aware of the marketing of technologies. In the realm of R&D, KIST was intended to make researchers problem-oriented and to impress the underwriters of such R&D with the importance of the implementation of the R&D results [3].

As industry grew, its technological requirements increased in level and diversity; as a result, laboratories, such as those in shipbuilding, petrochemicals, electronics, telecommunications, machinery and energy, which existed as integral parts of the institute, were no longer able to render sufficient technical support to industries growing at such a rapid rate. Independent research organisations specific to each industry and problem area became necessary. In order to address this formidable task, the government has made use of existing small laboratories at KIST as seeds and spun them off from the mother institute. Thus, they inherited not only accumulated experience but a workable management philosophy and a system all too often missing in a new organisation.

In order to operate these institutes effectively, the Daeduk Science Town designed to house research organisations, public and private, as well as higher educational institutions, and thus formed an intellectual complex to contribute to the development of science and technology [4]. It is expected that the Daeduk Science Town will develop as the cradle of Korea's burgeoning science and technology, eventually acting as the centre of national excellence.

Although there are several devices for inducing industries to perform R&D, the most essential one is to provide soft capital for technology development, due to industry's limited fund sources. In this respect, I will expand on Korean examples; the Korea Technology Advancement Corporation (K-TAC), Korea Technology Department Corporation (KTDC), Korea Development Investment Corporation (KDIC) and Korea Technology Finance Corporation (KTFC).

In 1974, K-TAC was organised to carry out the commercialisation of R&D results of KIST. Currently, K-TAC has 8 subsidiary companies and it will add 6 more subsidiaries.

KTDC was established in 1981 as an autonomous public enterprise. To promote the R&D projects of industry which inherently involve substantial risks coupled with expected high returns, KTDC is willing to share both the risk of failure as well as the benefits of success. To this end, the corporation offers three different types of financial support: long-term loans, conditional loans and equity investments. The major activities of KTDC include support for all aspects of the introduction, improvement and adaptation of advanced technology, particularly semi-developed technology from abroad, the commercialisation of R&D results, the development of new products and processes, and the support of plant engineering services. KTDC also provides special services in the area of technical advice, feasibility studies for R&D activities, technology transfer and management. Of the total funds approved during 1985, 37% were approved for 77 projects of the metal and machinery industry, 30% for 61 projects of the electric and electronics industry, and 33% for 50 projects of the chemical and other industries. The breakdown of projects showed that 86% of the total amount was provided for R&D activities and the commercialisation of R&D results, 12% for technology imports and training, and 2% for the purchase of R&D equipment.

In another case, KDIC was incorporated in December of 1982 by the seven Seoul-based short-term finance companies in Korea. As a limited liability venture capital company, KDIC is designed to foster and strengthen the technology oriented small and medium industries in Korea through equity investment and/or equity-type investment. In addition to the provision of financial support, KDIC expects to support the management of its portfolio companies through its participation on the Board of Directors of these companies, as well as through the provision of business advisory services. Since the establishment of KDIC, 98 projects have been invested in. The investments include common stocks, preferred stocks as minority shareholders, convertible debentures and debentures with warrants. As of the end of June, 1986, the portfolio of KDIC by industrial classification was as follows; 32% for 38 electronics projects, 14.4% for 11 metal fabrication projects, 14.6% for 7 non-metallic mineral products projects; 13.2% for 18 food projects and 2.3% for 2 miscellaneous projects. KDIC purchases debentures and makes short-term loan for working capital only to its portfolio companies.

Lastly, KTFC was established in October 1984 by the Korea Development Bank as a venture capital company. During 1985, KTFC approved 71 projects of 58 firms and supplied funds in the form of equity investments, acquisition debentures, credit loans and conventional loans. The financial support during the year consisted of 24% to R&D activities,

45% to the initial commercialisation of new technology, and 31% to improvements in the manufacturing process.

To meet the urgent need of industry for manpower trained to handle high level technology, the Korea Advanced Institute of Science (KAIS) and the Changwon Technicians College were established. KAIS provides post-graduate programmes in applied science and engineering in selected fields to educate a sufficient number of high calibre scientists and engineers to meet the emerging needs of Korean industry. This institution is trying to build a "centre of relevance" to the nation's economic development rather than merely creating a "centre of excellence" in academic pursuits. The Changwon Technicians College was established to guarantee a new social status based on professional pride in the skilled worker's career. This institution makes it possible for a skilled worker to become a master foreman through education in needed theoretical background and administrative skills, and later to become a manager or supervisor with the same social standing as a regular college graduate [6].

In the early 1970's, the growth of Korea's economy surpassed that of a semi-developed country and approached that of a highly industrialised nation. As a result, nurturing the potential of science and technology has become an immediate need. The trend at this stage of development is naturally to turn an eye to the development of basic science as well.

Research activities for basic sciences had to be supported on a national scale as the source for industrial technology. The support of basic research activities in universities and the establishment of the Korea Science and Engineering Foundation were a step forward in strengthening basic research fields. In addition, systematic and mutual cooperation between government, universities, industry and research institutes was recognised as a necessity.

While enterprises in industry are encouraged to finance their own R&D activities through the Law for the Promotion of Technology Development, those lacking in their own facilities and research personnel are induced to consign R&D tasks to "specialised research institutes" either entirely or on a cooperative basis. In addition to this way of promoting the co-operation between industry and academia, plans have been drawn up to establish an integrated research system to include basic, applied and development research.

The government also finances R&D activities jointly with private enterprises. The benefits of these joint research ventures naturally accrue to the enterprises involved, and have led to the creation of a number of laboratories financed jointly by private enterprises and the government.

Future government policy must focus on securing funds and research personnel for these institutions. As stated already, the lack of investment for R&D is one of the constraints to technology development, particularly in the private sector. Even enterprises with money for technology development may not know which organisation might be best able to solve their techonological problems. On the other hand, even if a research institute has developed a promising technology, it is not always an easy task to find the right client to use it. An intermediary to bridge academia and industry with sufficient funding capacity might offer a ready solution to these problems. It is suggested here that a specially equipped financial institution could act as such an intermediary. An industry, a research institution and a financial intermediary could form a tripartite system of cooperation to aid technology development. To create this tripartite system a financial institution which will play an effective intermediary function must be founded. Consequently, it will be imperative to strengthen the maximum utilisation of existing institutions and to expedite the establishment of Technology Development Bank to ensure a smooth flow of money for R&D activities[7].

Legal Back-up

The Ministry of Science and Technology spear-headed the enactment of several very important laws for the development of science and technology. They include: the Science and Technology Advancement Law of 1967 which defines the basic commitment of the government to support science and technology and to provide policy leadership; the Law for the Promotion of Technology Development of 1972 to provide, among others, fiscal and financial incentives to private industries for technology development; the Engineering Services Promotion Law of 1973 to promote local engineering firms by assuring markets on one hand and performance standards on the other; the National Technical Qualification Law of 1973 which, through a system of examination and certification, promotes the enhancement of status for professionals in technical fields, particularly for those who practice skills; the Assistance Law for Designated Research Organisations of 1973 which provides incentives in legal, financial and fiscal terms for research institutes in specialised fields where the government and private industry place particular emphasis, such as shipbuilding, electronics, communication, mechanical and material engineering, and energy and related area; and the Law for the Korea Science and Engineering Foundation of 1976 which provides a legal basis for the establishment of the Foundation to act as the prime agent for strengthening research in basic and applied sciences, as well as in engineering, centered chiefly around universities, and to facilitate more rapid application of science and engineering to national needs.

Among these measures, I shall draw special attention to the Law for the Promotion of Technology Development of 1972. The law was passed to encourage the private sector to adapt and improve imported technology, and to develop domestic technology through the R&D activities of government subsidised laboratories. Subsequently, various tax and financial incentives have been provided. As a result, an ever-increasing number of enterprises have been allocating funds for R&D projects, and long-term low interest loans have been granted to those enterprises seeking to utilise newly developed technologies for commercial purposes. Encouraged by the government policy, many firms in the private sector are now showing a keen interest in establishing their own laboratories and equipping them with necessary facilities and qualified staff.

The government took a follow-up step in 1977 by amending the aforementioned law to 1) extend the tax and financial incentives to a wider range of industries, while making R&D activities mandatory for strategic industries; 2) take protective measures to create demand for products embodying newly developed domestic technologies; and 3) organise the Industrial Technologies Research Association to search for solutions to the problems facing small and medium enterprises and provide them with guidance on technology development.

Creation of a Favourable Science and Technology Climate

Science and technology development gains momentum when a suitable environment for its popularisation is created. The creation and promotion of such an environment is a prerequiste for science and technology development, particularly in a country where social and economic patterns and customs are bound by tradition.

Korea has launched a movement for the popularisation of science and technology as an integral part of its long-range science and technology development plan. The movement aims to motivate a universal desire for scientific innovation in every one in all aspects of their lives. It has been led by the Ministry of Science and Technology, the Korea Science Promotion Foundation, and the Saemaul Technical Service Corps in cooperation with concerned government agencies, industry, academic circles, and the mass-communication media. The basic goal of this movement is a reorientation of the public's attitudes. This movement is in no way conceived as the special province of scientists and engineers, although this group can provide key support and resources in view of its pertinent talent and knowledge. It is not intended to focus attention solely on major scientific or technological advances, but rather a vast number of small advances made by people in every segment of society. In all aspects of the movement, primary emphasis is given to rationality, creativity, and workability.

It is necessary to develop a rational and scientific way of thinking among the Korean people and to discard passive attitudes and practices. As the first objective of this movement, they must comprehend the importance of science and technology in economic development and must develop the habit of applying elementary technical knowledge to everyday life. The second objective of this movement is to encourage everyone to acquire technical skills. Third, this movement is targeted at the strategic development and expansion of the economy will require increasing scientific and technical abilities. Korea, like the highly industrialised countries of the world, must have all the resources of science and technology effectively at its command. This can only be possible if the spirit of every individual is oriented towards the basic values and methodologies of science and technology.

International Technical Cooperation

In the present context, we can see clearly the necessity of international cooperation of increasing global interdependence, in science and technology for the benefit of both developed and developing countries. As one cannot expect development of science and technology in closed or isolated societies, modern science strongly demands active international interchange and mutual cooperation. From this point of view, the efficient scientific and technological cooperation with all other nations is perhaps the essential factor that may determine whether or not a country can sustain development past a certain level. It is more so when a developing country depends inevitably on the transfer of the science and technology.

The term, "technical cooperation", is often used when discussing the problems of developing countries, but its origin can be linked to the term, "economic and technical aid", that was used in the 1940s. In the 1960s, the term, "economic and technical aid" was changed to "technical cooperation" as the donor, who offered the aid, and the recipient, who received it, took more seriously the mutual cooperation and the supplemental nature of this cooperation in order to increase the effectiveness of technical aid.

From the early 1970s, two new dimensions in technical cooperation evolved: i) the country programme approach that was the heart of the new movement by the UNDP, being influenced by Jackson's report, and ii) mutual technical cooperation among developing countries (TCDC) which is accomplished through the new international economic order that was decided in the Sixth Special Session of the UN General Assembly.

The basic concept of the country programme approach was to orient the conventional, segmented and fragmental technical cooperation activities into recipient-sided cooperation which are more suited to the need of the recipient's country based on its long-term development plan. The establishment of a long range plan supported by the UNDP with links to individual nation's long term development plans enables developing countries to adapt technical cooperation to their needs. The truth is that there is strong desire to apply the country programme approach together with bilateral and multilateral technical cooperation to a large majority of developing countries. Recently in the international society, the basic spirit of mutual technical cooperation among developing countries, which is becoming more prominent, seeks to cast off the methods of technical cooperation that produced the former master-servant relationship by pursuing actions of the modern developing countries using the massive human and natural resources which they possess. All participants, without discriminating donor nations from recipients, are perceived with the dual roles of donor and recipient.

The process of technology-pace industrial development starts with the importation of foreign technology and proceeds to a final goal of technological self-reliance. The process needs a catalyst of technological cooperation to succeed.

The current international cooperation in technology has been more in form than in substance. The technological gap between advanced and less developed countries is too great and the understanding of mutual conditions and interests too meagre, resulting in technologies offered by donor countries proving to be inadequate or difficult to adapt to the capacities of recipient countries. Therefore, there is a need to design a new mechanism to render international technical cooperation, which is flexible and effective.

Furthermore, a review of the previous features of international cooperation in technology and economic development, indicates the following points to be considered for increasing the efficiency and effectiveness in utilising technical aids from advanced countries.

1)   The appropriateness of transferred technologies must be determined to suit specific conditions and basic needs of less developed countries (LDCs) under the "Country Programme" concept.

2)   It is recommended that techno-economic feasibility study must be performed prior to the formulation of projects, by the mobilisation of "in-country talents" supplemented by foreign experts.

3)   Utilising the absorbed and digested technology of the newly industrialised countries (NICs) with the accumulated experience of adaptation in their economic development might provide a helpful guide for the technology development of LDCs, minimising the trials and errors.

4)   Evaluation of project proposal and implementation plan offered by the donor countries should be done by the expert of third country which is not related to the supplier of equipment and services.

Under these circumstances, a tripartite cooperation system may be desirable to assist the LDCs to develop and adapt more effectively the technology needed for their economic development. The so-called "Tripartite Technical Cooperation" scheme consisting of the NICs, jointly with the advanced countries or relevant international organisations is believed to have a special merit for the technology development of LDCs to fill the gap between the recipient countries' capability of implementing the recommended development strategy and the donor countries' understanding of complex particular situations of the LDCs.

In this respect the Asian and Pacific Centre for Transfer of Technology (APCTT) of the UNESCAP has already initiated a few pioneering projects, such as the Technology Atlas project and the Technical Human Resources Development project. In these projects, APCTT and Korea together constitute the catalytic party for meaningful technical cooperation between the two conventional parties — donor and the recipient — and thus resulting in a tripartite cooperation.

Guidelines for Future Development

Once a country has reached the take-off stage in its development, some kind of boost may be necessary to ensure that the momentum of development is maintained; however, it is equally imperative that the industrialisation process be put back on a normal track as soon as possible. To this end, a unique industrial structure and direction for the industrialisation process must be established on the basis of the actual conditions in a given country.

In view of the Korea's conditions, it would seem that what is needed is not the blind pursuit of ever-increasing scale; instead, the industrial structure should stress the manufacture of products with a high added value stemming from the asset of a high quality labour force combined with a sparing use of natural resources and energy. In this way, it will be possible to develop strategically specialised industries which emphasize technology and brain-power. If kept small in scale, they may not be bogged down by huge infrastructures which in turn require immense capital investments. Thus, it should be possible to achieve stable prosperity while avoiding unnecessary competition in the international division of labour.

In making this argument, it is not my intention to minimise the significance of large-scale industries. Rather, the development of these basic industries should be pursued with some restraint in order to free resources necessary to support the minimum demands of the more specialised industries which produce high value-added products. After all, ensuring a stable supply of the major raw materials and semi-processed products is a prerequisite for a final product which will successfully compete in international markets. In other words, basic industries must be developed as a foundation for industrialisation, but the scale of these industries has to be determined in terms of what is appropriate at a given stage and in terms of the goals being pursued. Moreover, it is necessary to achieve a balance between quantitative and qualitative production as well as between facilities and technology.

It is, therefore, quite evident that, in a country like Korea with its limited territory, scarce natural resources and high population density, it is skill and brainpower which provide the base for national development. Consequently, while we are laboring to foster the needed manpower, we must also search for a technological development strategy which will employ this superior manpower within an industrial structure which makes the most of technology and brainpower. To place emphasis exclusively on those industries which require a huge infrastructure would mean prevailing instability with the concomitant loss of the opportunity to join the ranks of the devloped countries. Taking this perspective, it is clear that our efforts must be bent toward achieving that "small but advanced" type of development which is exemplified by such European countries as Switzerland, Belgium, the Netherlands, Denmark and Sweden.

To realise the technology-intensive industrial structure, it is necessary to 1) foster the development of strategically specialised industries; 2) optimise the social and industrial system; and 3) promote the quest for a high technology society[8]. As was pointed out earlier, strategically specialised industries will have to be characterised by a propensity to economise resources and create employment opportunities while requiring minimal capital investment and producing little environmental pollution. Furthermore, a country has to minimise its spending on social overhead to compete successfully with the fully industrialised nations and resource-rich countries. For this reason, optimisation of the social and industrial systems is a very important strategic goal.

Looking at the situation from another angle, science and technology, especially technology based on science developed during the second half of this century, has exerted a great global impact on mankind, resulting in the apex of the so-called "industrialised society". This impact has become greater and greater in recent years, leading to a societal transition. Such a societal change due to recent, rapid technological progress will transform our present society into a post-industrial or information society.

As can be easily observed, advanced countries are now tending to switch their industry-oriented development strategies to information-oriented ones. Developing countries are bound to be affected by this trend, and thus are required to turn their eyes toward a new information-oriented development strategy in the forseeable future. In order to meet this demand, they have to seek and adopt the basic concept of an information-oriented society. The first step towards this new concept is to establish a system for the settlement of an information-oriented society. Since the most pressing demands in this regard will be effective utilisation of computer systems, we have to become prepared to face the upcoming challenges.

Concluding Remarks

Now, let me make a few closing remarks. First, the notion that industrialisation in a developing country does not create enough employment to make it worthwhile, has limited validity. In the case of Korea, industry has provided at least one-third of all jobs created since 1962. Second, the idea that developing countries do not require high technology if they set their targets towards agriculture rather than industry is not completely valid either, especially when there is limited arable land, which must support a large population. Agriculture does require a considerable array of what might be called high technology, as in the development of high yield crops suitable for particular ecological and environmental conditions. Third, the argument that developing countries do not require domestic R&D, but require injections of technology from developed countries, is not sound. Domestic R&D is a prerequisite in enhancing the technological literacy necessary to make it possible to take advantage of foreign technologies. That is, industrialisation in a developing country with high selectivity in terms of sector, size, and degree of capital and technological intensities can bring about many essential improvements. The problems that need to be solved in a developing economy often require high technology, which can set development in motion to overcome insurmountable obstacles. Fourth, the developing countries have to become prepared to face the upcoming challenges in an age of technological change. It is asserted that the future society is one in which science and technology will determine the direction of socio-economic changes. The speed of such a societal change is accelerating and the area of impact is broadening, while the nature of change is even more sophisticated. Lastly, but probably most importantly, developing countries should not be swayed by the prevalent notion that the generation of technology in developing countries is not economically sound, if not impossible. On the contrary, I believe that there is a vast scope, and an absolute need in developing countries for the generation of technologies by those countries themselves or perhaps in collaboration with developed countries. To accomplish this end requires highly qualified people more than anything else; they are the only ones who can change the methods and the milieu.

References

1. Hyung Sup Choi, "The role of various stages of technology relevant to developing countries", Proceedings of Third Inter-Congress, The Pacific Science Association, Bali Indonesia, July 1977.
2. Franklin A. Long and Alexandra Oleson, *Appropriate Technology and Social Values — A Critical Appraisal* (Ballinger Publishing Company, Cambridge, Massachusetts, 1981).
3. Hyung Sup Choi, *Industrial Research in the Less Developed Countries*, Chapter 7 (Regional Center for Technology Transfer, ESCAP, Bangaloré, India, 1984).
4. Hyung Sup Choi, *Technology Development in Developing Countries*, Chapter 3 (Asian Productivity Organization, Tokyo, Japan, 1986).
5. Hyung Sup Choi, "Mobilization of financial resources for technology development", *Technology Forecasting and Social Change*, Vol. 31, No. 4, July 1987.
6. Hyung Sup Choi, *Bases for Science and Technology Promotion in Developing Countries*, Chapter 8 (Asian Productivity Organization, Tokyo, Japan 1983).
7. Center for Policy Alternatives at M.I.T., "A proposal for the establishment of the Korean technology development organization", Korea Institute of Science and Technology, Seoul, Korea, 1981.
8. Hyung Sup Choi, *Policy and Strategy for Science and Technology in Less Developed Countries*, Volume III, Chapter 7 (Korea Advanced Institute of Science and Technology, Seoul, Korea, 1981).

Appendix III

The International Centre for Theoretical Physics

Note by
Dr. A.M. Hamende
Scientific Information Officer

1. *Objectives*

The International Centre for Theoretical Physics is a multi-disciplinary institution for research and training-for-research. It was founded in 1964 and is part of the International Atomic Energy Agency (Vienna) and of UNESCO (Paris). Professor Abdus Salam, Nobel Laureate for Physics in 1979, suggested the Centre's creation. Its annual regular budget is 16.6 million US$, ninety percent of which comes from the Government of Italy while IAEA and UNESCO contribute for the remainder.

The ICTP was created in view of reaching several objectives which are:

(*a*)   to help in fostering the growth of advanced studies and research in physical and mathematical sciences, especially in the developing countries;

(*b*)   to provide an international forum for scientific contacts between scientists from all countries; and

(*c*)   to provide facilities to conduct original research to its visitors, associates and fellows, principally from developing countries.

2. *Range of Scientific Disciplines*

The programmes of the ICTP encompass a large spectrum of scientific disciplines from the most sophisticated subjects like the ultimate structure of elementary particles, down to more practical domains like remote sensing or telematics. Table I shows the range of the scientific disciplines which are or have been dealt with by the ICTP.

TABLE I

ICTP Fields of Research and Training-for-Research

Fundamental Physics	High Energy and Particle Physics Relativistic, Cosmology and Astrophysics
Physics of Condensed Matter	Condensed Matter Physics and related Atomic, Molecular Materials Science Surfaces and interfaces Liquids and statistical mechanics
Mathematics	Applicable Mathematics System Analysis, Mathematics of Development, Mathematics in Industry Algebra Geometry Topology Differential equations Analysis Mathematical Physics
Physics and Energy	Nuclear Physics and Fission Plasma Physics and Nuclear Fusion Non-conventional Energy (Solar, Wind and other)
Physics and Environment	Geophysics Soil Physics Climatology and Meteorology Physics of the Oceans Physics of Desertification Physics of the Atmosphere, Troposphere, Magnetosphere and Aeronomy
Physics Teaching	
Physics of the Living State	Neurophysics Biophysics Medical Physics
Applied Physics	Physics in Industry Microprocessors Communications Instrumentation Synchrotron Radiation Lasers Computational Physics Space Physics

3. Activities

The activities of the ICTP include several components, i.e. (a) research, (b) high-level training courses, (c) training in Italian laboratories, (d) external activities, (e) book and scientific equipment donation programme, and (f) training laboratories.

3.1 *Research* is carried out throughout the year in fundamental physics, physics of condensed matter and mathematics. A small permanent international staff, full professors of the Department of Theoretical Physics of the University of Trieste and of the International School for Advanced Studies (ISAS) and senior visiting scientists provide guidance to younger and less experienced physicists and mathematicians invited for periods ranging from one to twelve months and coming from all over the world. The ICTP also welcomes postdoctoral fellows for one or two years.

3.2 *High-level training courses and workshops, conferences and topical meetings* — Soon after the creation of the ICTP, it was realised that the scientific cadres of developing countries needed additional training for updating their research if they were to be competitive on the international scene. With this purpose in mind, high-level courses were instituted in condensed matter physics, nuclear physics, plasma physics and in mathematics during the first five years of existence of the ICTP. Many other subjects were added later (see Table I). High-level training courses have a duration of three to ten weeks and are attended by 70–90 participants mostly from developing countries.

Workshops, as a rule, differ from the courses in that they are more research-oriented and lectures are less numerous, leaving more time for discussion and research. In principle, they cater for already experienced scientists. In addition to them, the ICTP organises conferences and topical meetings on advanced subjects.

Between thirty and thirty-five courses, workshops, conference and other meetings are now held each year. As an illustration, Table II shows the programme for 1988.

3.3 *The programme for training and research in Italian laboratories* is the third component of the ICTP activities. It enables experimentalists from developing countries to participate in the research activities of laboratories belonging to universities or governmental and industrial

TABLE II

1988 Programme

College on Variational Analysis, 11 January – 5 February.
Spin and Polarization Dynamics in Nuclear and Particle Physics,
12 – 15 January.
Second School on Advanced Techniques of Computing in Physics,
18 January- 12 February.
Workshop on Functional-analytic Methods in Complex Analysis and
Applications to Partial Differential Equations, 8 – 19 February.
Workshop on Applied Nuclear Theory and Nuclear Technology Applications,
15 Feburary – 18 March.
Winter College on Laser Physics: Semiconductor Lasers and Integrated Optics,
22 February – 11 March.
Workshop on Optical Fibre Communication, 14 – 25 March.
Impact of Digital Microelectronics and Microprocessors on Particle Physics,
28 – 30 March.
Large-scale Structure and Motions of the Universe, 6 – 9 April.
Experimental Workshop on High-Temperature Superconductors, 11 – 22 April.
Spring School and Workshop on Superstrings, 11 – 22 April.
School on Non-accelerator Physics, 25 April – 5 May.
Spring College in Condensed Matter Physics: The Interaction of Atoms and
Molecules with Solid Surfaces, 25 April – 17 June.
Workshop on Modelling of the Atomospheric Flow Field, 16 – 20 May.
Course on Physical Climatology and Meteorology for Environmental
Applications, 23 May – 17 June.
Mini-Workshop on "Mechanisms of High-Temperature Superconductivity",
20 June – 29 July.
Summer School in High-Energy Physics and Cosmology, 27 June – 5 August.
Research Workshop in Condensed Matter, Atomic and Molecular Physics,
20 June – 30 September.
Unoccupied Electronic State, 21 – 24 June.
Computer Simulation Techniques for the Study of Microscopic Phenomena,
19 – 22 July.
Towards the Theoretical Understanding of High T_c Superconductors,
26 – 29 July.
Fifth Trieste Semiconductor Symposium (IUPAP); 4th International
Conference on Superlattices, Microstructures and Microdevices, 8 – 12 August.
Summer School on Dynamical Systems, 16 August – 9 September.
The Application of Lasers in Surface Science, 23 – 26 August.
Working Party on "Electron Transport in Small Systems",
29 August – 16 September.
Frontier Sources for Frontier Spectroscopy, 30 August – 2 September.
Summer Workshop on Dynamical Systems, 5 – 23 September.
Fourth Summer College in Biophysics, 12 September – 7 October.
Course on Ocean Waves and Tides, 26 September- 28 October.

TABLE II (Cont'd.)

College on Medical Physics, 10 October – 4 November.
First Autumn Workshop on Mathematical Ecology, 31 October – 18 November.
College on Neurophysics: "Development and Organisation of the Brain",
7 November – 2 December.
Workshop on Global Geophysical Informatics with Applications to Research
in Earthquake Predictions and Reduction of Seismic Risk,
15 November – 16 December.
College on Global Geometric and Topological Methods in Analysis,
21 November – 16 December.

institutions. Grants are given for periods ranging from a few months
to one year, depending on the conditions set by the host laboratory.

The programme was established in 1983, thanks to a grant from the
Dipartimento per la Cooperazione allo Sviluppo of the Italian Ministry of
Foreign Affairs and to the responsiveness of the Italian academic world,
the Consiglio Nazionale delle Ricerche (CNR), the Istituto Nazionale di
Fisica Nucleare (INFN) and the Ente Nazionale per le Energie Alternative
(ENEA) which provide grants for up to a total of 80/90 man/months per
year.

The network of host institutions include now more than two hundred
laboratories.

3.4   The fourth component is the *External Activities Programme*. The
rationale for this programme is that though the ICTP has been successful
in training for research several thousands of scientists in Trieste, little had
been done for building up communities of scientists in their own milieu.
It is true in the early seventies the ICTP had provided a modest financial
assistance to meetings, schools or conferences organised in the more
advanced developing countries. A more important effort had however
to be done if the investment was to have a lasting impact. Again, a grant
from the Dipartimento per la Cooperazione allo Sviluppo allowed the
ICTP to tackle the problem in a bigger way. An Office for External
Activities was created in 1985 and became operational in 1986. In a first
phase, the ICTP provided financial as well as intellectual assistance in five
programmes, i.e. *training activities, workshops, conferences, physics
and mathematics teaching, visiting scholars* who help research groups
wishing to embark on a new major project or introduce a new line of
research. It sponsored 79 activities in 33 countries.

In a second phase, the ICTP will give special attention to the formation of scientific networks and the establishment of centres.

3.5   Another important component of the ICTP is its *Book and Scientific Equipment Donation Programme.* The book donation programme was initiated at the ICTP some years ago to provide universities in developing countries with books, journals and proceedings. These publications are normally donated to the ICTP by individuals, libraries, publishing companies, international conferences and international organisations in industrialised countries for distribution among libraries in developing countries.

The ICTP receives unused surplus scientific equipment from laboratories such as CERN which are then shipped to institutions in developing countries once selected by a scientist from the recipient laboratory.

3.6   The *Microprocessor Laboratory* is the sixth component of the ICTP activities. It was created in 1985. It is jointly operated with the Istituto Nazionale di Fisica Nucleare of Italy and it is sponsored by the United Nations University (Tokyo, Japan). The laboratory helps scientists from developing countries to get acquainted with microprocessor technology and to develop projects of their own which they will use in their home countries. Fourteen such projects were carried out in 1987. It also provides technical support to other activities taking place at the ICTP or outside.

4.   *Networks — Associate and Federation Scheme*

4.1   One of the reasons for which the Centre was created, was to check the *brain-drain* which made the best scientists from the developing countries to emigrate to the advanced nations where they would find a congenial atmosphere for the progress of their research. Something had to be done to break the isolation of the scientists who had chosen to remain in their countries — an isolation due to the lack of opportunities to discuss with their colleagues or to attend international conferences and the nearly total absence of scientific journals and books in their libraries. The response of the ICTP to this necessity was the creation in 1964 of the *Associate Membership Scheme.* Associate Members are scientists from and working in developing countries who are appointed for a period of six years during which they are entitled to three research visits to the

ICTP. Each of such visits should not exceed three months but should last more than six weeks. During their stay at the ICTP, Associate Members work either independently or in collaboration with other scientists in residence.

In 1987, the list of appointed Associate Members included 319 scientists from 62 nations.

Some of the former Associates who have acquired international reputation or have distinguished themselves as *entrepreneurs* in their home countries in research or education, may be appointed as *Senior Associates* for six years. A fund of 4,000 US$ is reserved for each of them from where they may draw for their travel and subsistence at the ICTP. In 1987, the ICTP list of Senior Associates included 36 names from 17 Member States.

For younger scientists, the ICTP has set up the *Junior Associateship,* a scheme which is mainly meant to help those working in institutions in developing countries with poor library facilities. Junior Associates are selected among participants in courses or workshops and during their four-year appointment they are entitled to order books through the ICTP or subscribe to scientific journals for their home libraries up to 350 US$ each year. In 1987, the ICTP counted 122 Junior Associates mostly from Asia and Africa.

4.2   In 1964 also, the ICTP devised a scheme for relatively nearby universities from Austria, Yugoslavia and Hungary to have regular access to its activities. This was the *Federation Scheme.* Again, this scheme proved to be the genuine response to a widespread need. Federated institution are entitled to send junior scientists to the ICTP for a total number of days which may vary from 40 to 120, depending on the geo-graphical location of the institution. The subsistence expenses of the visitor are borne by the ICTP, while, as a rule, the federated institution bears the cost of travel. The ICTP may however contribute partially to the travel costs. There are also special arrangements with the Kuwait Foundation for Science, Kuwait University (for nationals from Arabic and Islamic countries), the Islamic Republic of Iran, the University of Qatar, the Government of Argentina, the Brazilian National Research Council (CNPq) and the Arab Bureau of Education in Saudi Arabia which contribute fixed sums each year in support of their nationals. Three hundred and thirty-four institutions are federated with the ICTP this year (see Table III).

TABLE III

Federation agreements in 1988

Geographical area	Number of agreements	Countries
Africa	103	Algeria, Benin, Burundi, Cameroon, Congo, Côte d'Ivoire, Egypt, Ethiopia, Gabon, Ghana, Guinea, Kenya, Liberia, Libya, Madagascar, Mali, Mauritania, Morocco, Nigeria, Rwanda, Senegal, Sierra Leone, Somalia, Sudan, Swaziland, Tunisia, Zaire, Zambia
Asia	138	Bangladesh, P.R. China, India, Iran, Iraq, Israel, Jordan, Korea Rep., Kuwait, Lebanon, Malaysia, Mongolia, Nepal, Pakistan, Philippines, Qatar, Saudi Arabia, Sri Lanka, Syria, Thailand, Vietnam, Israel-West Bank, Yemen P.D.R., Yemen A.R.
Europe	61	Austria, Bulgaria, Czechoslovakia, German Dem. Rep., Greece, Hungary, Poland, Portugal, Romania, Spain, Turkey, USSR, Yugoslavia
North & Central America	17	Cuba, Honduras, Jamaica, Mexico, Puerto Rico, Trinidad
South America	15	Argentina, Brazil, Chile, Ecuador, Guyana

In 1987, 606 scientists came to Trieste under the Federation Scheme for a total of 509 man/months.

Tables IV and V show summary statistics on the 1987 programme and the period 1970–1987 respectively.

5. *Facilities*

The ICTP programmes are carried out in several buildings located in Miramare, 7 km from the town. The main building — 3000 sq. m — houses the Library (30,000 books and 600 scientific periodicals), a 280-seat lecture hall, a 100-seat lecture room, offices for scientists and part of the secretariat, the computer facilities and a cafeteria. A new building, equal in size to the main one and adjoining it, is being completed and will be inaugurated in October 1989. It will provide more room for the Library and two additional lecture rooms. A second building, named after Galileo Galilei, is the first guest house of the ICTP. It has 40 double bedrooms and a lecture room for sixty people.

To cope with the dramatic increase in its activities since 1983, the ICTP rents a hotel, now name Adriatico Guest House, with 172 beds, two lecture halls and several meeting rooms as well as offices for scientists and the secretariat. The high-temperature superconductivity laboratory, part of the library collections and the printing shop are housed in that building.

The Microprocessor Laboratory is housed in a former elementary school rented to the Municipality of Trieste. All buildings are within walking distance of each other.

TABLE IV

Statistical summary on activities held at and outside the ICTP in 1987

Activities	Number of Visitors			Number of Man/months		
	Dev.	Ind.	Total	Dev.	Ind.	Total
1. At the ICTP: (a) Research:						
Total	464	178	642	1137.05	229.05	1366.10
(b) Training for research (courses, workshops and conferences)						
Total	1886	1418	3304	1313.97	419.09	1733.06
2. Outside activities: (a) Italian laboratories	108	–	108	730.10	–	730.10
(b) 3 major activities in dev. countries[1]	121	11	132	66.15	3.87	70.02
Total	229	11	240	796.25	3.87	800.12
GRAND TOTAL[2]	2579	1607	4186	3247.27	652.01	3899.28

Figures on research include long- and short-term scientists as well as Associate Members, some scientists from Federated Institutes and seminar lecturers.

[1] As regards to the activities held outside the ICTP, the 79 courses sponsored but not organised by ICTP are not included.

[2] Actual numbers of visits (some scientists took part in more than one activity). The actual number of scientists is

1987						
	2171	1529	3700	3247.27	652.01	3899.28

TABLE V

Overview of ICTP activities: scientists and scientific preprints since 1970

Year	No. of Scientists		No. of Man/months		No. of preprints		No. of Member States represented	
	Total	from Dev. countries	Total	from Dev. countries	Total	from Dev. countries	Total	from Dev. countries
1970	582	218	864	389	154	81	53	35
1971	885	338	533	323	160	125	68	37
1972	888	407	1214	697	161	108	71	53
1973	878	352	1258	738	194	142	64	47
1974	862	329	854	588	141	104	65	48
1975	928	399	1018	664	172	141	82	62
1976	962	387	820	563	127	102	71	54
1977	1331	644	1080	776	158	108	92	71
1978	1327	655	1079	791	160	116	91	70
1979	1470	619	961	608	167	108	90	68
1980	1461	615	1296	991	183	148	93	72
1981	1933	960	1533	1148	239	159	90	70
1982	2139	978	1749	1278	236	179	83	63
1983	2188	1160	1810	1397	238	186	99	79
1984	2082	1086	1870	1425	249	210	96	76
1985	2720	1671	2669	2179	313	266	109	89
1986	3651	2180	3820	3149	401	323	109	86
1987	3700	2171	3899	3247	421	358	120	91

Appendix IV

The Third World Academy of Sciences

Note by
Dr. M.H.A. Hassan
Executive Secretary

1. *Foundation of the Third World Academy of Sciences*

The idea of setting up a Third World Academy of Sciences was conceived by Professor Abdus Salam of Pakistan on the occasion of a general meeting of the Pontifical Academy of Sciences of the Vatican in Rome on 6 October, 1981. After discussing the idea with Members of the Pontifical Academy from the Third World, a Memorandum was drawn up in support of the initiative, with the aim of exploring the possibility of creating such an organisation.

The Foundation Meeting of the Academy took place in Duino Castle and at the University of Trieste, Italy, during the period 10 – 11 November, 1983, under the sponsorship of the Trieste International Foundation.

On 5 July, 1985, the Third World Academy of Sciences (TWAS) was officially launched by the United Nations Secretary General, Mr. J. Perez de Cuellar, in Trieste, Italy, on the occasion of the opening of a Conference on "South-South and South-North Cooperation in Sciences", which was organised by the Academy. The Conference, which marked the Academy's real birth, was attended by 250 delegates, representing Academies and Research Councils from the South and the North, and representatives of International Organisations.

The Academy is the first international forum to unite distinguished men and women of science from the Third World, with the objective of promoting basic and applied sciences in the Third World through nurturing excellence and fostering future generations of promising scientists from developing countries.

The Academy is a non-governmental, non-political and non-profit making organisation, whose objectives are to recognise and promote high calibre scientific research carried out by scientists from developing countries, to facilitate their mutual contacts, strengthen their scientific research work and foster it for the development of the Third World and in the service of mankind.

The Academy became a Scientific Associate of the International Council of Scientific Unions (ICSU) in 1984 and was granted official NGO status by the United Nations Economic and Social Council in 1985. It is presently located on the premises of the International Centre for Theoretical Physics at Miramare, Trieste, Italy, a Centre sponsored by the International Atomic Energy Agency (IAEA) and the United Nations Educational, Scientific and Cultural Organisation (UNESCO).

2.   *Objectives*

The principal objectives of the Academy are:

(*a*)   To recognise and support excellence in scientific research performed by individual scientists from the Third World;

(*b*)   To provide promising scientists in the developing countries of the South with the conditions necessary for the advancement of their work;

(*c*)   To promote contacts between research workers in developing countries of the South among themselves and with the world scientific community;

(*d*)   To provide information on and support for scientific awareness and understanding in the Third World;

(*e*)   To encourage scientific research on major Third World problems.

3.   *The Membership*

The Membership of the Academy consists of Fellows, Associate Fellows and Corresponding Fellows. Fellows are elected from among scientists of developing countries who have made outstanding contributions to their respective field of science. Associate Fellows are elected from among citizens of industrialised countries who either have their origin in developing countries or have distinguished themselves in the context of Third World science, and who have attained highest international standards. Corresponding Fellows are elected from among promising scientists of developing countries.

At present, there are 96 Fellows, coming from 42 developing countries, 39 Associate Fellows and 3 Corresponding Fellows. Out of these 138 Members, 78 are also members of nine of the world's prestigious Academies and 10 are Nobel Laureates of Third World origin.

4.   *Promoting South-South and South-North Collaboration*

South-South Fellowships

The aim of the Fellowship Programme is to facilitate and promote mutual contacts between research scientists in the Third World and to further relations between their scientific institutions. Fellowships covering travel costs are awarded for visits to scientific institutions within the Third World for a minimum period of six weeks. Living expenses are borne by local sources. The Fellowships are offered to nationals of developing countries, normally with research experience and with positions in universities or research institutions in those countries.

Governments and scientific organisations in Argentina, Brazil, Chile, China, Colombia, Ghana, India, Iran, Kenya, Madagascar, Mexico, Vietnam and Zaire have so far agreed to provide local hospitality for a total of over 250 annual visits under this programme.

TWAS/ICIPE Associateship Scheme

The Academy and the International Centre of Insect Physiology and Ecology (ICIPE) in Nairobi, Kenya, have instituted a joint associateship programme for visits to ICIPE.

Associates are selected from among senior researchers in various aspects of insect science, who afe nationals of developing countries and are working and living in those countries. They are appointed for a fixed period of six years during which they are entitled to visit ICIPE three times, each for a minimum period of three weeks and a maximum of three months. The Academy covers travel expenses, while ICIPE provides local support. A number of 7 Associates have so far been appointed.

The Academy plans to institute a similar programme with other centres of excellence in the Third World.

Support for Scientific Meetings

The Academy encourages the organisation of scientific meetings in biological, chemical and geological sciences in Third World countries by providing financial support in the form of travel grants for principal speakers from abroad and/or participants from the region. Special consideration is given to those meetings which are likely to benefit the scientific community in the Third World and to promote regional and international cooperation in developing science and its applications to the problems of the Third World.

Fellowships for Research and Training in Italian Laboratories

The Academy supports visits by Third World scientists to laboratories in Italy active in the fields of biological, chemical and geological science, for the purpose of pursuing research or training. The Fellowship covers living expenses in Italy for a period of six months to one year.

ICSU/TWAS Programme of Lecturers

The International Council of Scientific Unions (ICSU) and the Academy have recently launched a joint Lectureship Programme. The general objective of this programme is to provide scientists in developing countries with the opportunity for discussions and scientific collaboration with colleagues from other countries who have made outstanding contributions to the advancement of science.

Action on Drought, Desertification and Food Deficit (DDFD) in Africa

The Project is a joint venture initiated by the Academy, the African Academy of Sciences (AAS) and the National Academy of Sciences of the USA. It is supported by the Academy, the World Bank and the MacArthur Foundation. The aim of the project is to apply Science and Technology in overcoming Drought, Desertification and Food Deficit in Africa. A Task Force of internationally renowned scientists was set up in July 1985, and its first meeting was held in Trieste in December 1985. The DDFD Project was officially launched at an international conference in Nairobi in June 1986. An international fact finding mission on biotechnology and long-term soil and water management was organised by the three academies, which visited a number of African countries during 1987.

TWAS Awards

The Academy awards prizes to individual scientists from developing countries who, in the opinion of the Council, have made outstanding contributions to the advancement of science. Consideration is given to prove achievements judged particularly by their national and international impact.

5.   *Relations with Third World and International Scientific Organisations*

In recognition of the importance of establishing close ties with leading scientific organisations in the Third World, the Academy has signed agreements of cooperation with over 60 Academies and Research Councils in 46 developing countries. The Academy has also signed a Memorandum of Understanding on Scientific Cooperation with the African Academy of Sciences (AAS), the Latin American Academy of Sciences (ACAL) and the Federation of Asian Scientific Academies and Societies (FASAS). Through this Memorandum, the Academy provides some financial assistance to each of the three regional Academies in support of their regional activities.

In the same spirit, the Academy has initiated the establishment of a Network of Scientific Organisations in the Third World. Approximately eighty Academies, Research Councils and Ministries of Science, Technology and Higher Education have so far agreed to join the Network. In addition to strengthening cooperation between its members, the Network will further the contribution of the South to global science projects and areas of today's frontier science and technology which are most likely to have a strong impact upon the economic and social development of the Third World.

6.   *Finance*

Support for South Fellowship Programme comes from Argentina, Brazil, Chile, China, Colombia, Ghana, India, Iran, Kenya, Madagascar, Mexico, Vietnam and Zaire.

Programmes and projects of the Academy, for 1988, are financed by the Kuwait Foundation for the Advancement of Sciences (KFAS), the OPEC Fund for International Development, the Consiglio Nazionale delle Ricerche (CNR), the Government of Sri Lanka and, generously, by the Italian Government (through the Direzione Generale per la Cooperazione allo Sviluppo) and the Canadian International Development Agency (CIDA).

Appendix V

The Third World Network of Scientific Organisations (TWNSO)

1. *Foundation of the Third World Network of Scientific Organisations*

During 1986, the Third World Academy of Sciences (TWAS) invited several national Science Academies and Research Councils in developing countries, with which TWAS had already established close links, to sign an agreement envisaging the strengthening of cooperative links between TWAS and these scientific bodies. An invitation was extended to some thirty Science Academies and Research Councils in the South, of whom twenty-three responded positively.

Encouraged by this positive response, the President of the Third World Academy of Sciences proposed at the opening of the TWAS Second General Conference in Beijing, China, in September 1987, that the TWAS initiative should be extended in scope and that a "Network" linking Science Academies, Research Councils and other leading scientific organisations in the South be formed with the full participation of Ministries of Science and Technology and Higher Education, in order to enhance communication and collaboration among them and to increase the effectiveness of science in the South. The South can then collectively have substantial input in frontier science programmes (such as biological studies of the human genome, space research and nuclear fusion), which may have particular significance to development in the Third World. The President's proposal was strongly supported by the representative of the Italian government attending the Conference, who in his opening address to the Conference pledged financial support for the "Network".

The "Network" proposal was subsequently discussed and endorsed by the participants of the Conference, among whom were Ministers of Science and Technology and Higher Education. Presidents of Science Academies and Chairpersons of Research Councils from more than 40 Third World countries. It was decided at the Meeting that an Ad-Hoc Committee under the chairmanship of Professor J. Aminu of Nigeria and the membership of Professors M.G.K. Menon of India and M. Roche of Venezuela, be set up to explore further the institution of the Network.

Academies, Research Councils and Ministries of Science, Technology and Higher Education in the South have subsequently been invited to join the Network as Members. It is hoped that this will make it possible for these bodies to apply the necessary political pressure, both within and outside the South, in support of the development of Science and Technology in the Third World. Scientific and technological organisations of expatriates of Third World origin living in industrialised countries have also been invited to become Associate Members of the Network. Ninety scientific organisations from fifty-nine Third World countries have agreed to become Members of the Network and four Expatriate Organisations have accepted to become Associate Members.

The Foundation Meeting of the Network was convened by the Third World Academy of Sciences at its Headquarters from 4 to 6 October 1988 and was attended by over ninety participants including 15 Ministers of Science and Technology and Higher Education, 12 Presidents of Academies and 17 Chairpersons of Scientific Research Councils from 36 Third World countries. The Meeting approved the Statutes for the Network and elected a President and an Executive Board for an interim period of one year. The Meeting also adopted a declaration referred to as "The Trieste Declaration on Science and Technology as an Instrument of Development in the South".

The office of the Network has been established within the Third World Academy of Sciences' Secretariat. In order to facilitate the operation of the Network, it has been decided to set up four regional offices at the location of the four Vice-Presidents, i.e. in Nigeria, Tunisia, Mexico and Malaysia.

The second meeting of the Network will be held in Bogota, Colombia, during the period 16–20 October 1989, in conjunction with the Third General Conference of the Third World Academy of Sciences.

The Network is a non-governmental, non-profit making and autonomous scientific organisation. Its objectives and Membership structure are set out below.

2. *Objectives of the Network*

The general objective of the Network is to promote South-South and South-North Cooperation in the development and application of Science and Technology in the Third World.

This may be achieved by:

(*a*) Furthering the contribution of the South to global projects of science *(such as Man and the Biosphere programme of UNESCO and the International Geosphere Biosphere Programme of ICSU);*

(*b*) Furthering the contribution of the South to areas of today's frontier science and technology which are most likely to have a strong impact upon the economic and social development of the Third World *(such as space science and technology, thermo-nuclear fusion, high technology and biotechnology);*

(*c*) South-South collaboration: Promoting and strengthening co-operation between Academies, Research Councils and scientific organisations in the Third World and enhancing their role in the development of the Third World, including information sharing through the setting-up of data banks.

(*d*) South-North collaboration: Furthering relations between scientific institutions and organisations in the South and with their counter-parts in the North through the development of bilateral links and exchange programmes.

(*e*) Encouraging Third World Governments to take appropriate political action to develop their scientific enterprise through self-reliance and adequate allocation of resources.

(*f*) Undertaking any other activities that will further the objectives of the Network.

3. *Membership of the Network*

The *Membership* of the Network is open to the following organisations:

1. Ministries responsible for Science and Technology in the Third World;

2. Ministries responsible for Higher Education in the Third World;

3. National Science Academies in the Third World;

4. National Research Councils responsible for Science and Technology in the Third World;

5.   Major Science and Technology foundations in the Third World;

6.   Regional, inter-governmental and international organisations and
      centres promoting science and technology in the Third World;

7.   Any other organisations promoting Science and Technology in the
      Third World which, in the judgement of the Council, are eligible
      for membership.

The *Associate Membership* of the Network is open to scientific and
technological organisations of Third World origin of scientists working
in industrialised countries, to Ministries and organisations corresponding
to 1–6 above, provided that each case is treated on its own merit.

Membership will be given to qualified organisations that request it
in writing and accept the Statutes of the Network.

Founding Member status will be given to organisations that have
already requested membership and which have not withdrawn their
request by 31 December 1988.

Members of the Network:

1.   National Academy of Exact, Physical and Natural Sciences
      Avenida Alvear 1711, 4°
      1014 Buenos Aires
      Argentina

2.   National Council of Scientific and Technical Research (CONICET)
      Rivadavia 1917
      1033 Buenos Aires
      Argentina

3.   Bangladesh Academy of Sciences
      3/8 Asad Ave
      Mohammadpur
      Dhaka 7
      Bangladesh

4.   Ministry of Secondary and Higher Education
      Cotonou
      Benin

5. Bolivian National Academy of Sciences
   Av. 16 de Julio No. 1732
   P.O. Box 5829
   La Paz
   Bolivia

6. Brazilian Academy of Sciences
   Rua Anfilófio de Carvalho 29
   C.P. 229
   20000 Rio de Janeiro, RJ
   Brazil

7. National Council for Scientific and Technological Development
   Av. W-3 Norte Quadra 507 Bl. B
   707040 Brasilia
   Brazil

8. Ministère de l'Enseignement Superieur, de l'Informatique et de la
   Recherche Scientifique
   c/o Central Post Office
   BP 1457
   Yaoundé
   Cameroon

9. Chilean Academy of Sciences
   Almirante Montt 453
   Clasificador 1349
   Santiago
   Chile

10. Ministry of Public Education
    Avda Libertador B
    O'Higgins 1371
    Santiago
    Chile

11. Academia Sinica
    Beijing
    The People's Repulic of China

12. State Scientific and Technological Commission
     of the People's Republic of China
     52 Sanlihe
     Fuxingmenwai
     Beijing
     China

13. Colombian Academy of Exact, Physical and Natural Sciences
     Carrera 3.A No. 17-34 Piso 30
     Apdo. Aereo 44.763
     Bogotá 1, DE
     Colombia

14. Fondo Colombiano de Investigaciones Científicas y Proyectos
     Especiales Francisco José de Caldas (Colciencias)
     Transversal 9A No. 133
     Apdo. Aereo 051580
     Bogotá
     Colombia

15. Consejo Nacional de Investigaciones Científicas y Tecnológicas
     (CONICIT)
     Apartado 10318
     San José
     Costa Rica

16. Ministry of Higher Education and Scientific Research
     4, Sharia Ibrahim Nagiv
     Garden City
     Cairo
     Egypt

17. National Research Centre
     Al-Tahrir St.
     Dokki
     Cairo
     Egypt

18. Ethiopian Science and Technology Commission
    Addis Ababa
    Ethiopia

19. Centre National de la Recherche Scientifique et Technologique
    (CENAREST)
    B.P. 842
    Libreville
    Gabon

20. Council for Scientific and Industrial Research
    P.O. Box M.32
    Accra
    Ghana

21. Ghana Academy of Arts and Sciences
    P.O. Box M.32
    Accra
    Ghana

22. Ministry of Industries, Science and Technology
    POB M39
    Accra
    Ghana

23. Academy of Medical, Physical and Natural Sciences
    Apdo Postal 569
    Guatemala, C.A.
    Guatemala

24. Instituto Nacional de Estudios e Pesquisas (INEP)
    Caixa Postal 112
    Bissau
    Guinea-Bissau

25. National Science Research Council
     Institute of Applied Science and Technology
     University Campus, Turkeyen
     P.O. Box 101050
     Greater Georgetown, S.A.
     Guyana

26. Academia Hondureña de la Lengua
     Tegucigalpa, D.C.
     Honduras

27. Council of Scientific and Industrial Research (CSIR)
     Rafi Marg
     New Delhi 110 001
     India

28. Federation of Asian Scientific Academies and Societies (FASAS)
     c/o Indian National Science Academy (INSA)
     Bahadur Zafar Marg
     New Delhi 110 002
     India

29. Indonesian National Research Council
     Lembaga Ilmu Pengetahuan Indonesia (LIPI)
     Jl. Jenderal Gatot Subroto No. 10
     P.O. Box 250
     Jakarta Selatan
     Indonesia

30. Atomic Energy Organization of Iran
     P.O. Box 14155-1339
     Teheran
     Iran

31. Ministry of Higher Education and Culture
     Tehran
     Iran

32. Federation of Arab Scientific Research Councils (FASRC)
    P.O. Box 13027
    Baghdad
    Iraq

33. Third World Academy of Sciences (TWAS)
    c/o International Centre for Theoretical Physics (ICTP)
    P.O. Box 586
    Strada Costiera 11
    34136 Trieste
    Italy

34. Scientific Research Council
    P.O. Box 350
    Kingston 6
    Jamaica

35. Ministry of Higher Education
    POB 1646
    Amman
    Jordan

36. Royal Scientific Society
    P.O. Box 925819
    Amman
    Jordan

37. African Academy of Sciences (AAS)
    P.O. Box 14798
    Nairobi
    Kenya

38. Kenya National Academy of Sciences
    P.O. Box 47288
    Nairobi
    Kenya

39. National Council for Science and Technology
    P.O. Box 30623
    Nairobi
    Kenya

40. Ministry of Science and Technology (MOST)
    77-6 Saechongro, Chongro-ku
    Seoul 100
    Korea

41. Academy of Sciences of D.P.R.K.
    Sosong District
    Pyongyang
    Democratic People's Republic of Korea

42. Kuwait Foundation for the Advancement of Sciences (KFAS)
    P.O. Box 25263
    Safat 13113
    Kuwait

43. General People's Committee for Education and Scientific Research
    Tripoli
    Libya

44. Académie Malgache
    BP 6217
    Tsimbazaza
    Antananarivo
    Madagascar

45. Ministry of Scientific and Technological Research for Development
    Box 694
    Anosy
    Antananarivo 101
    Madagascar

46. National Research Council of Malawi
    P.O. Box 902
    Blantyre
    Malawi

47. Malaysian Scientific Association
    c/o International Relations Committee
    38 Flat Melor
    Jalan Loke Yew
    55100 Kuala Lumpur
    Malaysia

48. Ministry of Science, Technology and Environment
    14th floor, MUI
    Plaza Wisma Sime Darby
    Jalan Raja Laut
    50662 Kuala Lumpur
    Malaysia

49. Centre National de la Recherche Scientifique et Technologique
    BP 3052
    Bamako
    Mali

50. Academia de la Investigacion Cientifica
    Av. San Jeronimo No. 260
    Jardines del Pedregal
    Mexico D.F. 04500
    Mexico

51. Académie du Royaume du Maroc
    Route des Zaërs
    B.P. 1380
    Rabat
    Morocco

52.   ISESCO
      Agdal
      Rabat
      Morocco

53.   National Council for Science and Technology
      Kirtipur
      Kathmandu
      Nepal

54.   Royal Nepal Academy of Science and Technology
      P.O. Box 3323
      1/48 Kopundole
      Lalitpur
      Nepal

55.   Federal Ministry of Education
      Ahmadu Bello Way
      Victoria Island
      Lagos
      Nigeria

56.   Federal Ministry of Science and Technology
      9 Kofo Abayomi Street
      Victoria Island
      Lagos
      Nigeria

57.   The Nigerian Academy of Sciences
      c/o Department of Computer Science
      Faculty of Science
      University of Lagos
      P.M.B. 1004, University of Lagos Post Office
      Lagos
      Nigeria

58. Ministry for Science, Technology and Petroleum and Natural
     Resources (MOST)
     Islamabad
     Pakistan

59. Pakistan Academy of Sciences
     5, Constitution Avenue
     Sector G/5
     Islamabad
     Pakistan

60. Pakistan Council for Science and Technology
     37, School Road, F-7/1
     Islamabad
     Pakistan

61. Consejo Nacional de Ciencia y Technologia (CONCYTEC)
     Camilo Carrillo 114
     Apdo Postal 1984, 9o piso
     Lima 11
     Peru

62. National Peruvian Academy of Sciences
     c/o Instituto Nacional de Cultura
     P.O. Box 5247
     Lima
     Peru

63. National Academy of Science and Technology (NAST)
     P.O. Box 3596 Manila
     Bicutan, Taguig
     Metro Manila
     Philippines

64. Academia de Artes y Ciencias de Puerto Rico
     Apartado 3308
     Hato Rey, P.R. 00919
     Puerto Rico

65.   Ministry of Higher Education and Scientific Research
      BP 624
      Kigali
      Rwanda

66.   Arab Bureau of Education for the Gulf States
      P.O. Box 3908
      Riyadh
      Saudi Arabia

67.   Centre National de Documentation Scientifique et Technique
      BP 411
      61 Boulevard Pinet-Laprade
      Dakar
      Senegal

68.   Ministry of Scientific and Technical Research
      BP 411
      61 Boulevard Pinet-Laprade
      Dakar
      Senegal

69.   Singapore National Academy of Science
      c/o Chemistry Department
      National University of Singapore
      Lower Kent Ridge Road
      Singapore 0511

70.   National Academy of Sciences of Sri Lanka
      120/10 Wijerama Mawatha
      Colombo 7
      Sri Lanka

71.   Natural Resources, Energy and Science Authority of Sri Lanka
      47/5 Maitland Place
      Colombo 7
      Sri Lanka

72. Ministry of Higher Education and Scientific Research
     Khartoum
     Sudan

73. National Council for Research
     P.O. Box 2404
     Khartoum
     Sudan

74. Centre for Scientific Research
     BP 4470
     Damascus
     Syria

75. Ministry of Higher Education
     Damascus
     Syria

76. Tanzania Commission for Science and Technology
     P.O. Box 4302
     Kivukoni Front
     Dar-es-Salaam
     Tanzania

77. National Research Council
     196 Phahonyothin Rd.
     Bangkhen
     10900 Bangkok
     Thailand

78. Institut National de la Recherche Scientifique
     B.P. 2240
     Lomé
     Togo

79. Ministry of Education
     POB 61
     Nuku'alofa

80. Ministry of Scientific Research and Higher Education
     1 Rue de Beja
     Tunis
     Tunisia

81. National Institute for Scientific and Technical Research
     Bordj-Cedria
     B.P. 95
     Hamman-Lif
     Tunisia

82. Scientific and Technical Research Council of Turkey (TÜBITAK)
     Emek Is Hani Kat 16
     06650 Kizilay
     Ankara
     Turkey

83. Uganda National Academy of Science and Technology
     P.O. Box 16606
     Kampala
     Uganda

84. Consejo Nacional de Investigaciones Cientificas
     y Tecnicas (CONICYT)
     Sarandi 444, Piso 4
     Casilla de Correo 1869
     Montevideo
     Uruguay

85. Academia de las Ciencias Fisicas, Matematicas y Naturales
     Palacio de las Academias, Avenida Universidad
     Bolsa a San Francisco
     Apdo, de Correos 1421
     Caracas 1010-A
     Venezuela

86. Latin American Academy of Sciences (ACAL)
     c/o Instiudo Internacional de Estudios Avanzados
     Centro de Biociencias, Parque Central
     Apartado 17606
     Caracas 1015-A
     Venezuela

87. National Council for Scientific and Technological Research
     (CONICYT)
     Apartado 70617
     Los Ruices
     Caracas
     Venezuela

88. National Centre for Scientific Research of Vietnam
     Nghia Do
     Tu Liem
     Hanoi
     Vietnam

89. Ministry of Higher Education
     Haile Selassie Ave
     POB 50464
     Lusaka
     Zambia

90. National Council for Scientific Research
     P.O. Box CH. 158
     Chelston
     Lusaka
     Zambia

Associate Members of the Network:

1. Canadians for the Promotion of Research and Education
    in Pakistan (CANPREP)
    35 Claudet Crescent
    Ottawa, Ontario
    KIG 4R4
    Canada

2. World Social Prospects Association (AMPS)
    5 Route de Morillons
    CH-1218 Grand Saconnex
    Geneva
    Switzerland

3. Third World Scientific and Technological Development Forum
    (STD Forum)
    9 Daleview Avenue
    Glasgow G12 OHE
    United Kingdom

4. Association of Scientists of Indian Origin in America, Inc. (ASIOA)
    c/o Department of Neurology
    University of Mississippi Medical Center
    Jackson, MS 39216-4505
    USA

Appendix VI

Basic Sciences

Here is the text of a letter from a Nobel Laureate, Max F. Perutz, who was responsible for the determination of the structure of hemoglobin. Notice the difference of approach between a scientist and a bureaucrat so far as the running of the enterprise of sciences is concerned. Writing to Professor Denis Noble, Secretary of the British Society for Basic Sciences, Max Perutz says:

"I am deeply disturbed about the ABRC Report on "A strategy for the Science Base" because it threatens to undermine the greatest strength of British science — its originality. Most original ideas spring from young people. There is no reason to believe that young British people are intrinsically more original than German or French ones, but British Universities have given them more scope. In Germany and France a young scientist has to work at his professor's bidding. Here a gifted young scientist who wins a university position is independent and can apply for a research grant to pursue his own ideas. If they are good, he is likely to receive support, no matter whether he is in Oxford or Aberystwyth. The ABRC Report first pays lip service to these ideas (para 1.19, p. 4) and then recommends that the policy that has made them flourish be discontinued (para 1.25, p. 5).

"Some of the most original recent ideas have sprung from small universities. Liquid crytals from Gray in Hull; hypervariable human DNA from Jeffreys at Leicester; vibrations spectroscopy of surfaces from Sheppard at East Anglia; metal ałkoxides and metal alkyls, keys to new gallium arsenide technology for solar panels, from Bradley at Queen Mary College; Multan programme for solving crystal structures automatically by direct methods from Woolfson in York; magic angle spinning NMR by Andrew in Banglor; solid state catalysts and a new order of complexity of crystalline solids from John Thomas at Abserystwyth; and many more.

These developments have great industrial potential, but they all arose from basic curiosity-motivated research.

"Under the three-tier system proposed in the ABRC Report, only those scientists finding appointments in R universities could be sure of receiving independent Research Council support. Those in the T would get "a small margin" (para 1.36, p. 8) to facilitate research in association with more advanced research centres, and those in type X may also be told that they can only have shared facilities in a university other than their own. Some scientists now perform experiments in powerful central facilities and work out the results at home, but for most people in the chemical and biological sciences experimental work is continuous and must be done at their home university.

"The likely impact of the three-tier scheme on young people can be gauged from the present distribution of 671 'New Blood' lecturers appointed in natural science and engineering. I have made a tentative list of the likely top 15 universities on the basis of their UGC ratings. According to this only about half the 'New Blood' lecturers now work in universities likely to be R's. The research careers of the other half will be in jeopardy: active research workers in type T universities could either apply for posts in the new interdisciplinary centres or emigrate: those in type X universities would be uncertain where they stand. Some may be lucky and get grants, others may be told that their institution does not qualify for research support. Those who find positions in one of the new interdisciplinary research centres envisaged by the Report would be subject to "more positive management than is generally the norm for university research at present", under a Director appointed by the Research Council in question. They might find better facilities there than at a small university but they will not have the same independence.

"The 'New Blood' lecturers comprise the best of the young generation of scientists. Any decision to deny a major, arbitrarily defined proportion of them substantial research support would be tragic for the individuals and a loss to Britain. Moreover, their fate will discourage the next generation from entering science. It is true that the Report recommends an expansion of the long-term support for young researchers now provided by Royal Society Research Fellowships, but their numbers will remain small compared to that of young university teachers who make up the bulk of Britain's research potential. Concentration of university resources required by financial stringency has been going on for a long time and could be continued on a departmental basis without depriving entire universities of the right to do substantial independent research.

"The authors of the Report seem to believe that most good work is now done by large teams, but that is true of only some subjects. Monoclonal antibodies were invented by two people. High temperature superconductivity was discovered by two men at IBM Zurich by research done on their own initiative. I understand that it was not part of the laboratory's "mission". W.F.H. Jarrett has developed a very promising vaccine against AIDS in the Veterinary School of the University of Glasgow; starting from small beginnings, it was done by a team of eight people in a department with about 30 graduates, not in a great interdisciplinary research centre. The entire Veterinary School is now under threat of closure by the UGC.

"My laboratory is often held up as a model of a centre of excellence, but this is not because I ever "managed" it. I tried to attract talented people by giving them independence, listening to them and taking an interest in their work, helping them to get what they needed for it and making sure that they got the credit for it afterwards. I also followed the tradition that I had learnt at the Cavendish Laboratory of letting young people publish their work independently, because I knew this to be one of the most important stimuli to originality. Had I tried to direct people's work, the mediocrities would have stayed and the talented ones would have left. The laboratory was never "mission-oriented". The National Institute for Medical Research was founded with the mission of developing antibiotics, but these have not been discovered there, while it has achieved great things in other fields. The FMBL Laboratory in Heidelberg is organised exactly as prescribed by the Report, but has achieved little commensurate with its enormous cost.

"The brilliance of British science is one of the country's greatest cultural achievements, if not the greatest, but it is a fragile flower as I know from Austria, my country of birth. Once destroyed by bad politics it cannot be restored."

Appendix VII

Approaches to Science in Industry

*by K. Alex Muller**

(He received the Nobel Prize in Physics in 1987, together with J. Georg Bednorz, for the discovery of High Temperature Superconductivity.)*

Projects in Industry, especially IBM

I should like to expose the way research and development (R&D) is done in IBM.

In IBM, R&D is carried out in both the research and production divisions. One distinguishes three categories: sciences, advanced projects, and development. In the sciences, efforts are undertaken to increase the knowledge and deepen the understanding in areas possibly, but not necessarily, of importance to the company; to name just some of them: mathematics, especially the applied branch, computer sciences, of course, physics, materials sciences, and also certain areas of chemistry ranging from anorganics to polymers. Results obtained in these areas are normally published in the appropriate journals and presented at conferences and symposia.

There is a clear-cut distinction between advanced projects and development. The former are of high-risk character and not time-limited, i.e. they are terminated by either success or in the opposite case if one can show that they have no future. The Josephson computer project is an example of the latter case, as the Zurich Laboratory group could establish that it was not possible technically to create the appropriate memory chip with this device and reasonable tolerances. Development projects, on the other hand, are of low risk and time-limited. They have a definite goal in the form of a product. Usually development is carried out in the Production Divisions at various levels of sophistication, whereas advanced projects and sciences are pursued in the Research Division, which at present totals 3500 employees.

The categorisations outlined above are however not at all strict. Sometimes development is carried out in the Research Division, and at other times quite fundamental insights are gained at a laboratory in a production facility. A recent example of the former is the local-area network called "Token Ring" developed at the Laboratory in Rüschlikon, Switzerland, which belongs to IBM's U.S. domestic Research Division. This Token Ring is now an IBM product. Of course, in such a case, the issue of technology transfer from research to production is posed. This is not specific to IBM, but can be found in all larger companies. Recalling from my years at Battelle Memorial Institute, novel solutions found under contract research for specific medium-size companies had to overcome the reservations of "not invented here" at middle-echelon levels of the respective company.

Under the leadership of Dr. R. Gomory, then the Director of Research, IBM adopted a policy which successfully helps in technology transfer and innovation: staff members of the production divisions spend longer periods of time in appropriate departments of the research division and, vice versa, individual people or whole groups from research are assigned to particular production divisions. This substantially reduces the barriers in the transfer process. To assess the quality of research, for a number of years, a committee led by the IBM Chief Scientist visited about half a dozen IBM locations each year. The committee consists of two to three IBMers, usually Fellows of the company, and half a dozen well-known scientists from universities.

To decouple research to a certain degree from the pressures inherent in product manufacture, the director of research usually reports to the highest executive of the company, until recently to the president, now to the senior vice president and chief scientist, who in turn reports to the president. However, this decoupling is not of a strict nature. In case of an alarming situation in the company, a substantial number of the research division staff may become involved in understanding and remedying the difficulties, partially at the production line, as was the case when the yield of the newly introduced multilayer ceramics was very low. Multilayer ceramics are used to interconnect up to a hundred VLSI chips. Another example was the sensitivity of the electronic components to radioactivity, and other such situations may arise in the future.

Industrial, University and Government Laboratories

In the electronics industry, companies comparable in size to IBM all invest substantially in research and development, be they in Europe, the USA or Japan. Although the designation of the efforts may vary nominally, the scope and emphasis as well as the amount spent for research in comparison to that for development will not differ too much. In stating this, I have in mind vital and healthy R&D. However, it cannot be overlooked that in both Europe and the USA there are companies which once were leaders in R&D but are now receding in this respect. Partly they abandoned sciences or they lack in innovative spirit. To be vital in the latter is a matter of utmost importance and of substantial concern to management. The approaches and evaluation of success in research vary from company to company. From this also the self-understanding and what I would call the "local" culture, taste and ethics of the staff members, their behaviour towards their colleagues and in their groups are distinctly different.

In the above, I alluded to the fact that even larger companies prefer supporting the advanced-project, or even only the development, over the science aspect. On not too long a timescale, this does not preclude innovativeness to generate competitive products on the marketplace. Small companies are usually restricted to such a policy owing to their limited R&D budgets. Help in this respect can come from government contracts or institutes like the Battelle Memorial which have expensive equipment from their own resources at their disposal. However, only a sizeable science department leads to breakthroughs like the transistor or the maser/laser, which after development times of the order of decades changed technology altogether.

From what I have just said, one may think that a substantial manpower effort is needed to initiate such impressive developments. Here, one has to be considerably more differentiating: the important innovations are achieved by one person or a small group of two or three. This was true for the transistor, the maser principle, or, much earlier, even the electron microscope, the radar idea, and, more recently, the tunneling microscope and the high-T_c superconductive oxides. However, I have to add immediately that in all these cases, even with the radar klystron at Stanford, the environment in which these achievements came about was very important! With this I mean the high scientific level and expertise of a research centre or laboratory, for which the hiring of excellent young staff members is crucial. Their interaction as well as that with visitors for shorter and longer periods of time and relations to high-class university institutions are essential.

Examples of centres with excellence are, as just mentioned, major well-known universities in the USA, Japan and Europe. Others are national and supranational centres, like the high-field magnetic laboratory in Grenoble, where the quantum Hall effect was discovered. Media people have frequently asked whether in materials research there was a difference of approaches between universities and industry. This is certainly not the case with regard to quality as such, when looking at the best laboratories. However, in the science laboratories of larger industrial companies the structure is more "amorphous", at least in our company, and in a way also more liberal. Therefore interactions between scientists of different departments are rather easy and creative. For example, Dr. Bednorz and I were not in the same research group when we worked together on the copper-oxide ceramics. Despite the democratisation since 1968, university groups are more dominated by the professor leading the group, and thus tend to be tied to a specific field. This, of course, has to be so because one of the goals is the instruction of the students in order to lead them to academic maturity, i.e. a degree. So it is quite natural that the theme of a thesis will be chosen without too high a risk, from the field of interest of the research group.

The facilities dedicated to materials research, such as the magnet laboratory at Grenoble or reactor and synchrotron facilities to carry out neutron or X-ray diffraction experiments, are usually at the disposal of university staff people. Thus, they represent a possibility to counter-balance the relatively more closed university structure as compared to industry. At such centres, here I mention the facilities in Grenoble and Desy in Hamburg in Europe, or in the USA, the Brookhaven National Laboratory, it is exactly this immense opportunity to meet and discuss with colleagues and experts from other fields and cultures, which makes them so valuable and attractive. This is comparable to the large research centres of industry. One also has to include the "offsprings" of former nuclear research centres, such as Los Alamos in the USA or Jülich in Germany, which have recently been redirected more towards materials research.

The specific projects in such government facilities dedicated to materials sciences are comparable to what is spent at industrial centres. In total budget and manpower, their size is comparable to that of the research divisions of large companies. This is also true for the high-energy centres until recently. But the emphasis of the latter is quite different, especially the specific experiments are undertaken at costs larger by a factor of one or two orders of magnitude; a ratio which is also reflected by the number of people involved. Often several university groups cooperate in such experiments, which requires planning and coordination over several years. Thus, the personal freedom of the staff involved can become quite restricted. From this, there appears an affinity of such undertakings with the larger development efforts in industry. Therefore, it is not surprising that staff members from say, CERN, transfer to industry and vice versa, or that the directors of industrial research and of high-energy centres meet and consult.

Appendix VIII

A State's Duties towards Science

1. In the context of a State's duties towards Science, permit me to present the proclamation 5461 of 17 April 1986 of the President of the United States of America. *I wish similar proclamations could be made by the Presidents of the developing countries.* This proclamation reads:

"Since the time of its beginnings in Egypt and Mesopotamia some 5,000 years ago, progress in mathematical understanding has been a key ingredient of progress in science, commerce, and the arts. We have made astounding strides since from the theorems of Pythagoras to the set theory of George Cantor. In the era of the computer, more than ever before, mathematical knowledge and reasoning are essential to our increasingly technological world.

"Despite the increasing importance of mathematics to the progress of our economy and society, enrolment in mathematics programmes has been declining at all levels of the American educational system. Yet the application of mathematics is indispensable in such diverse fields as medicine, computer sciences, space exploration, the skilled trades, business, defence, and government. To help encourage the study and utilisation of mathematics, it is appropriate that all Americans be reminded of the importance of this basic branch of science to our daily lives.

"The Congress, by Senate Joint Resolution 2261, has designated the week of April 14 through April 20 1986 as "National Mathematics Awareness Week" and authorised and requested the President to issue a proclamation in observance of this event.

"NOW, THEREFORE, I RONALD REAGAN, President of the United States of America, do hereby proclaim the week of April 14 through April 20 1986, as National Mathematics Awareness Week, and I urge all Americans to participate in appropriate ceremonies and activities that demonstrate the importance of mathematics and mathematical education to the United States.

"IN WITNESS WHEREOF, I have hereinto set my hand this seventeenth day of April, in the year of our Lord nineteen hundred and eighty-six, and of the independence of the United States of America the two hundred and tenth".

2. A similar and more recent proclamation has been made by President Reagan in respect of superconductivity. In view of this topicality, I repeat it here.

"THE PRESIDENT'S SUPERCONDUCTIVITY INITIATIVE.

"The President has announced an eleven-point initiative to promote further work in the field of superconductivity and ensure US readiness in commercialising technologies resulting from recent and anticipated scientific advances.

"The US has been a leader for years in the field of superconductivity — the phenomenon of conducting electricity without resistance. US private and Government researchers have also been at the forefront of recent laboratory discoveries allowing superconductivity to occur at higher temperatures and with greater current-carrying capacity than was previously possible.

". . . The Federal Government is currently spending approximately $55 million on superconductivity research, with more than one half of that reallocated within the last six months.

"The President's initiative reflects his belief that it is critical that the US translate our leadership in science into leadership in commerce. While the US private sector must take the lead, the Administration is taking important actions to facilitate and speed the process, including increasing funding for basic research and removing impediments to procompetitive collaboration on generic research and production and to the swift transfer of technology and technical information from the Government to the private sector.

". . . The Superconductivity Initiative includes both legislative and administrative proposals. . . . The major components of the Initiative are: . . . *"Administrative"*

1.   "Establishing a "Wise Men" Advisory Group on Superconductivity under the auspices of the White House Science Council. . . .

2.   "Establishing a number of "Superconductivity Research Centres" . . . that would: i) conduct important basic research in super-conductivity; and ii) serve as repositories of information to be disseminated throughout the scientific community.

3.   ". . . The National Science Foundation (NSF) will augment its support for research in high temperature superconductivity pro-

grammes at three of its materials research laboratories. In addition, NSF is initiating a series of "quick start" grants for research into processing superconducting materials into useful forms including wires, rods, tubes, films and ribbons.

4. "The Department of Defence is developing a multi-layer plan to ensure use of superconductivity[an] technologies in military systems as soon as possible. The Department of Defence will spend nearly $150 million over three years. . . ."

[an] High temperature superconductivity is a subject to which fundamental contributions were made by Chinese physicists; Zhau Zhong Xian was awarded the annual $10,000 Physics Prize of the Third World Academy of Sciences on 14 September 1987. Apparently any nation may join in this potentially rich and, fortunately, still open quest if it can afford just thirty thousand dollars for equipment and for the physicists.

Appendix IX

Special Academy Lecture

Technology and the Future of Europe

by Hubert Curien

*(From the **Bulletin of the American Academy of Sciences**, Vol. XLI, No. 8, May 1988)*

Last fall, Hubert Curien, former President of the Europen Science Foundation and the European Space Agency, presented a special lecture at the House of the Academy on the relationship between the structure of the European economy and the advancement of scientific research in Europe. Mr Curien has also served as Director General of the Centre National de la Recherche Scientifique (CNRS), as President of the National Space Center and of the Air and Space Academy, and as French Minister for Research and Technology. The following is a summary of his remarks.

Many of you have visited Europe, and are aware of the difficulties which Europe faces in coordinating its economic and scientific activities. I would like to present a short analysis of this problem with a particular emphasis on the implications of scientific research for the economy of the European continent as a whole.

To strengthen the bonds between scientific and economic progress, Europe must deal with a number of short-comings. Among the most serious is the lack of an effective common market. It is true that what we term the "Common Market" will be formally established in 1992, but for the present we must accept the fact that our activities in technological development are redundant. The same kinds of products and systems are being developed independently in every major country in Europe. If we are to make our science and industry more efficient, we must begin to move toward greater cooperation. Europe must also find ways to encourage innovation within its small to middle-sized industries. The common practice in Europe is for large industries to subcontract to smaller firms, indicating exactly the type of inventions and products they want. Consequently there is little need for smaller firms to employ good scientists and engineers, and innovation is circumscribed within a few enterprises.

On the financial side, the European use of capital has not been directed toward support of innovative practices. Some positive changes are being made in this regard; for example, we have collected a number of legal rulings which should stimulate venture capital, and large banks throughout the continent have sought the advice of scientists and engineers on possible investments. Again, it will take time to change the attitudes of a citizenry, and especially a banking community, that has traditionally demonstrated little interest in innovation. Related to the limited availability of private capital is the fact that Europe has no institution analogous to the United States Department of Defense, which invests a large percentage of its enormous financial resources in research carried out by American universities and industry. In Europe each country supports its own defense effort, and the percentage of funds allocated for research is much smaller. Of the federal funds presently dedicated to research in the United States, some 72 percent comes from the Department of Defense as compared to 50 percent in the United Kingdom, 30 percent in France and 10 percent in Germany. Although military funding may be controversial, diminishing it would not increase the budget for research in civilian ministries, and greater allocations from the defense budgets would enable European countries to expand their research efforts significantly.

There is the further issue of the role of research within European industries and universities. The leaders of European industries tend to be oriented toward marketing and production rather than research. On the other hand, academic research is not closely linked with industrial needs. To help reverse this situation, an effort has been made to include a "Mister Research" on industrial boards of directors. The reasons for the separation of industry and academia vary from one country to another. In France, for example, it is the engineers who generally become the industrial leaders of the nation, yet they are educated in schools which allow them only very limited access to laboratory facilities, and thus they bring virtually no research experience to their influential positions. Research scientists, in turn, are trained at the universities which have excellent laboratory facilities, but these scientists tend to become teachers or researchers as such and are seldom found in important decision-making positions in industry. Here again we are faced with traditions that must be changed. Ten years ago, we built a new Ecole Polytechnique with excellent laboratory facilities to be used by both teachers and students; thus far, however, the curriculum has been so tight that few students have had the opportunity to actually work in the labs. There is a further point to be made about the career choices of both scientists and engineers. When a student leaves a school or university, he does not think of the whole of Europe as his job market but rather only his home country, be it France, Germany, or England. In the United States, there is, for example, a nationwide American Physical Society with annual meetings and services which enable its members to seek out job opportunities throughout the country. Few organisations of this kind now exist in Europe; some fifteen years ago, we founded the European Physical Society and the European Mineralogical Society. If the strength of our scientists and engineers is to be fully realised, they must regard themselves as members of a European community. . . .

Now let us examine what we have achieved in Europe, first in terms of such "big programmes" as aeronautics, space, and atomic energy. I was fortunate to have been in a position of responsibility for space activities in France and Europe during a particularly exciting period. With the development of one launcher, Ariane, we now hold more than one-half of the market for orbiting satellites — a considerable success for Europe. The story of Europe's involvement in space is an interesting one because we started very early and very badly — just after Sputnik, some thirty years ago. Charles De Gaulle was the President of France, and he had decided that the French must have a part in space activities. We tried to convince our neighbors to launch a common European effort, but the result was not what we expected. A three-stage launcher was conceived, with Britain producing the first stage, France the second, and Germany the third. Although there were no problems with the individual stages, they failed to work together because there was no unity of conception. The project was an example of juxtaposition rather than integration, and it epitomised Europe's inability to bring about significant scientific advances.

In the 1970s, Europe achieved some very real success with the formation of the European Space Research Organisation (ESRO) which built seven scientific satellites to study the magnetosphere and the ionosphere. However, the European governments felt that the future of satellites lay not in the areas developed by ESRO but in telecommunications and observation, and thus ESRO was disbanded.

In its place, we established the European Space Agency with the two-fold purpose of "encouraging strong scientific activities" and developing satellites and launchers for specific applications, including telecommunications and observation. Support of the scientific programmes is mandatory, with the contribution of each of the eleven participating countries determined by its GNP. Support of programmes focused on specific applications is termed "a la carte". For example, France chose to take the lead in the development of the launcher, Ariane, providing 66 percent of the funds while Germany contributed 10 percent, Italy 10 percent, and Britain about 3 percent, whereas Ariane is a completely European endeavor, a second project, Space Lab, is a European-built laboratory to be launched on an American shuttle. In this case, Germany has taken the lead and has chosen to involve the United States because it maintains that the goal of advancing fundamental research in this area cannot be accomplished without the expertise and active participation of America. Space Lab appears to be a magnificent facility which works very well, but given the problems and priorities of NASA, there remains the question of when it can actually be launched.

Ariane and the Space Lab raise the question of the extent to which European space activities should be autonomous. In the late 1960s and early 1970s, France and Germany built two telecommunications satellites under a programme called "Symphonie". But since Europe lacked its own launching capabilities, it had to request that the United States put the satellites into orbit — a request honoured by the United States on a commission basis. This was, of course, a fair arrangement for a commercial venture, but it points up the fact that without autonomy in its launching facilities, Europe can never be commercially active in space.

Nonetheless, by the mid 1980s, a number of successful European space programmes were in place — Ariane, Space Lab, telecommunications and observation satellites. At that point, the principal issue was whether to advance gradually by building a more sophisticated version of Ariane or to take a giant leap by adopting a British plan for the construction of one of the world's first horizontal take-off launchers (Hotol). Given the uncertainty over the feasibility of "jumping over" an entire generation of launchers, it was decided that the French would proceed to design and build a larger; more reliable Ariane, known as Ariane 5, while the development of the Hotol system remained under study. ...

In all of these undertakings, the critical link between European science and the economy must not be overlooked. The European Economic Community, when created, had not stated responsibility for scientific matters except in the area of atomic energy. The Treaty of Rome has now been reformulated to incorporate the Community's direct responsibility for support of scientific research and technological development. At present, the amount of that support is minimal; of the total funds spent on European research and technology, the Community provides only $2-3$ percent. Yet that limited contribution has been put to good use as a catalyst in establishing joint industry-university programmes in telecommunications and information technology, in material sciences, and in biotechnology. Such programmes are an important step toward developing "networks" of university laboratories and industries which are seeking ways to integrate their related functions. The Community's support is intended to initiate basic research and extends only as far as what might be called the "pre-competitive" stage.

To stimulate networking in market-oriented enterprises, a programme entitled Eureka has been formed. Under Eureka, a product with a projected development time of $5-10$ years is identified and analysed, with particular emphasis on the approach needed to make it marketable. Eureka's first principle is to avoid the redundancy which has hindered both innovation and efficiency in Europe's technological growth. The second principle is "variable geometry" — a practice that is difficult to achieve in an organisation such as the European Economic Community which tends to promote the proportional participation of every country involved. Since each nation has the right of veto, quick decisions are impossible to obtain. The main purpose of "variable geometry" is to avoid the right of veto for each project by considering the programme as a whole and by granting each country reasonable participation in the total effort. The third principle is to work from the "bottom up" as opposed to the "top down", to search for the best ideas and encourage initiative from individual countries and industries rather than to issue directives from above and stifle, or fail to respond to, the most inventive proposals.

In sum, Europe is striving to advance both its science and its economy through integration and innovation. Yet in seeking to become more autonomous, we are not trying to distance ourselves from the United States. Quite to the contrary; it is our firm conviction that a strong Europe will be best for the United States as well.

Appendix X

The Key to World Domination

by Kurt Mendelssohn

(Chapter One of Science and Western Domination published by Thames and Hudson, London, 1976.)

Five hundred years ago, when the ships of Prince Henry the Navigator edged their way along the coast of Africa in the first step towards dominating the world, Europe was desperately poor in comparison with the great civilisations of the East. Soon she was going to be made even poorer by wars and plagues. By then, however, the foundation of Europe's future wealth had been laid. Its roots lay in a single philosophical idea which today is called science, but whose old name, 'natural philosophy', provides a rather better description. ...

What is so spectacular is the rapid rise of science, and its fairly negligible effect on man's moral behaviour. Admittedly, science has offered an enhanced scope for the destruction of life, but it has done the same for its preservation and for the alleviation of suffering. In fact, for what it is worth, the world population has increased. ...

The object of this book is to trace the essential steps in thought that have led along the path of science to white domination over the rest of the world. It is neither a history of science nor an account of scientific achievement, but a history of those concepts and ideas of natural philosophy which have inexorably forced the West into a position of economic superiority. It was an enterprise in which greed came only second to the spirit of adventure. ... Natural philosophy in the modern sense could not be reborn because it had not existed before. Instead, it came into the world as something entirely new, a revolution that has shaped the lives of more people than any previous idea, the great religions included. Science has provided mankind with a new dimension which is both terrifying and exhilarating.

Before we can embark on our history of scientific ideas, something has to be said about the reason why natural philosophy has had this enormous impact on human affairs, which far exceeds the success of any other philosophical method. Its immense strength lies in its power of accurate prediction. There is a minimum of fumbling. Trial-and-error methods are reduced to a level which was quite unknown in former phases of human activity. An aero-engineer who, applying scientific principles, designs a new aircraft not only knows, even in the drawing-board stage, that his aircraft will be capable of flying, but he also can forecast, with a fair degree of accuracy, its performance. A political group, planning a new election campaign, does not necessarily meet with the same degree of success. This disparity has nothing to do with the complexity of the problem since the aircraft is bound to be infinitely more complex than the party's manifesto. A similar difference in certainty exists when we consider operation. The aircraft is far more likely to deliver its occupants via a predetermined route to the required destination than does an elected government.

. . . The remarkable progress of science is thus merely a relative phenomenon, and it is a matter of taste whether we say that technological development has advanced or that human society has lagged behind. What is remarkable, on the other hand, is that the development of the scientific method should have been limited to European man. The amount of political and sociological trial and error in the great oriental civilisations is of the same order of magnitude as in Europe, and in all other respects, too, these civilisations compare well with the West.

The fact that Asia has now chosen to deviate from its traditional pattern and to follow the technological road demonstrates, better than any other argument, that the philosophical method of science is the most outstanding contribution which has been made to human progress in the last millennium. Since, for reasons unknown to us, scientific progress has not been matched by a corresponding development in morality, the white race has used this powerful method to dominate the globe. It seems that those who apologise for the latter fact tend to forget the achievement of having developed science in the first place. There is, moreover, no reason to believe that, if another civilisation had developed science, it would have desisted from using it for exactly that purpose.

The fundamental belief on which all science is based is that of an integral creation. It is tacitly assumed that the universe in all its component parts is interdependent, and that the interdependence should be unique. Applying loose, but for the moment convenient terminology, this means that a particular set of causes can have only one particular set of effects. The object of science is to discover the nature of this interdependence. Its strength and success lie in making use of such fragments of inter-dependence as have already been established. From where we stand today we can see that this knowledge is very fragmentary indeed.

What really interests the scientists, and what we will be mainly concerned with from now on, are the formulations of the laws of nature and the changes in formulation which have taken place as science has developed. The basic belief in the unique interdependence has gradually grown out of the observation of regularities. Some of these are so obvious that their mention must appear trite, such as the regularity with which day follows night, and the annual periodicity with which the day lengthens and shortens. A quite different and even more obvious regularity is that, whichever way you hold it, the level of water in a jug will always be parallel to the horizon.

The primary object of scientific inquiry is to establish as many of these regularities as possible, and then to reduce their number by trying to find a common explanation for some of them. In most cases this is only partially possible. For instance, while in both the regularities already mentioned the concept of universal gravitation is relevant, it does not suffice. The two laws to which these and many other regularities have been reduced, that of universal gravitation and that of conservation of momentum, were both enunciated by Newton. He was not, however, able to find the ultimate connection which these two laws must have if the belief in unique interdependence is correct. Such a connection was only proposed 250 years later by Einstein, through the concept of general relativity, but we have as yet not sufficient evidence to accept this unifying concept with the same confidence as the two laws of Newton.

There thus exists a secondary object of science, which introduces an element of control by testing the law to which the regularities have been reduced. This can be done either by making observations on a great number of similar regularities or, even better, by investigating corollaries of the law. In the case of general relativity, it is a necessary corollary that a beam of light should deviate from its straight path when passing close by a massive object. Using the sun as the largest mass near at hand, and waiting for an eclipse to reduce its glare, we find that the apparent shift in the position of a star seen close to the sun can just be detected by our most accurate methods of observation. This is the reason why expeditions are sent to remote and usually inconvenient places wherever the rare total eclipses of the sun are due to occur. The evidence which they have brought back from those lucky occasions when they were not foiled by cloud, is as yet inconclusive.

We therefore have two grades of predictability in science. The more direct one deals with cases which have been investigated and used without doubt for safe prediction. If the circuit of an electric torch is closed by pressing the switch, a current will flow and the bulb will light up. This is accepted as quite certain even for a newly assembled, and as yet unused, torch. If it does not light up at the first attempt, the shop assistant will look for a manufacturing fault rather than doubt the laws of nature pertaining to electric currents. The same is true for a new power station.

The regularities concerning electric currents and their connection with magnetism were thoroughly investigated in the first half of the nineteenth century, and the laws of nature containing these regularities were formulated in 1865 by James Clerk Maxwell in a set of differential equations. These equations were able to account for all known observations and uses of electric currents but, as Maxwell noticed immediately, they allowed for other phenomena which had never been observed. From his equations Maxwell predicted the existence of electromagnetic vibrations, and such waves were eventually produced and detected a quarter of a century later by Heinrich Hertz. They form the basis of the whole technique of radio communication, including television.

Maxwell's prediction of radio waves is of a higher grade than the switching on of the new torch, or the functioning of a new power station. It goes further than a repetition of known regularities. Both sets of regularities, however, the known and the newly predicted, are intimately connected by the set of differential equations. Provided that Maxwell's formulation of the laws of electromagnetism was correct, the existence of radio waves had become as inevitable by this formulation as the switching on of the torch.

It must be noted that in all these stages of using the scientific method of philosophy, observation plays a most important role. It is the key to the whole operation, be it in the primary gathering of regularities, or in their testing, or in the proof of the formulated law. It is thus quite justifiable that some of the early university chairs of physics were established as professorships of experimental philosophy, a term which expresses the occupation of the holders better than the later designation. The key position of the experiment in science is indispensable in that economy of effort which has made science so successful.

In the first place, the experiment sees to it that speculation cannot run riot. When a set of observed regularities is combined into a theory, the next thing to do is to test this theory by experiment. Normally, the next step in speculation is only taken if the experiment has provided confirmation for the correctness of the initial step. Failure to confirm the theoretical prediction requires that the theory be discarded or modified. . . .

These are the rules of the game, but only too often do they have to be broken legitimately. The theoretician may not be able to wait for the checking experiment for the simple reason that the experiment is too ·difficult, or too expensive. Wasted theoretical effort, may, in fact, be very much cheaper than the experiment. Nevertheless, the experiment will ultimately have to be performed, whether the taxpayer likes it or not. The interesting state has not been reached when, as is the case with space research or particle bombardment, the taxpayers of several large countries have to combine their resources in order to get the experiment performed.

The other important function of the experiment is the discovery of new facts, such as the discovery of America which foiled Columbus's experiment to reach the Indies. Here we mean by 'new fact' not one of the suspected regularities, as with Maxwell and the radio waves, but some completely unsuspected phenomena. For instance, at the turn of the century there existed two rival theories according to which the electrical resistance of a metal should either drop gradually to zero, or rise gradually to an infinitely high value as the absolute zero of temperature is approached. A decision has to wait until a method for the close approach of absolute zero had been found. When in 1908 Kamerlingh Onnes in Leyden succeeded in doing this, he immediately performed the experiment and found that the first alternative was almost, if not quite, the correct one. In addition, however, he had discovered that most metals will suddenly and completely lose their resistance a few degrees *above* absolute zero. This 'superconductivity' was an entirely new phenomenon which had no place in the earlier predictions. In fact, for a long time it remained unexplained, i.e. it was completely unconnected with any of the known regularities. When such a very isolated new fact is discovered, it always means that we have entered into a new field of natural phenomena of which, up to then, we had no knowledge whatever. At the time of their discovery, nobody could suspect, and therefore predict the existence of either America or superconductivity. . . .

. . . The white man's first venture over the uncharged oceans, besides bringing knowledge of new continents, also gave direct proof that the earth is a globe. In our time the exploration of the land surface has been brought virtually to a close, but a sphere holds within its surface a content, and of this content we know next to nothing. Emanating from the interior of the earth there is volcanic heat and a magnetic field of whose cause we are as yet largely ignorant. Little is known about the beds of the oceans and until quite recently we had not been able to penetrate very far into the atomosphere. Then a new window was opened on that day in October 1957 when the first satellite was launched into outer space. It is indeed fortunate for mankind that such strategic and political importance is being attached to the occupancy of extra-terrestrial space, because the opening of this new window is quite expensive, and weighty reasons have to be advanced to justify it. The mere thirst for knowledge would certainly cut no ice with those who prefer to see the money spent rather for the good of humanity, which means to provide them with three Sundays each week instead of two. . . .

The experiment requires instruments, and the instruments of science form a most important chapter in this branch of philosophy. The invention of new methods of exploration and measurement is an integral part of the peculiar intellectual adventure which gave the white man command of the earth. The conception and design of a new probe is a feat in science which in importance often parallels the concept of a new abstract idea. It is for this reason that one is inclined to deplore the disappearance of the old term 'experimental philosophy', since the invention of a new probe often goes far beyond the solution of how it can be done. It always also involves the problem of what can be done, which means asking questions. And asking questions, preferably of such a kind as to admit of an answer, is the business of philosophers. When in 1932 Cockroft and Walton built the first particle accelerator, it was a superb exercise in experimental ingenuity, ranging from electrical engineering to high vacuum technique. Yet more important than all this was the guiding idea of building *any* machine which would make atomic nuclei react with each other, a process which does not normally take place in the physical world of our experience.

... The question has often been asked why the great civilisations of the East did not develop science and technology. The answer to this question was once given by Einstein, who pointed out that it is the wrong question. The miracle, he said, was not that the East failed to create experimental philosophy, but that the West did. However, why this happened is a riddle that has never been solved. To be candid, we simply don't know. It would indeed be rash to suggest that it was developed merely out of a desire to enslave others. But whether it was intended or not, this is what in the end happened. In addition to greed for riches and domination, the white man became possessed suddenly of a strange spirit of adventure, of an insatiable intellectual curiosity that has driven him on for the past five hundred years. It is this curiosity and this spirit of adventure, in thought as well as in action, to which we will now turn. For the first scene the stage will be set in a little town on the windswept southwestern corner of Europe where nothing remains today — except the name.

No ideas, not even scientific ones, grow in the test tube. They germinate and develop in men's minds, and these minds are conditioned by their environment. Beyond the discoverer's character there is the world into which he is born and bred, and whatever his brain thinks up will be shaped and sharpened by this world. The man and the mould into which he has been cast form an integral part of any new thought which he may contribute. Science is that essential ingredient of Western civilisation which distinguishes it from the cultures of others.

Original scientific ideas and concepts are basically the work of individuals rather than the collective effort of many, and the success of our world has for centuries now depended on the achievements of geniuses. That the execution of their ideas may often have to rely on the work of thousands of other people does not alter this fact. We thus have come to regard the leading role of the genius as a necessary and desirable feature of the world, and the names of men like Copernicus, Newton, Faraday and Einstein are deeply imprinted on the mind of the school-child. It is here perhaps, more than in any other respect, that the West differs from the civilisations of the East.

In China or India, regard for the achievement of the individual has been bestowed much more sparingly by society. It has been limited to the rare phenomenon of a great religious leader or a great philosopher. To Eastern peoples, harmony of mankind in general, both with the forces of nature and with one and other, has been the main aim of life; and attitude that fights shy of drastic innovation in an accepted pattern of existence. The belief that this pattern, which had been achieved in centuries of patient evolution and self-discipline long ago, must be maintained, contrasts profoundly with the West's obsession with change and progress. The men whose life and work form the subject of this book have never existed in the East because the East would have given them no encouragement.

Now, after rejecting consistently in the centuries past the blandishments of Western science, the Eastern nations have changed their minds and are absorbing, in a somewhat undigested form, the so-called blessings of Western science, from electronics to automatic weapons. Most of them are quite willing to build on this foundation without concerning themselves particularly with its origin. It rarely occurs to Easterners that the West too, might have a history, not only of wars but also of ideas. I know my colleagues in Delhi, Peking and Tokyo well enough to realise that after six p.m., or whenever their day's work is finished, they long to re-immerse themselves into the quiet solace of their own heritage. They simply have no time to bother about the West's. To them this book is dedicated.

Appendix XI

Infinite in All Directions

by Freeman J. Dyson

(Chapter Three, Manchester and Athens)

Thirty years before he became Prime Minister, Disraeli wrote a novel called *Coningsby.* . . . About halfway through the story, Disraeli's hero spends a few days in Manchester, and here are the thoughts which the Manchester of 1844 called to his mind:

> *A great city, whose image dwells in the memory of man,*
> *is the type of some great idea. Rome represents conquest;*
> *Faith hovers over the towers of Jerusalem; and Athens*
> *embodies the pre-eminent quality of the antique world,*
> *Art. . . . What Art was to the ancient world, Science is to*
> *the modern; the distinctive faculty. In the minds of men*
> *the useful has succeeded to the beautiful. Instead of the*
> *city of the Violet Crown, a Lancashire village has expanded*
> *into a mightly region of factories and warehouses. Yet,*
> *rightly understood, Manchester is as great a human*
> *exploit as Athens. . . .*

What was so exciting about Manchester? Disraeli with his acute political and historical instinct understood that Manchester had done something unique and revolutionary. Only he was wrong to call it science. What Manchester had done was to invent the Industrial Revolution, a new style of life and work which began in that little country town about two hundred years ago and inexorably grew and spread out from there until it had turned the whole world upside down. Disraeli was the first politician to take the Industrial Revolution seriously, seeing it in its historical context as a social awakening as important as the intellectual awakening that occurred in Athens 2,300 years earlier. . . .

Science did flourish in Manchester during the crucial formative years of the Industrial Revolution, but the relations between science and industry were not at all in accordance with Disraeli's ideas or with the ideas of later Marxist historians. Science did not arise in response to the needs of industrial production. The driving forces of the Manchester scientific renaissance were not technological and utilitarian; they were cultural and aesthetic. . . .

So the anti-academic, anti-establishment brashness of Manchester made a fertile ground for the growth of science. And the science which grew in that northern soil had a style different from the science of Athens, just as two hundred years later the music of the Beatles growing up in nearby Liverpool had a style different from the music of Mozart. The science of Athens emphasises ideas and theories; it tries to find unifying concepts which tie the universe together. The science of Manchester emphasises facts and things; it tries to explore and extend our knowledge of nature's diversity. . . .

Manchester brought science out of the academies and gave it to the people. Manchester insolently repudiated the ancient prohibition, "Let nobody ignorant of geometry enter here," which Plato is said to have inscribed over the door of his academy in Athens. . . .

The discoveries of recent decades in particle physics had led us to place great emphasis on the concept of broken symmetry. The development of the universe from its earliest beginnings is regarded as a succession of symmetry-breakings. As it emerges from the moment of creation in the Big Bang, the universe is completely symmetrical and featureless. As it cools to lower and lower temperatures, it breaks one symmetry after another, allowing more and more diversity of structure to come into existence. The phenomenon of life also fits naturally into this picture. Life too is a symmetry-breaking. In the beginning a homogeneous ocean somehow differentiated itself into cells and animalcules, predators and prey. Later on, a homogeneous population of apes differentiated itself into languages and cultures, arts and sciences and religions. Every time a symmetry is broken, new levels of diversity and creativity become possible. It may be that the nature of our universe and the nature of life are such that this process of diversification will have no end.

If this view of the universe as a steady progression of symmetry-breakings is valid, then Athens and Manchester fit in a natural way into the picture. The science of Athens, the science of Einstein, tries to find the underlying unifying principles of the universe by looking for hidden symmetries. Einstein's general relativity showed for the first time the enormous power of mathematical symmetry as a tool of discovery. Now we have reason to believe that the symmetry of the universe breaks explicitly and the laws of its behaviour become unified if we go back far enough into the past. Particle physics is at the moment at the threshold of a big new step in this direction with the construction of Grand Unified models of the strong and weak interactions. The details of the Grand Unified models are worked out by studying the dynamics and composition of the universe as it is presumed to have existed for an unimaginably small fraction of a second after its beginning. Whether or not the particular models now proposed turn out to be correct, there is no doubt that the concept of unifying physics by going back to a simpler and more symmetric past is a fruitful one. The science of Athens, what Einstein called the ancient dream that pure thought can grasp reality, is then nothing else than the exploration of our remotest past, "A la recherche du temps perdu" on a bolder scale than Proust ever imagined.

In a similar fashion, the science of Manchester and of Rutherford, the science of the diversifiers, is an exploration of the universe oriented toward the future. The further we go into the future, the more diversity of natural structures we shall discover, and the more diversity of technological artifice we shall create. It is then easy to understand why we have two kinds of scientists, the unifiers looking inward and backward into the past, the diversifiers looking outward and forward into the future. Unifiers are people whose driving passion is to find general principles which will explain everything. They are happy if they can leave the universe looking a little simpler than they found it. Diversifiers are people whose passion is to explore details. They are in love with the heterogeneity of nature and they agree with the saying, "Le bon Dieu aime les détails." They are happy if they leave the universe a little more complicated than they found it.

Now it is generally true that the very greatest scientists in each discipline are unifiers. This is especially true in physics. Newton and Einstein were supreme as unifiers. The great triumphs of physics have been triumphs of unification. We almost take it for granted that the road of progress in physics will be wider and wider unification bringing more and more phenomena within the scope of a few fundamental principles. Einstein was so confident of the correctness of this road of unification that at the end of his life he took almost no interest in the experimental discoveries which were then beginning to make the world of physics more complicated. It is difficult to find among physicists any serious voices in opposition to unification. . . .

In biology the roles are reversed. Very few of the greatest biologists are unifiers. Darwin was a unifier, consciously seeing himself as achieving for biology the unification which Newton had achieved for physics. Darwin succeeded in encompassing the entire organic world within his theory of evolution. But the organic world remains fundamentally diverse. Diversity is the essence of life, and the essential achievement of Darwin's theory was to give intellectual coherence to that diversity. . . . Darwin had no peer and no successor.

Or perhaps I should say, Darwin has only one successor and his name is Francis Crick. In saying this, I am not expressing a judgement of the greatness of Crick as compared with other contemporary biologists, and still less am I expressing a judgement of the importance of molecular biology as compared with botany and zoology. I am merely saying, Crick is a unifier of biology in the style and tradition of Darwin. . . .

Fortunately, the recent successes of particle physics and of cosmology do not exclude the possibility that the world of physics is truly inexhaustible, that Michael Polanyi was right when he said: "This universe is still dead, but it already has the capacity of coming to life," that John Wheeler is right when he says: "The universe is a self-excited circuit," that Emil Wiechert was right when he said: "The universe is infinite in all directions."

Technology and Pakistan's Attack on Poverty

Address by Professor Abdus Salam at the XIII Annual All Pakistan Science Conference, Dacca, 11 January 1961.

I wish to begin by offering my sincerest thanks to my colleagues for the honour they have done me in electing me as General President. I feel doubly proud because our meeting takes place in this historic city of Dacca. In my experience there is no part of Pakistan where scholarship in its own right carries more esteem, and where a scholar receives more personal affection than East Pakistan. This unfortunately is a dying tradition elsewhere but one which lives in Dacca and I would like to begin by paying a tribute to this.

In my address today I would have liked to speak about the scientific field I have been privileged to work on, about the elementary particles of physics — those ultimate constituents of which all matter and all energy in the Universe is composed. I would have liked to explore with you the frontiers of our knowledge and our ignorance, to tell you of some of the concepts the physicist has created to comprehend God's design. I would have liked to show you that with all his pragmatism, the modern physicist possesses at once the attributes of a mystic as well as the sensitivity of an artist. I would have liked to convey to you some of the wonder, some of the fascination, as well as some of the heartbreaks of the physicist's craft.

But I shall not do this. In electing to speak on a general subject like Technology in relation to Pakistan's Attack on Poverty, rather than on Elementary Particle Physics, I am following the illustrious tradition of my predecessors in this office. More particularly I have in mind the eloquent Presidential address on 'Technology and World Advancement' delivered by Professor P.M.S. Blackett to the Dublin meeting of the

British Association for Advancement of Science in 1957. If I speak part of the time about the laws of economics rather than the laws of quantum physics, it is because like Blackett I interpret technology not in its narrow industrial sense but as something embracing the scientific organisation of most modern life. There are times when, in all humility, a mere scientist may also express himself on ideological matters, not because he has new insights to reveal but because there are things he believes passionately in, which need saying and cannot be said often enough.

We, in Pakistan, are very poor. This poverty we share with the majority of the human race, with some one thousand million people in about a hundred countries. Fifty per cent of us in Pakistan earn and live on less than eight annas a day; seventy-five per cent live on less than a rupee. This rupee a day includes the two daily meals, clothing, shelter and education, if any. In contrast some four hundred million inhabitants of Europe and North America live on an average daily income of fifteen rupees.

It is important to realise that this uneven distribution of wealth is of a relatively recent origin. Three hundred and fifty years back Akbar's India and Shah Abbas's Iran compared favourably in living standards with Elizabeth's England. Soon after, however, the Western growth started. It coincided with a great technological advancement in agriculture and manufacturing methods. Now technical advances on a limited scale have occurred from time to time in the history of human societies. These advances have always led to increased prosperity. What, however, distinguished the nineteenth century technological revolution was the fact that it was firmly based on a scientific mastery of natural law. This gave man so much power, and it has led to so great an increase in production, that for the first time in human history, there is no physical reason for the existence of hunger and want for any part of the human race.

The realisation that hunger, ceaseless toil and early death can be eliminated for whole societies, and not merely for parts of societies is something new. The last hundred years have seen nation after nation start with something like our conditions and crash through the poverty barrier. The laws governing this type of transformation are now well understood. First, a society must acquire the requisite technological skills; secondly, it must save and re-invest more than 5% of its national income in productive enterprises. This minimum of five per cent just about offsets the depreciation of existing wealth. To double the standard of living in forty years needs an investment rate of 10−15%; to double it in a decade, a nation needs to invest about 25% of its national income.

Skills and capital — these then are the two prerequisites for building up a self-reinforcing economic growth. Nation after nation has achieved this in the last two centuries, each nation leaving the imprint of its own peculiar experience. Four of these experiences — those of Britain, Japan, Russia and China — however, stand out clearly. The British were the first to show that the poverty barrier can be crashed through if skills and capital are available. The Japanese showed that technology is communicable; that it is easy to learn and acquire. Having been conditioned for years to look with misty, uncomprehending eyes at the engineering miracle of an airliner, I still remember the shock of my life when I first visited the De Haviland Aircraft Factory at Hatfield. Instead of an organised assembly line where I expected to see molten aluminium being poured in at one end and a Comet Airliner coming out at the other, all I saw was something like an overgrown metalsmith's workshop in rural Pakistan. And when two women in overalls lifted a couple of aluminium sheets while a third started welding them together with a manually operated welder to make part of the fuselage, I am afraid I lost my respect for the mysteries of the manufacturing craft.

I do not for one moment wish to suggest that all technology is electrical welding. There is the other part of the story — the aerodynamic design of the Comet where the high-level scientific talent comes in. But the Japanese experience forced home the moral that technological competence is not a hereditary characteristic; that it can be acquired and in fact acquired rather quickly.

The third important lesson came from Russia. It showed that transition to sustained growth need not take a century or longer. It can be telescoped into the span of one man's life provided heavy industry receives top priorities. And then finally there is the Chinese experience underlining that cheap labour is itself a form of capital.

Summarising the economic part of our argument, skills and sufficient capital rightly invested are the major ingredients of self-reinforcing growth. On the road to achieving sustained, compounded growth of this type all nations have left the imprints of their peculiar experience, but four stand out clearly: the British experience, showing that it can be done; the Japanese experience, that technology is easy to acquire; the Russian experience, that priorities on heavy industry accelerate the growth, and the Chinese experience that cheap labour is itself a form of capital.

From this brief and highly idealised economic summary, let us turn to the realities of the situation in Pakistan.

The facts of our poverty are obvious enough and I am not going to mince words about it. You can go out in the streets and see it all around you. I am not referring now to the obviously shelterless, the obviously needy. I have in mind, more, the uncomplaining millions, with their suppressed hunger; the millions who, and I speak from experience, seldom get the two regular meals of the day; the millions who must often choose between buying badly needed food or a book for their schooling child. We live with a crushing poverty of the sort which Europe or America have not seen since the day when Dickens wrote. The marvel to my mind always is that the human spirit does not break and that most of the needy are still able to keep a dignified exterior.

The sense of what can be achieved, on the other hand, hits you most when you visit an affluent society like that of the United States. You just cannot believe the plenty — the plenty not for the few but for everyone. Everytime I am privileged to visit that great country, I have to remind myself afresh that it is indeed possible to produce so much for so many.

I do not say all this in any spirit of envy. This prosperity is due to an organisation of society where scientific knowledge is fully exploited to increase national productivity. This prosperity is a portent of hope; hope that possibly within our life-time, using the same methods, we in Pakistan may also achieve the same.

Our poverty raises not merely material but also spiritual issues. The Holy Prophet, may peace and blessings of Allah be upon Him, said "It is near that poverty may become synonymous with *Kufr.*" . يَكَادُاَنْ يَكُونَ الْفَقْرُكُفْرَا I shall not attempt to translate *Kufr* into English; the nearest equivalents, apostasy or unbelief, can never convey the connotations which *Kufr* has for a Muslim audience. Let me say with all the vehemence at my command that I would like to see this saying of the Prophet on the door-piece of every religious seminary in Pakistan. There may be other criteria of *Kufr* as well, but in the conditions of the twentieth century, in my opinion the most relevant criterion of *Kufr* is the passive toleration of poverty without the national will to eradicate it.

I have mentioned technological skills and capital as the two prerequisites before a pre-industrial society like ours can crash through the proverty barrier. Actually there is a third and even more important prerequisite. And that is the National Resolve to do so. In Professor Rostow's words "a nation's take-off into sustained growth awaits not only the build-up of social overhead capital — capital invested in communication networks, schools, technical institutes — it not only awaits a surge of technological development in agriculture and industry, but it also needs the emergence to political power of a group prepared to regard the modernisation of the economy as a serious high order political business." Such was the case in Germany with the revolution of 1848, such was the case with Japan with the Meiji restoration of 1868; such was the case with the Russian and Chinese revolutions. Our independence in 1947 could have provided us with the necessary stimulus. Unhappily this was not the case. Our independence did not — definitely did not — coincide with the emergence of a political class which made economic growth the centre piece of state policy. I can still recall the interminable arguments, conducted in private and public, in the early years of Pakistan, about its ideology. Never in these discussions did I hear the mention of total eradication of poverty as one of the primary ideological functions of our new state.

True enough the country registered commendable progress in the manufacture of consumer goods — though, one must not forget, with appalling suffering to the consumer himself. True enough the establishment of the Pakistan Industrial Development Corporation was a triumph. But at no time was this development purposefully designed to achieve the breakthrough we have talked about. The first five-year plan was commissioned in 1955, full eight years after independence. It did not receive formal approval of the Government till 1957. During these years there was a total neglect of the primary sector of our economy — agriculture; we squandered the windfall surpluses of the Korean war boom in buying, on Open General Licences, European cosmetics and radiograms. It is not that we failed to develop basic heavy industries. We did not even make any provisions for their future establishment; not even to the extent of starting to get our men trained in basic technologies. And, lastly, we completely neglected the exploitation of our minerals. Not even a survey was undertaken.

It would be right to date our progress to the "take-off" from the assumption of power by the present Government. I believe when the future history of Pakistan is written the greatest significance of the revolution of 1958 will come to be recognised as the resolve for the first time of Pakistan's Government to achieve the breakthrough within five years. This resolve is reflected first by a recognition of the need for bold planning, for agricultural development, for exploitation of minerals and most important of all, for heavy industry. Secondly, it is reflected in the recognition that a liberal provision for developing technological and scientific skills is the wisest investment a nation can make.

Take our new five-year plan first. It is a sagacious plan though perhaps not as audacious as I would like to see. It aims at achieving the crucial 10–15% investment level. It places due emphasis on our primary sector of agriculture. It envisages the beginning of a basic heavy industrial complex, particularly the steel industry. And, most important, it sets about exploiting our one industrial source material — the Sui and Sylhet gas — to set up a petrochemical industry.

Quite often one hears abroad the rather sneering statement that underdeveloped countries look upon steel mills as national monuments. I personally confess to this complex and for very good economic reasons. Without a heavy industrial base nothing is possible in the long run. To take one pertinent example given by Professor Mahalanobis — the great Indian statistician — let us consider the problems of providing 700,000 tons of extra grain needed for the five million annual increase of Indian population. There are four ways of getting this extra grain; buy the grain; buy fertiliser to grow the grain; buy plant to make fertiliser or finally build heavy engineering capacity capable of making fertiliser plants. The cost of buying grain works out at 300 million pounds; the cost of buying fertiliser is one-third of this, and the cost of fertiliser plant about one-fourth. But the real saving comes if one sets up heavy manufacturing capacity to make fertiliser plants. The cost then is just some 10 million pounds. If the last alternative is chosen, one must however start planning some eight or ten years in advance of the season in which the fertiliser would be used.

It is gratifying that so far as fertilisers are concerned, our planners have chosen the third alternative. We are not planning to buy fertiliser but we shall make it in the country. Personally, of course, I would strongly favour Mahalanobis' last alternative — to set up the heavy manufacturing capacity within the country to make fertiliser machinery. The Second Five Year Plan has made a beginning towards this by contemplating steel production of 400,000 tons. As steel producers and consumers it will put us in the same world class as the Republic of Chile and though I cannot say I feel satisfied with this, it is at least a beginning.

Turning again to the question of 10 – 15% capital investment needed for achieving the economic breakthrough, there is a vital 3% of which must be provided for in Foreign Exchange to buy foreign goods, foreign machinery and foreign know-how. It is this crucial 3 – 4% which must come from the advanced countries either in the form of long-term loans or outright gifts. During 1957 – 58 over the world, some two and a half billion dollars were provided by U.S.A., U.K., U.S.S.R. and France as aid to underdeveloped countries. Let us make no bones about it; this gift entails sacrifices for ordinary people like ourselves in the donor countries. In the United States, stores always show prices without the Federal Tax. The tax is added on at the counter so that one is highly conscious of the extra imposition by the time the purchase is completed. Thus shopping around whenever I have had to pay the ten cents of Federal Tax, the thought that at least a quarter of a cent was going into foreign aid lightened the burden for me. It also gave me added respect and admiration for all those who constantly make this sacrifice.

The economists have estimated that in order that this aid brings its fullest impact, it must be stepped up from two to at least three billion dollars annually and kept at this level — with a guaranteed continuity — for a very long time to come. To get the scale right, it is perhaps worth mentioning that the Marshall aid to Europe just after the War ran to about twice this figure, though of course the rapid recovery of Europe made its continuance unnecessary after three years.

As I said before, aid is a gift and it necessarily entails sacrifices and there is very little we can offer in return — at least for a very long time. Whether it will or will not be forthcoming is in the end a moral and spiritual question. I can only quote sages like Rostow who has spoken of "the resonances of spirit, will, and insight which the West needs, quite as much as steel and electric gadgets, to do the jobs which extend not only to missile arsenals and the further diffusion of welfare at home" but to the Five Years Plans of the nations abroad. I can only quote Blackett when he speaks about "the uneven division of wealth and comfort among the nations of mankind, which is the source of discord in the modern world, its major challenge, and unrelieved its moral doom." I do not know if a future historian will find it ironic that in the 1960's three billion dollars of aid were not easy to find while 60 billion dollars were annually spent stocking the world arsenals with atomic weapons, missiles and rockets. And I find it strange that during 1957 – 58 while underdeveloped countries received 2.4 billion dollars in aid, they lost 2 billion dollars in import capacity — in getting paid less for the commodities like jute and cotton which they sell and in paying more for the industrial goods which they buy. Paul Hoffman calls this a "subsidy or contribution by the underdeveloped to the industrialised countries" — a subsidy which almost washed away the entire sums received in aid. And as a physicist, I find it the height of hypocrisy to pretend that the man-made satellites orbiting in space and each costing at least as much as the entire yearly budget of Pakistan have been sent up only to collect data on cosmic rays. All this makes no sense. It all points to one thing; the bankruptcy of world statesmanship in dealing with problems of hunger and want. Dare I say what the world needs today is a great successor to Keynes to preach on a global scale that the raising of living standards of any depressed region is a collective world responsibility. Dare I say that we need a great successor to Roosevelt to give a New Deal not simply to one part of the United States but to a large part of the human family.

I have talked so far about our plans and the position regarding capital. I now wish to turn to the question of the provision of technical skills. And this is where we as scientists directly come in.

Nowhere more than in this respect can we see the force of my remarks regarding the recent change of climate in Pakistan. This change is embodied particularly in the world of the Education and the Scientific Commissions.

Consider first the category of technicians who understand the scientific foundations of their craft. It is an awful fact, but nevertheless true, that in the entire liberal arts dominated educational history of British India, there never was anything analogous to the British National or Higher National Certificates in Technology. I could not believe it when I was first told that Great Britain has 300 colleges of Technology spread all over the country training 30,000 technicians every year. One of the most far-reaching recommendations of the Education Commission is the provision to establish enough technical schools and polytechnics to produce 7,000 technicians a year. Our major problem is the staffing of these technical institutions. I purred with pride last when Sir John Cockcroft spoke to me about the excellence of our army technical schools and their technical instructors. I am sure it will not be impossible to tap this reservoir for providing teachers in the early stages.

We, in Pakistan, are disposed to think of Scotland as a prosperous state within the British Commonwealth. I was startled the other day to read an article by Dr. J.M.A. Lenihan, entitled "What is wrong with Scotland?" After painting a rather gloomy picture of consistent economic decline Dr. Lenihan concludes that this decline stems entirely from lack of trained technologists. To the objection that if there is no industry in Scotland, there is no need for technical college, Dr. Lenihan counters by remarking "The scientist, the technologist and the technician are, in the main products of the educational system, not of the industrial system in which they hope to work. A coherent demand for technical education facilities will not rise from an assortment of industries but the existence of technically trained people will facilitate the growth of new industries."

Dr. Lenihan's viewpoint about skills coming before industrialisation has of course a peculiar relevance to our situation in Pakistan. Some ten days back I heard a similar comment from Professor S. Tomonaga, the great Japanese physicist, now President of Tokyo University. Speaking of the spectacular rise of the Japanese transistor industry, Professor Tomonaga attributed it to a careful cultivation of the art of calligraphy. Every Japanese child must spend years learning the calligraphic arts at school; this develops a sensitivity of touch, a nimbleness of fingers, peculiarity suited, as they have now discovered, to transistor assembly and development. Clearly no skills or special talents a nation may cultivate are ever wasted when the spark of industrialisation comes along.

There is one other passage from Lenihan's address which I would like to quote. After listing a number of difficulties which face the Scottish economy he goes on to say "Many of the difficulties that have been mentioned are the natural consequences of living in a country ... where science is not taken seriously enough. How else can we describe a country which, fighting for economic survival in a world dominated by technology, allows the basic sciences of physics and chemistry only the status of half subjects in the school curriculum." There is perhaps, in Dr. Lenihan's remarks, a considerable moral for Pakistan's secondary education.

Perhaps the most depressed community, till recently, among technicians in Pakistan was — and so far as University teachers are concerned still is — the community of scientific workers. All scientific research institutes in Pakistan have been run under the uncomprehending bureaucratic control of Government Ministries. And when I say control, I mean control. We never seem to have recognised that in a science-dominated world there never could be any tasks for the Pakistani scientists. The official attitude towards Science has at best been one of reluctant indulgence; somewhat like the attitude of the learned divines in the worst and most intolerant days of the Bukhara Emirate towards the local clock-maker who was a Christian. He was permitted to enter the mosque for repairing the tower clock only on a plea that after-all in the matter of technical usefulness he was on par with the donkeys which carried the stone-slabs into the mosque in the first place. Why should the clock-maker suffer a greater social disability? Not only did our bureaucracy adopt the divines' attitude to the clockmaker but also if possible the clockmaker was hired from abroad.

One aspect of this neglect is the awful fact that there are so few of us in the country. According to the statistics collected by the Scientific Commission there is a total of sixty trained physicists in Pakistan. To get the scale right, this is roughly the number of correspondingly trained men you may find in any one London College. In scientific research, unfortunately, it is no longer possible for a single person to achieve his individual breakthrough. Before science can flourish and a scientific tradition can develop, there has to be a critical size, a critical number of trained scientists at one place. Once the critical number is reached, the chain reaction starts; the group becomes self-reacting. Otherwise it simply withers and dies away.

I have great hopes that all this is going to change. As you know a Scientific Commission was appointed by the Government last year, and it has presented its report. From the manner the Government has reacted to reports of its previous Commissions, I venture to predict that 1961 may be the beginning of a new era for scientific research in respect of its organisation, in respect of the massive training programmes which may be initiated and in respect of the calls which the nation may make on its scientific talent. In fact, a feeling that the boot may soon be on the other foot. I only hope we, as scientists, can rise to the challenge and are not found waiting and unprepared.

What exactly are the tasks where we as scientists can make an immediate contribution? One could list a number, ranging from problems of low productivity in agriculture, problems of flood control and water-logging, to the optimum use of Sui-gas. To take one concrete example, a new method for gas-reduction of low grade iron ore has been developed in Mexico. Most steel producing countries are of course not interested in gas-reduction for they possess plentiful supplies of coke . . . The Mexican process is producing one million tons of steel annually. Our situation in Pakistan is similar to that of Mexico. We possess gas as well as low-grade iron ore. It is gratifying that our Department of Scientific and Industrial Research has independently started a small development project for the process. If successful, it may revolutionise our steel economy. Would you not agree that the project needs topmost blessing and the highest priorities?

I would like to end by briefly reiterating some of my remarks. In hoping to achieve the breakthrough to national prosperity, we, like most other poor countries, depend considerably on numerous factors beyond our national control. But there are a number of internal prerequisites the nation must satisfy before the transformation of our society will take place. The first and foremost of these is the firing of the entire nation and harnessing of its spiritual energy, to the objective of eradication of poverty within one generation. This will need constant reiteration of the economic objectives; in particular this will need convincing the nation that the economic policies are designed to enrich the whole society and not merely a part of it. I do not know how the youths of Dacca spend their evenings but as a measure of the nation's consciousness, I shall feel happy when Lahore, for example, makes a transition from its present literary to a technological culture and instead of love-lyrics in the Mall Cafes the discussions range freely and fiercely — at least part of the time — over the targets of the Five Year Plan.

Let us be absolutely clear about the nature of the revolution we are trying to usher in. It is technological and scientific revolution and thus it is imperative that topmost priorities are given to the massive development of the nation's scientific and technological skills. And finally let us as scientists face and live up to the challenge thrown up by Pakistan's poverty. Let future historians record that the fifth important lesson in economic transition to prosperity was taught by Pakistan in achieving a rate of growth as rapid as the Russian and the Chinese but without the corresponding cost in human suffering.

Let me end quoting from the Holy Quran.

إِنَّ اللَّهَ لَا يُغَيِّرُ مَا بِقَوْمٍ حَتَّى يُغَيِّرُوا مَا بِأَنْفُسِهِمْ

*The Lord changeth not what is with a people
until the people change what is in themselves.*

Highlights of Science for Turkey

*Speech delivered by Abdus Salam at the Symposium on "**Turkey in the Year of 2000**", Istanbul, 5 – 7 November 1986.*

أَشْهَدُ أَنْ لَاإِلَهَ إِلاَّ اللهُ وَأَشْهَدُ أَنَّ مُحَمَّدًا عَبْدُهُ وَرَسُولُه

أَعُوذُ بِاللهِ مِنَ الشَّيْطَانِ الرَّجِيمِ

بِسْمِ اللهِ الرَّحْمَنِ الرَّحِيمِ.

1. It is an honour and a privilege for me to have been invited to speak on Physics in the Year 2000, in Turkey and in the Islamic world, at this gathering convened by the prestigious journal, the *Insan ve Kainat* and in the presence of those who run this great country.

My experience of dealing with development-related sciences derives from directing and running a United Nations International Centre for high level physics research located in Trieste, since 1964. Since its inception, this Centre has had the privilege of welcoming of the order of 29,200 experimental and theoretical physicists, more than half of them working in research institutes and universities in developing countries. I am also privileged to be responsible for the Third World Academy of Sciences which comprises prestigious scientists from our countries and as such a purview of the year 2000 in sciences falls under the Academy's mandate. Professor Feza Gursey whom you all know is one of the distinguished Fellows who graces this Academy.

2. Now first and foremost one must realise that Turkey represents a great nation which has contributed from the earliest times to Islamic scholarship and science. The recently published Encyclopedia of Islamic Science, edited by Saban Dogen, is a witness to the preponderance of Turkish names among those who created science in Islam.

Today Turkey has a population level nearly that of Great Britain and France and some five times larger than that of Sweden. There is no reason why Turkey should not be a leader of sciences, certainly by the year 2025, if the right priorities are allocated to science. So far as physics is concerned, during the last fifteen years, the Centre at Trieste alone has been privileged to welcome 434 visits of physicists from Turkey — such is the enthusiasm of Turks for this subject. In this context it is wise to remember that physics is an incredibly rich discipline: it not only provides us with the basic understanding of the laws of nature, it also is the basis of most of modern high technology. Thus physics is the "science of wealth creation" par excellence. This situation may well change in the 21st century, but this is true today. This is even in contrast to chemistry and biology which together provide the "survival basis" of food production as well as of pharmaceutical expertise. Physics takes over at the next level of sophistication. If a nation wants to become wealthy, it must acquire a high degree of expertise in physics, both pure and applied.

3. The Third World as a whole is slowly waking up to the realisation that — in the last analysis — science and technology are what distinguish the South from the North. On science and technology depend the standards of living of a nation and its defense standing. The widening gap between nations of the North and the South is basically the science gap. To see this gap, just turn over the pages of a multidisciplinary science journal — like *Nature*. Not more than 1% of the papers originate in the South, and if there are any occasional Southern names among other contributions, these are either Ph.D. students or Southern researchers who have migrated to the North.

Now, while the South is making some purposeful efforts to acquire technology, very few of us have yet woken up to the need of acquiring science as well. I shall therefore spend the bulk of my time speaking of science transfer.

Let me begin by calling to your minds the year 1799 in Turkey: Against the opposition of the Ulema — and surprisingly even of a section of the military establishment — in that year Sultan Selim III did introduce the subjects of algebra, trigonometry, mechanics, ballistics and metallurgy into Turkey. He imported French and Swedish teachers for teaching these disciplines. His purpose was to rival European advances in gun-founding. Since there was no corresponding emphasis on research in these subjects, and particularly, in materials research, Turkey could not keep up with the newer advances being made elsewhere. The result was predictable: Turkey did not succeed. Then, as now, technology, unsupported by science, will not flourish.

As my second example, take the situation in Egypt at the time of Muhammad Ali, thirty years after the episode with Selim III I have just recounted. Muhammad Ali in Egypt had his men trained in the arts of surveying and prospecting for coal and gold, in Egypt. This attempt was unsuccessful but it did not strike him, nor his successors, to train Egyptians on a long term basis in the sciences of geology or of related environmental sciences. Thus, till this day, there is not one high-level desertification research institute in the entire sub-continents of North Africa or the Middle East (except in Israel). When we recently organised a course on the physics and mathematics of the desertification process, we had to import teachers from Denmark — with their experience with the wastes of Greenland!

My third example is again from Egypt, where, I am told, three million dollars was spent in setting up a factory for the manufacture of thermionic valves. The factory was built in the same year that transistors were perfected and began to invade the world markets. The recommendation to set up the thermionic valve factory was, of course, naturally made by foreign consultants. It was, however, accepted by Egyptian officials who were not particularly perceptive of the way science was advancing, and who presumably never consulted the competent physicists in their own country.

4. Why do we neglect science and technology based on science so far as development is concerned? First and foremost, there is the question of national ambition. Let me say it unambiguously. Countries of the size of Turkey, or Egypt, or my own country, Pakistan, have no science communities geared to development because we do not want such communities. We suffer from a lack of ambition towards acquiring science, a feeling of inferiority towards it, bordering sometimes even on hostility.

In respect of ambition, let me illustrate what I mean by the example of Japan at the end of the last century, when the new Meiji constitution was promulgated. The Meiji Emperor took five oaths — one of these set out a national policy towards science — "Knowledge will be sought and acquired from any source with all means at our disposal, for the greatness and security of Japan". And what comprised "knowledge"? Listen to the Japanese physicist, Hantaro Nagaoka, specialising in magnetism — a discipline to which the Japanese have contributed importantly, both experimentally and theoretically since. Writing in 1888 from Glasgow — where he had been sent by the Imperial Government — to his Professor, Tanakadate, he expressed himself thus: "We must work actively with an open eye, keen sense, and ready understanding, indefatigably and not a moment stopping . . . There is no reason why the Europeans shall be so supreme in everything. As you say, . . . we shall . . . beat those *yattya bottya* (pompous) people (in science) in the course of ten or twenty years".

The same happened in the Soviet Union sixty years ago when the Soviet Academy of Sciences, founded by Peter the Great, was asked to expand its numbers and was set the ambition of excelling in all sciences. Today it numbers a self-governing community of quarter of a million scientists working in its institutes, with priorities and privileges accorded to them in the Soviet system that others envy. According to Academician Malcev, this principally came about in 1945, at a time when the Soviet economy lay shattered by the war. Stalin decided at that time to increase emphasis on sciences. Without consulting anyone else, he apparently decided to increase the emoluments of all scientists and technicians connected with the Soviet Academy, by a factor of three hundred per cent. He wanted bright young men and bright young women to enter massively the profession of scientific research and he succeeded.

5. Among the developing countries today, from experience at Trieste, we perceive just five which do value science and science-based technology, whatever else be their hang-ups. These countries are: Argentina and Brazil in Latin America, and China, India and South Korea in Asia. Barring these five, the Third World, despite its realisation that science and technology are the sustenance, and its major hope for economic betterment, has taken to science as only a marginal activity. This is, unfortunately, also true of the aid-giving agencies of the richer countries and also of the agencies of the United Nations.

Assuming that you agree with me that science has a role for development, why am I insistent that science in developing countries has been treated as a marginal activity? Two reasons:

First: policy makers, prestigious commissions (even the Brandt Commission), as well as aid-givers, speak uniformly of problems of technology transfer to the developing countries as if that is all that is involved. It is hard to believe but true that the word "science" does not figure in the Brandt Commission report. Very few within the developing world appear to stress that for *long term effectiveness, technology transfer must always be accompanied by science transfer*; that the science of today is the technology of tomorrow and that when we speak of science it must be broad-based in order to be effective for applications. I would even go so far as to say: if one was being machiavellian, one might discern sinister motives among those who try to sell to us the idea of technology transfer without science transfer. There is nothing which has hurt us in the Third World more than the recent slogan *in the richer countries* of "Relevant Science". Regretfully this slogan was parroted in our countries unthinkingly to justify stifling the growth of *all* science.

Second: Science transfer is effected by and to communities of scientists. Such communities need building up to a critical size in their human resources and infrastructure. This building up calls for wise science policies with four cardinal ingredients — long-term commitment, generous patronage, self-governance of .the scientific community and free international contacts. Turkey, I am told, has a total of 7,000 non-clinical Ph.D.s — of these around 1,000 are basic scientists — 400 being physicists working in the universities and research organisations. On US, Japanese, or European norms, the numbers of physicists should be about *4000, a factor of ten larger.* But this is not the whole problem. The real problem is that, in our countries, the high-level scientist has not been allowed to play a role in nation-building *as an equal partner to the professional planner, the economist and the technologist* — and this has gone on so long that the scientist has even forgotten that he should be claiming a stake towards development. Few developing countries have promulgated such policies for science; few aid agencies have taken it as their mandate to encourage and help with the building up of the scientific infrastructure, with a view to the scientists' deployment in development, and few scientists can or do fight for their community's share of such tasks.

6. *Why Science Transfer?* What is the infrastructure of sciences I am speaking about and why? First and foremost, we need scientific literacy and science teaching at all levels, and particularly at the higher levels — at least for the sake of the engineers and technologists. This calls for inspiring teachers, and no one can be an inspiring teacher of science, unless he has experienced and created at least some modicum of living science during some part of his career. This calls for well-equipped teaching laboratories and (in the present era of fast moving science), the provision of the newest journals and books. This is the minimum of scientific infrastructure any country of any size must provide for.

Next should come demands on their own scientific communities from the developing country government agencies and their nascent industries, for discriminatory advice regarding which technologies should be acquired.

Still next, for a minority of the developing countries, there is the need for indigenous scientists to help with their colleagues' applied research work. For any society, the problems of its agriculture, of its local pests and diseases, of its local materials base, must be solved locally. One needs an underpinning from a first-class base in sciences to carry applied research in these areas through. The craft of applied science in a developing country is made harder, simply because one does not have available next door, or at the other end of the telephone line, men who can tell you what one needs to know of the basic principles, relevant to one's applied work.

I spoke earlier of indifference towards science. When I was recently consulting my Turkish colleagues, I was told that this came sometimes even from the engineering community — a community which, in Turkey, enjoys reputation and status. I was surprised by this, for many reasons. Firstly, in Pakistan, my experience is that a lack of appreciation of the possible role of scientists stems from the shortsightedness of planners and economists and not engineers. (The same remark was made to me incidentally by Brazilian scientists).

Secondly, I was surprised because in the history of recent fundamental advances of physics, a crucial role has been played by engineers. Thus, for example, Y. Nishina, the man who first brought high-level physics to Japan and who was the teacher of the two Japanese Nobel Laureates in physics, H. Yukawa and S. Tomonaga, was an electrical engineer by profession. P.A.M. Dirac, the creator of quantum mechanics who, in my opinion, is the greatest figure in physics of the 20th century, was trained as an electrical engineer.   Eugene Wigner, who won a Nobel Prize for physics, started life as a chemical engineer.

To reinforce my remarks, let me recall that in 1961, I attended the centennial celebrations of the founding of the Massachusetts Institute of Technology, perhaps the most important technological school of the United States.   To my surprise, it was the engineers at this school who wanted the modicum of science to be increased in their curricula.

7. The remarks I have just made are a repetition of what I had the privilege to say at a UNDP meeting held here on 2 September 1985 in Istanbul. In my speech at that time, I mentioned that Greece had joined the Centre for Nuclear Research at Geneva, the largest and the most prestigious European organisation for particle physics research, with an annual budget of a quarter of a million dollars.   Greece displayed the ambition of joining the big league in science and one can visibly see as a result the maturity which Greek physics has acquired and its transformation year after year. How this maturity will reflect itself in the area of development, will, of course, depend on the policies which Greece will pursue in employing these men.   But the physicists will be there, at any rate.

I am very glad that my voice did not go unheeded and that Turkey is proposing to join the European Organisation for Nuclear Research (CERN) like Greece has already done — at least as an associate member. If properly used, this could be a source of high technology besides high science — particularly as CERN has wisely decided to grant a ten year remission to a payment of the full quantum of membership dues (a few million dollars) for new applications like Portugal, Spain and Turkey. In the first year of joining, 9/10 of these funds will be spent on building up technical infrastructure — computers, data analysis, microprocessors, detection-device technology, vacuum science and technology — *within* Turkey.   The following year this share will be 8/10's, the following year 7/10's within Turkey and so on.   As I said, if it is wisely planned, Turkey could build up high science as well as high technology through this enforced spending within Turkey.

While I am speaking of high technology, I am assuming, of course, that Turkey is going to follow the same pattern for its development as a country equal to it in size, so far as population is concerned.   This is South Korea.   South Korea had the advantage over Turkey of 100%

literacy — as against 52%[a] — when it started on this forward course 15 years ago. As you are aware, South Korea has increased its GNP from $100 per capita to $2000 over this period. They had access to a number of Koreans trained in the US. Turkey has a similar advantage today of having had a part of its working population trained in the technologies of metallurgy, automobile fabrication and the like, in Germany and other European countries. Just now, South Korea has embarked on a new phase of wishing to develop its basic sciences, in addition to excelling in fibre optics, microchip manufacture, and microprocessors. So much so that when I visited that country six weeks ago, I was asked to appear on a television interview lasting two and a half hours, together with Korean physicists. They told me that South Korea had made it a national objective to win Nobel Prizes and they wanted me to give advice to the young people on how to achieve this. Such an objective may seem rather strange to some people, but not to the South Koreans.

8. The second area which may be exploited by the year 2000 is the area of photon physics. In its weekly section on science and technology, the London *Economist* three weeks ago (18 October, 1986), writing under the caption "The Future Belongs to the Photon", had this to say:

"Electronics has been the main engine of innovation since the invention of the transistor forty years ago. Most of tomorrow's interesting technologies will work by manipulating light, not electricity. The electronics revolution is not young. The electron was identified less than a century ago and the microchip, on which today's information-technology industry utterly depends, has been around for fewer than twenty. The successes crammed into these two hectic decades have created the impression that electronics is a technology capable of limitless improvement.

---

[a]One factor which may have affected literacy in our countries is the authorities' and the Ulema's opposition to printing. Though Turkey was the first among Islamic countries to authorise printing — as early as 1727 — when Ibrahim Muteferrica received authorisation through an imperial edict to publish the first edition of a 1583 manuscript on the "newly discovered America" — Ibrahim Muteferrica's presses were stopped at his death, and printing did not resume in Turkey till the middle of the nineteenth century. The authorisation to print the Holy Book (but in Arabic text only) was granted only in 1874 — a full 320 years after Gutenberg's Bible!

"It is not. Electronics will give way to a superior technology based not on electricity but on light. Physicists did not realise until early in this century that light came in the separate packets they now call photons. But science has made startling progress in manipulating photons. A photonics revolution is already in the making.

". . . Why is the switch (to photons) worth making? Because photons travel faster than electrons; because they have no mass; because (unlike electrons, which interfere with each other) photons can be made to pass through each other unperturbed . . .

"Moreover, electronics is discovering its limits . . . Electronics has not reached that limit yet, but it is drawing close enough to worry engineers.

"The customary way to make computers cheaper and faster is to squeeze electronic components closer together. The number that can be fitted on a single chip has grown from about a dozen twenty years ago to two million today. But miniaturisation, too, is bumping against limits . . . And when components get too close, the chips are plagued by "cross-talk" — the leakage of charges from one component to another.

". . . The case for a photonic solution is compelling. Sending several electric currents through one chip at the same time risks cross-talk and disaster. Not so with beams of light: a chip could process several at once without their interfering with each other . . . Consider how rapidly light has nudged electronics out of two pillars of information technology: telecommunications and the storage of information.

"In communications, telephone companies are tearing out their copper cables as quickly as they can afford to and replacing them with hair-thin optical fibres made of glass. Light is a better messenger than electricity.

". . . One way or another, light looks like the wave of the future".

This is an area where the future is just beginning. A nation can join at the ground floor. As I was told in Japan, companies like Fujitsu and Sony and Hitachi do not rely on their superb technologists alone to excel. It is their Ph.D.s in physics, etc. — men who know photons (and electrons) intimately — who are responsible in the last analysis for adding innovative quality to their products. Turkish craftsmen have demonstrated their predilection for miniaturisation technology — writing whole suras of the Holy Quran on a grain of rice. Could they not excel equally in the newly emerging field of photon physics?

9. The third area I wish to stress for the year 2000 for Turkey is biotechnology. I wish to stress this since, partly on account of my urging, the Italian government has accepted, together with the UNIDO organisation, to create in Trieste an international centre on biotechnology for developing countries, like mine for physics.

As is well-known, the modern advances in genetics started with the unravelling of the genetic code by Watson and Crick.

This great discovery in biology — one of the most synthesising discoveries of the 20th century, possibly of all times — was made at Cambridge in April 1953 by two contemporaries of mine, one American, the other British — working at the world-famous Cavendish Laboratory which specialises in *basic physics*. An American pupil of mine for Ph.D. in theoretical physics — Walter Gilbert — with whom I published a paper on dispersion phenomena — was a neighbour of the genetic code's American co-discoverer, J.D. Watson. When Gilbert left me in 1956 after his Ph.D., both he and Watson went back to Harvard. The next time I saw my good pupil, Gilbert, was in 1961 in the US. Assuming that he was still working on some problem of theoretical physics I asked him what he was up to. He was somewhat sheepish; he said, "I am sorry, you will be ashamed of me; I am spending my time growing bacteria." Watson had seduced Gilbert for genetics.

Gilbert soon discovered a most elegant technique for deciphering the genetic code. For this work, he received *the Nobel Prize in chemistry in 1980*. In 1981 he left his chair at Harvard to found a company which exploits techniques of genetic manipulation to manufacture, among other medications, human insulin. This company is Biogen and is registered in Switzerland. It went public recently. Apparently, Gilbert's first investment in the company (of which he was President — he has subsequently returned to science at Harvard) was of US$4,000; this, I am told, is currently worth more than 14 million dollars.

Notice the mutuality of science and technology. Notice that the greatest discovery in molecular biology is made in a laboratory for physics, by men trained in the use of X-rays with fairly modest equipment. Notice Gilbert's transition from research in theoretical physics to fundamental genetics and then to practical genetic engineering. The point I am trying to make is twofold: first, science and technology go hand in hand in modern times; second there is a premium placed on excellence and brain power in civilisations other than ours. We must ask ourselves: do we provide like opportunities for our best men, nurturing their talents for our civilisation, or do we leave them to wither away, or if they are strongly committed to science, to migrate and to enrich the countries of Europe and America with their talents and their contributions?

Perhaps my examples appear too distant for comfort. Perhaps the intervening centuries of neglect of sciences have lured into us a feeling that we can never catch up in the creation of sciences, and that we need not even try. Is this true?

Biotechnology is one of the newer sciences. Like physics today, biotechnology's applications are expected to dominate the 21st century — in agriculture, in energy, in medicine. My next excerpt describes the obstacles which the developing world (including the World of Islam), faces in building up expertise in this subject. This is a quote from a guest editorial from the journal *Biotechnology:* "Biotechnology thrives on new knowledge generated by molecular biology, genetics and microbiology, but these disciplines are weak, often nonexistent, in the under-developed world. Biotechnology springs from universities and other research institutions, centers that generate the basic knowledge needed to solve practical problems posed by society. *But the universities of the underdeveloped world are not research centers . . .* And the few creative research groups operate in a social vacuum; their results might be useful abroad, but are not locally . . . Biotechnology needs dynamic interactions among the relevant industries. These interactions, however, are weak in countries *in which science is perceived as an ornament, not as a necessity . . .* Biotechnology requires many highly skilled professionals, but . . . underdeveloped nations lack sufficient people well trained in the pertinent disciplines . . . Economic scarcity and political discrimination induce professionals and graduate students to emigrate or abandon science altogether".

The writer goes on to ask "What can be done" and his answer is: "First of all, *underdeveloped countries must understand that they need to reform their universities . . .* ·They must recognise that molecular biology is not just another branch of biology, but the one and only tool available for understanding biological structure and function . . . Success in biotechnology depends on the conquest and consolidation of the moving frontiers of cell biology and medicine".

I am sure India, China, Argentina, Brazil, South Korea, among the developing countries, will take heed of this call. The question is shall

Turkey and the Muslim nations take heed also or shall we lose out in this new race to master and utilise biotechnology?

10. Let me summarise: we must ensure that we do not lose out in new physics, nor in physics-based high technology, nor in biotechnology. That is if we wish to live honourably in the 21st century; and wish to defend our culture and our civilisation. We must ensure that our scientific enterprise is of first class quality in these disciplines — like the South Koreans have ensured — and that it maintains living contact with international science. After all, science is at present being created — and at a furious pace[b] — *outside* the confines of our countries. At present, very few of us, if living and working in our own countries, can travel to scientific institutions and meetings abroad. Such travel, as a rule, is considered a wasteful luxury. In some of our countries, incredibly, it needs authorisation from the highest authority in the land!

So far as biotechnology is concerned, as I said, the Italian local community at Trieste, in consonance with the government in Rome, has donated funds to found an international institute of biotechnology. The local funds from the Government of the Trieste Region are of the order of 40 million dollars. Many other client countries will benefit — Turkey also. But I cannot understand why Turkey — the Region of Istanbul — should not found an international institute of biotechnology at the same level with its own manpower, which, if planned properly, can be built up in a short time.

For physics, I know from personal experience of working with them, that Turkish physicists are some of the most imaginative. They undertake difficult problems in physics consciously — and this is something I respect. I had the privilege of visiting this great country a short while back when I was honoured to be received by President Kenan Evren. I suggested to him that, in my opinion, what Turkey needed in its national priorities and plans was something analogous to the Bell Telephone Laboratories in the United States in the field of physics of communications. The Bell

[b]To stress this furious pace, let me recall that last week at Trieste, during his inaugural lecture on the frontier subject of Brain Research, the speaker started with the remark: "The last ten years have seen more accumulation of knowledge (and more books being made obsolete) in brain research than in the entire history of mankind".

Laboratories have produced six Nobel Laureates who have contributed to basic physics, besides including transistors in their roster of inventions. I estimate that the Turkish (or the Egyptian or the Pakistani) analogue to the Bell Laboratories for *Communication* physics would cost 40 million dollars to build and around four million dollars yearly to run. I believe it can be done with the highest level of quality, and that one can find those who could create it in Turkey.

In the context of a State's duties towards science, permit me, before I conclude, to present to you the proclamation 5461 of 17 April 1986 of the President of the United States of America. This proclamation reads:

> *"Now, therefore, I, Ronald Reagan, President of the United States of America, do hereby proclaim the week of April 14 through April 20 1986, as National Mathematics Awareness Week, and I urge all Americans to participate in appropriate ceremonies and activities that demonstrate the importance of mathematics and mathematical education to the United States.*
>
> *"In witness whereof, I have hereunto set my hand this seventeenth day of April, in the year of our Lord nineteen hundred and eighty-six, and of the independence of the United States of America the two hundred and tenth".*

Here is an example, for our nations and our rulers to emulate, of what it means to be aware of basic sciences and their importance. At the bottom of this page is a chart which gives the figures for the *voluntary* contributions from the public, raised in the US (52 million dollars during 1986) for bringing mathematics to American consciousness.

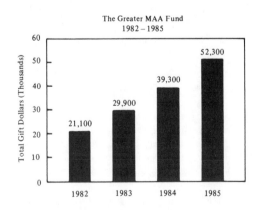

The Greater MAA Fund
1982 – 1985

11. I was asked recently by the Education Minister of Kuwait to write on the "Future of Science in Islam" for the Islamic Summit Conference, which is going to be held in Kuwait on 26 January 1987. Among other recommendations, in this paper I have also pleaded for the creation of foundations of science and prizes for science. These foundations and prizes should come from both Islamic Governments as well as from private donations. I have asked the question why 1/8th of the Auqaf Funds in our countries should not be devoted to sciences, in keeping with the emphasis on sciences in the Holy Book of Islam. Let us make no mistake about it; in contrast to 250 verses which are legislative, some 750 verses of the Holy Quran — almost one-eighth of it — exhort the believers "to study Nature — to reflect, to make the best use of reason and to make the scientific enterprise an integral part of the community's life". I have been asking Muslim divines in India and Pakistan if they were devoting one out of every eight of their Friday sermons to stressing these aspects of sciences. In reply, they said that they would have liked to but they do not know enough science themselves. Has the time not come for them to learn and to speak of the fundamental forces of Nature and their unification, of the structure of the earth, of the fascination and magic of the genetic code as marvels revealed by contemporary science, and to bring these to the consciousness of the believers, as the Holy Book enjoins us to do?

Let me reiterate; I miss an attitude towards science which considers it as being an integral part of our lives. This attitude has been absent in our countries from the 15th century onwards. May I suggest that the moment has come when our courts of State should be adorned again with scientists. I am reminded of the court of King Arthur of legendary fame; at that court there was a court magician; his name was Merlin. It was his function to use magic for forging steel for swords and to provide medicinal potions. We the scientists are the Merlins of today. We can perform feats of magic undreamt of by Merlins of yesteryears. We can even transform society. But in our Third World countries, the Merlins have no place in the courts of State. Should they not be invited back? Some will say — and perhaps rightly — that the Merlins in developing countries are amateurs, they do not know their applied craft. They choose to live in their own ivory towers, and our societies are thereby forced to import the real Merlins from the West. This may be true, but why is this? Could this emasculation have come about through the fact that our Merlins are so few in numbers, and even these few have never been permitted to make a contribution to development of their own countries? Not even by their colleagues — the economists — who in this metaphor are the High Priests of State. Only experience can teach the Merlin-Scientist the craft of developmental problem solving, even if he knew his science. This vicious cycle of lack of mutual trust must be broken, I hope before the year 2000. From all the portents I can see — including the holding of this Conference — the year 2000 will be a glorious year of distinction in sciences and their purposeful application towards the problems of this great land of Turkey. Inshallah.

رَبَّنَا وَآتِنَا مَا وَعَدْتَنَا عَلَى رُسُلِكَ وَلَا تُخْزِنَا يَوْمَ الْقِيَامَةِ إِنَّكَ لَا تُخْلِفُ الْمِيعَادُ .

The Isolation of the Scientist in Developing Countries

By Abdus Salam, reprinted from Minerva, Vol. IV, No. 4, Summer 1966.

Metropolis and Province in the Scientific World

Five hundred years ago, around 1470 A.D., Saif-ud-din Salman, a young astronomer from Kandhar, working then at the celebrated observatory of Ulugh Beg at Samarkand, wrote an anguished letter to his father. In eloquent words Salman recounted the dilemmas, the heartbreaks of an advanced research career in a poor, developing country:

"Admonish me not, my beloved father, for forsaking you thus in your old age and sojourning here at Samarkand. It's not that I covet the musk-melons and the grapes and the pomegranates of Samarkand; it's not the shade of the orchards on the banks of Zar-Afshan that keeps me here. I love my native Kandhar and its tree-lined avenues even more and I pine to return. But forgive me, my exalted father, for my passion for knowledge. In Kandhar there are no scholars, no libraries, no quadrants, no astrolabes. My star-gazing excites nothing but ridicule and scorn. My countrymen care more for the glitter of the sword than for the quill of the scholar. In my own town I am a sad, a pathetic misfit.

"It is true, my respected father, so far from home men do not rise from their seats to pay me homage when I ride into the bazaar. But some day soon all Samarkand will rise in respect when your son will emulate Biruni and Tusi in learning and you too will feel proud."

Saif-ud-din Salman never did attain the greatness of his masters, Biruni and Tusi, in astronomy. But this cry from his heart has an aptness for our present times. For Samarkand of 1470 read Berkeley or Cambridge; for quadrants read high-energy accelerators; for Kandhar read Delhi or Lahore and we have the situation of advanced scientific research and its dilemmas in the developing world of today as seen by those who feel in themselves that they could, given the opportunity, make a fundamental contribution to knowledge.

But there is one profound change from 1470. Whereas the emirate of Kandhar did not have a conscious policy for the development of science and technology — it boasted of no ministers for science, it had no councils for scientific research — the present day governments of most developing countries would like to foster, if they could, scientific research, even advanced scientific research. Unfortunately, research is costly. Most countries do not yet feel that it carries a high priority among competing claims for their resources. Not even indigenous *applied* research can command priority over straightforward projects for development. The feeling among administrators — perhaps rightly — is that it is by and large cheaper and perhaps more reliable to buy applied science on the world market. The resultant picture, so far as advanced research is concerned, remains in practice almost as bleak as at Kandhar.

Why Advanced Research Lags in Underdeveloped Countries

First and foremost among the factors that affect advanced scientific research is the supply of towering individuals, the tribal leaders, around whom great institutes are built. These are perhaps 2−3 per cent of all men who are trained for research. What is being done in the under-developed world to ensure their supply? Most developing countries are doing practically nothing. Quite the contrary, with all the obstacles and hazards which beset a poor society, it is almost miraculous that any talent at all is saved for science. These hazards are, first, the very poor quality of education; second, the higher or administrative grades of the civil service — in India, the Indian administration service, and in Pakistan, its analogue, the civil service of Pakistan — which skim off the very top of the sub-continent's intellect; third, the poor chances for a promising young research student to learn to do research as an apprentice to a master scientist. The greatest obstacle of all lies in the very low probability of having the opportunity to work with the few men — in the case of India and Pakistan, the Siddiquis, the Usmanis, the Menons, the Sarabhais, the Seshachars — at the few centres of excellence, who appreciate at all the demands of a research career and who run laboratories which are reasonably well equipped. There are just too few scientists who retain the creativity of which they gave promise when young and there are therefore too few to train younger scientists through a fruitful master-apprentice relationship. It remains a sad fact that, though India and Pakistan may have built specialised institutes outside the university system where advanced research is carried out, by and large their vast university systems remain weak, static and uninspired. It is not part of their tradition to make a place for advanced research or even for research at all. The colleges which provide a very large proportion of under-graduate education in India and Pakistan have grown up in a tradition of concentrating such resources as they have on the instruction and moral formation of undergraduates. I shall always remember my first interview with the head of the premier college in Pakistan, which I joined after a spell of theoretical work in high-energy physics at Cambridge and Princeton. My chief said: "We all want research men here, but never forget we are looking more for good, honest teachers and good honest college men. This college has proud traditions to uphold. We must all help. Now for any spare time you may have after your teaching duties, I can offer you a choice of three college jobs; you can take on wardenship of the college hostel; or be chief treasurer of its accounts; or if you like, become president of its football club". As it was, I was fortunate to get the football club.

Admittedly, this was 12 years ago. I should be ungrateful if I did not mention that this same college today is contending with the Atomic Energy Commission of Pakistan for the control of a high-tension laboratory with a 2.5 Mev Cockcroft-Walton set. This is a measure of the change brought about by the heroic efforts of the Pakistan Government since 1958. Things have changed. Nonetheless the situation of advanced research in underdeveloped countries still remains greatly in need of help.

In a number of fields, advanced scientific research in developing countries is beginning to reach the stage of maturity in which first-rate work can be done. Indigenous resources are being skilfully employed but there is still a desperate need for international help. The truth is that, irrespective of a man's talent, there are in science, as in other spheres, the classes of haves and have-nots; those who enjoy the physical facilities and the personal stimulus for the furtherance of their work, and those who do not, depending on which part of the world they live in. This distinction must go. The time has come when the international community of scientists should begin to recognise its direct moral responsibility, its direct involvement, its direct participation in advanced science in developing countries, not only through helping to organise institutions but by providing the personal face-to-face stimulation necessary for the first-rate individual working in these countries.

In advanced scientific research, it is the personal element that counts much more than the institutional. If, through meaningful international action, allied with national action, we could build the morale of the active research worker and persuade him not to make himself an exile, we shall have won a real battle for the establishment of a creative scientific life in the developing countries.

Breaking the Barrier of Isolation

As an example of what is needed, I shall take the science with which I am personally associated. Theoretical physics happens to be one of the few scientific disciplines which, together with mathematics, is ideally suited to development in a developing country. The reason is that no costly equipment is involved. It is inevitably one of the first sciences to be developed at the highest possible level; this was the case in Japan, in India, in Pakistan, in Brazil, in Lebanon, in Turkey, in Korea, in Argentina. Gifted men from these countries work in advanced centres in the West or the Soviet Union. They then go back to build their own indigenous schools. In the past, when these men went back to the universities in their home countries, they were perhaps completely alone; the groups of which they formed a part were too small to form a critical mass; there were no good libraries, there was no communication with groups abroad. There was no criticism of what they were doing; new ideas reached them too slowly; their work fell back within the grooves of what they were doing before they left the stimulating environments of the institutions at which they had studied in the West or the Soviet Union. These men were isolated, and isolation in theoretical physics as in most fields of intellectual work — is death. This was the pattern when I became associated with Lahore University; this is still the pattern in Chile, in Argentina, in Korea.

In India and Pakistan we have been more fortunate than most other underdeveloped countries in the last decade. A number of specialised institutes have grown up for advanced work in theoretical physics — the Tata Institute at Bombay, the Institute of Mathematical Sciences at Madras and the Atomic Energy Centres at Lahore and Dacca — where a fair concentration of good men exists. But this is not enough. These institutes are still small oases. They are too small for the fertilisation of the area around them. They are also in continuous danger of being dried up because the area around them is too arid and they still do not have vigorous contacts with the world community. Tata and Madras have partly solved their problem; they have funds to invite visitors — they have fewer funds to send Indian physicists abroad, mainly because of the serious shortage of foreign exchange.

It was with this type of problem in mind that the idea of setting up an International Centre for Theoretical Physics[a] was mooted in 1960. The idea was to establish a truly international centre, run by the United Nations family of organisations, for advanced research in theoretical physics. It was planned with two objectives in view: first, to bring physicists from the East and the West together; second, and even more important, to provide extremely liberal facilities for senior active physicists from developing countries.

The International Centre tries to deal with the problem of isolation in a number of ways. We have ordinary fellowships which are given mostly to those from developing countries. In addition, the International Centre has instituted an associateship scheme. A number of carefully selected senior active physicists from developing countries are given the privilege of coming for a period of one to four months every year to the centre with no prior formalities other than a letter to the director announcing their arrival. The centre pays for their transportation and their maintenance. The aim is to have eventually, at any one time, a group of about 50 senior active physicists from developing countries who possess this privilege.

Looking back on my own period of work in Lahore, as I said, I felt terribly isolated. If at that time someone had said to me, we shall give you the opportunity every year to travel to an active centre in Europe or the United States for three months of your vacation to work with your peers; would you then be happy to stay the remaining nine months at Lahore. I would have said yes. No one made the offer. I felt then and I feel now that this is one way of halting the brain drain, of keeping active men happy and contented within their own countries. They must

[a]*Cf. Minerva*, III, 4 (Summer, 1965), pp. 533 536.

be kept there to build for the future, but their scientific integrity must also be preserved. By providing them with this guaranteed opportunity for remaining in contact with their peers, we believe we are making a contribution to solving the problem of isolation.

Ideally the associateship scheme should be wide enough to cover nearly every active physicist in developing countries. It should be well publicised: every first-rate research worker should know and feel confident that he could almost, as it were, demand its privileges if he were living in a developing country. Unfortunately, the International Centre at Trieste does not possess funds to do this. Yet the scheme is not very costly. Since it pays no salaries — only the fare and a *per diem* allowance — it costs us something like $100,000. Since the associateship scheme seems thus far to be the most fruitful of all the available ways for breaking the isolation which kills the creativity of creative scientists, it should be extended.

Universities and institutions with the wealth and scientific eminence of Princeton, Harvard, Cambridge, All Souls, Rockefeller University, New York State University, the Imperial College in London and others should seriously consider the establishment of their own associateship schemes. It ought to be considered not only for theoretical physics but for other subjects too. Rockefeller University, for example, might extend the privilege of giving its freedom not only to a scientist of the distinction of Professor Seshehar, but also to other active microbiologists in most developing countries. The European Organisation for Nuclear Research at Geneva has already started a scheme similar to our own, which, I believe, covers both experimental and theoretical physics. It is designed of course only for less-developed countries within Europe (Greece and Spain).

If every active, first-rate worker in the developing countries could be covered, we would go very far towards the removal of one of the curses of being a scientist in a developing land.

II.2
INTERNATIONAL CENTRE
FOR THEORETICAL PHYSICS

Need for an International Centre for Theoretical Physics

Speech delivered by Abdus Salam at the IAEA Annual Conference, Vienna, 1962, moving the resolution for the creation of the International Centre for Theoretical Physics.

Two years back in September 1960 the Pakistan delegation had the privilege to co-sponsor the first resolution requesting the Agency to create an International Centre for Theoretical Physics. During these two years the idea has progressed; firstly by the generous financial offers from the Danish and the Italian Governments to help in setting up such a Centre and secondly by the enthusiastic support the idea has received from the world community of physicists. The panel of physicists which was convened by the Director General in March 1961 (whose report was circulated to member Governments) enthusiastically endorsed the idea of the Centre; they defined its scope and objective and discussed the manner in which it could be set up and run. I shall be referring frequently to the conclusions of this panel.

In considering the question of whether the Agency should set up a Centre, there are three questions we should ask ourselves:-

(1) Does research in theoretical physics fall within the scope of the Agency's activities?

(2) Do physicists from the emerging countries really need and desire such a Centre?

(3) If the Centre is desirable, can it be created and can the Agency afford it?

Let us examine the issue in the light of these three questions. First; *Does research in theoretical physics fall within the scope of the Agency's activities?*

The submission of the sponsors of the resolution is that there are indeed fewer sciences which have contributed more to the coming of the atomic age than theoretical physics. Even if we ignore the fact that Einstein was the first scientist to dream of equivalence of mass and energy and to create the whole basis of our science, even if we forget that two of the world's leading theoretical physicists, Fermi and Wigner, actually built the world's first atomic reactor, we should not forget that there are still uncharted areas in theoretical plasma physics which are vital to the tapping of fusion power. We must not forget that in spite of all our advances in nuclear physics, we still do not know the theoretical expression for the law of force between two nucleons. These indeed are areas in theoretical physics of direct and immediate concern to the Agency; with research in them its major responsibility. Let me name these areas again; these are reactor theory, plasma theory, low energy nuclear theory, theoretical high energy physics. This last may sound on the very speculative side of theoretical physics. But I sometimes wonder what reply an Agency like ours may have given to a request of a young and unknown theoretical physicist, Albert Einstein, in 1904, if he had made an application for a Fellowship to follow his theoretical speculations on the nature of space and time. Who among us would have dreamt that he would come up also in the same paper with the relationship of energy and mass? Who among us dare say today that a theoretical speculation on the muon fusion may not be relevant to the energy problems of tomorrow?

My remarks about theoretical physics and its relevance to the Agency's activities are not to deny the role of other scientific disciplines nor to deny the Agency's responsibilities to them. But as we shall see, the claim theoretical physics has rests on one thing peculiar to it; theoretical physics needs no apparatus, it is the least costly of all sciences. The return relative to cost for this discipline is the highest. If the Agency has to choose among fundamental sciences on a tight budget, theoretical physics is undoubtedly the subject of choice.

The case for the Agency to support theoretical physics gets stronger when we consider this subject from the point of view of the emerging countries. First and foremost let us not forget that young scientists in the under-developed world feel the urge to meet the challenge of fundamental science as much as anyone else. Among the fundamental sciences, theoretical physics has a peculiar fascination for them.

(*i*)First, no costly apparatus is needed.

(*ii*) Second, in this field individual initiative — rather than collaborative effort — can still lead to a breakthrough. Almost invariably theoretical physics is the first science in smaller countries, which gets developed at the advanced level. History bears this out; this was the case in Japan with Yukawa and Tomonaga; this was the case with India, in Brazil, Turkey, Lebanon, and Argentina. No one can reverse this historical process of the order in which science grows in rich or poor soils. But in spite of the native ability, in spite of the ambitions of these scientists, they, in common with other scientists in their countries, suffer from one fatal disability — isolation. After an initial period of brilliant work at some active centre, they are faced with a cruel choice; either to leave their countries or to ossify and become scientific administrators. Unlike other scientists whose disabilities may include lack of costly equipment and apparatus, the theoretical physicist can be helped at a very small cost, by making frequent contacts possible, and by awarding him frequent visiting Fellowships to live for periods at active centres.

Boiled down to its essentials, we are therefore talking of a visiting Fellowship programme. I believe up to this point a large number of delegates are with us. I think the misgivings of some of the delegates start at this point, regarding the manner in which such a Fellowship programme may be administered. The simplest way to do this as suggested by SAC is to have a number of Fellowships available at a selected number of regional or national centres like CERN, Copenhagen, Dubna and Princeton. (If these Fellowships are given by IAEA, in view of the fact that the normal time for processing a Fellowship by the Agency is 17 months, one would have to place these particular Fellowships at the disposal of the Centres we are talking about.)

Now such a Fellowship programme is fine. But its weakness is this; there are not enough places in the existing centres to meet the demands. This was emphasised by the panel of distinguished physicists set up by the Director General — some of the panelists were from the centres I have mentioned. This has been re-emphasised by the letters the Director General has received from CERN and Copenhagen.

I have mentioned the work of the panel of the theoretical physicists. This panel was attended also by a representative of UNESCO. He told us

that while UNESCO is deeply interested in such a Centre, by its charter
it is barred from creating it under its own auspices. An important question
often asked is, will such a Centre lure the scientists from the emerging
countries away from their homes? Our answer is an emphatic "No";
on the contrary, the Centre will help in stemming the torrent of migration.
If one is certain of being able to spend periods of a few months working
in an active Centre, the incentive to become a permanent exile considerably
decreases.

Finally *do the physicists from the underdeveloped countries want
such a Centre?* We have before us a document signed by 53 of the
participants who attended the Seminar at Trieste. Let me read from
this document:

"While important cross-fertilisation of ideas has taken place during
the six weeks of the Seminar, the Seminar has served to underline the
need for a Centre — in order for important joint work to be done, contact
for more than six weeks of a summer seminar is necessary —"

Let us turn now to the third point:-

Is it possible to create such a Centre? Can one find three or four
distinguished physicists who feel that they would make a more important
contribution by joining this Centre as permanent staff? Would other
distinguished men come to this Centre as visiting Professors? We humbly
submit that the answer to these questions cannot be given by a debate
at a meeting of the Board of Governors. The answer depends on whether
the world community of physicsts is enthusiastic about such a Centre.
I can only give my personal impressions. I know for a fact that men
like Niels Bohr, Hideki Yukawa, Hans Bethe, Robert Oppenheimer,
Victor Weisskopf, Robert Marshak, Julian Schwinger, Abraham Pais,
Leopold Infeld, to name but a few, are strongly in favour of an International Centre. No such Centre now exists; once it is created there
are scores of activities it can engage in, like arrangements of seminars
in developing countries. Inevitably the Centre will become the clearing
house for new ideas. With it is associated at present the idealism of the
world community of theoretical physicists.

Gentlemen, let us project to twenty years from now. The world
is moving closer, economically, intellectually, scientifically. In twenty
years, there will be international research centres not only for theoretical
physics but for most fundamental sciences. The world trend is in this
direction and nothing can stop it. It is possible for us in this Agency

to take the initiative in forwarding this movement. I do hope very much we shall. With these words I commend to you the resolution in front of us —

Trieste — World Rendezvous for Physicists

by Dan Behrman

Reprinted from **The UNESCO Courier,** *May 1971.*

Every year, some five hundred of the world's brightest young minds in science are exposed to the International Centre for Theoretical Physics, a rather unusual United Nations institution on the Adriatic coast of Italy just outside Trieste. Most of these scientists come from developing countries and, under ordinary circumstances, they would be likely candidates for the brain drain.

This is precisely why the Trieste centre is in operation with the joint support of two UN agencies, the International Atomic Energy Agency and UNESCO, and the Italian government. As a way out of the intellectual isolation that drives young scientists to emigrate, it offers them training, an opportunity to do research at regular intervals and, most of all, a place to think, talk and work.

From this scientific centre where chalks, blackboards and desks are the only visible apparatus, come more than 130 papers every year in the basic fields of elementary particles, high energy physics, field theory, nuclear physics, solid state physics and plasma physics.

The centre serves to link east and west as well as the developed and the developing worlds. Research workshops have brought together the top people in the United States and the USSR on many topics and particularly plasma physics where problems related to the domestication of the thermonuclear energy of the hydrogen bomb are being studied. If they can be solved, the world will be presented with a new source of power, pollution-free and well-nigh inexhaustible.

Yet the pursuit of theoretical physics cannot be justified in terms of its immediate applications. It is the most philosophical of sciences for it is concerned with the study of the very nature of matter. As such, it attracts the most talented brains of the developing world, the Einsteins, the Fermis, the Niels Bohrs of tomorrow and the day after. They will not devise ways to build better mousetraps but they learn to think in terms of original solutions. If they are not given the chance to work in contact with others at their own level, they languish . . . and they leave.

Such was the experience of the founder and director of the Trieste centre, Prof. Abdus Salam. It might even be said that it sprang from his own life, from the isolation that he himself suffered when, after taking his doctorate at Cambridge and conducting research in Princeton, he returned home to Pakistan to teach in 1951.

"I was the only theoretical physicist in the country at that time," he told me as I sat in his office sharing his lunch of sesame seeds. "The nearest one was in Bombay. You have no idea of what that can be like. A theoretical physicist has got to be able to talk, to discuss, to shout if need be.

"I remember, I received a cable one day from Wolfgang Pauli, the Nobel Laureate from Zurich, who was in Bombay. He said he was alone and he wanted me to come to talk to him. So I took a plane to Bombay and a taxi to his hotel. I went up to his room, I knocked on the door.

"He told me to come in and then, without a word of greeting, he said to me:

"'The problem is, if we have derivative terms in Schwinger's action principle . . .'"

Prof. Salam was called into an adjoining office for a moment and I had a chance to take in his surroundings. On one wall hung a framed 16th century prayer in Persian which, he had told me, invoked the name of Allah to ask for a miracle. A typewritten notice had been slipped under the glass top of his desk:

"Reminder: Mornings to be spent on physics: No visitors — No phone calls — No mail (except personal) before noon — Administrative matters and visitors during the period after lunch until 4 p.m. only. Remaining time to be spent on physics."

Also under glass on a wall to the right of the desk was a quotation: "We have all of us to preserve our competence in our own professions, to preserve what we know intimately, to preserve our mastery. This is in fact, our only anchor in honesty."

Prof. Salam might have written that himself, but it was signed by the late Robert Oppenheimer, one of the earliest supporters of the Trieste centre. "The day that a director of a research centre like this one stops being a scientist, he's useless," Prof. Salam remarked. "Any fool can administer. People forget that they were made heads of centres because they were doing good science. So they lose their competence, they become manipulators of men just to keep themselves in power."

The entire full-time professional staff of the centre could fit into a small Fiat or a short sentence: Prof. Salam, the director; Prof. Paolo Budini from Italy, deputy director; and Dr. Andre Hamende, a Belgian, who is everything else. At Trieste, Parkinson's Law has been repealed. The administrative staff has actually dwindled from five to three since the centre opened in 1964 but the number of scientists it reaches every year has more than quintupled.

The International Centre for Theoretical Physics does all this on a budget of no more than $600,000 a year. Of this figure, the biggest chunk is a generous $250,000 grant from the Italian government which also financed the construction of its $2 million building. Then the International Atomic Energy Agency and UNESCO each gives $150,000. The remainder is made up mainly of contributions from the Swedish International Development Authority and the Ford Foundation.

This sum covers virtually all expenses from fellowships and publications to heating and administration: included is the operation of the centre's library with 6,000 volumes and an up-to-date reference section of journals. The output in physics is so great today that one American journal alone runs to eighteen volumes in a single year.

This all began in 1960 when Prof. Salam was a member of the Pakistani delegation to the General Conference of the International Atomic Energy Agency in Vienna. He has always had a great gift for doing a number of things all at once: even today, he is still science adviser to the President of Pakistan and professor of theoretical physics at the Imperial College of Science and Technology in London along with his tasks in Trieste. Commuting between such jobs would numb an ordinary man, but Prof. Salam claims it enhances his productivity.

As a delegate in Vienna, he put forth the idea of an international centre for theoretical physics. "I was naive then, I wouldn't dare do it today. People took it half-jokingly and many delegations abstained on the vote when it was approved for a preliminary study. I found out that the idea interested the poor countries. What I wanted to do was to give the poor a place of their own where they would not have to beg anybody. Why shouldn't a bright youngster in Pakistan have the right to receive the same stimulating atmosphere as an Englishman or an American, provided he deserves it?"

His proposal got over the first hurdle in 1960. It was helped over the succeeding ones by Prof. Salam's fortuitous meeting with Prof. Budini at a symposium in Trieste on elementary particle interactions.

Prof. Budini was also seeking a way out of isolation, in this case the geographic predicament of Trieste in a cul-de-sac at the far corner of Italy. Nationalism did not make much sense to this physics professor at the University of Trieste whose birthplace, an island that once belonged to Venice, has changed flags three times in his own lifetime. He dreamt of a Trieste lying instead in the centre of Europe, a pole of attraction to fellow physicists from the world over. He and Prof. Salam had no trouble putting their dreams together.

Money was put up by a local bank, the Cassa di Risparmio di Trieste. An offer of land, later converted to money, came from Prince Raimondo di Torre e Tasso whose nearby castle at Duino has played host to Liszt, Mark Twain, Rilke and, most recently, to the 1970 Pugwash Conference. The prince said: "Trieste is my daughter and this is my dowry."

In 1962, the General Conference of the International Atomic Energy Agency approved the creation of a centre. "That was the most momentous day of my life," Prof. Salam told me. "I seldom smoke, but I must have smoked fifty cigarettes that day and I went through a kilo of grapes. At the end of the debate, sixty hands went up in favour — and we had won."

The following year, the Italian government's offer of Trieste as a site was accepted and, in 1964, Prof. Salam and his staff moved into temporary quarters in the heart of the city. Four years later, they were in their present building at Miramare, a long double-decker sandwich in concrete with two rows of wood-framed windows as the filling.

On the grounds of the building, there is a small house where Prof. Salam lives while he is at the centre. It is only twenty yards or so from his office window but he can spend two weeks at a stretch seeing nothing more of the outside world than those twenty yards. He has one group working in Trieste, another at Imperial College in London. In the centre, he and the collaborator, John Strathdee, share an office decorated principally with blackboards and equations.

Prof. Salam told me they were endeavouring to put under one unified scheme the micro and macro universes inside the nucleus and outside in galactic space to cover the frontiers between the behaviour of elementary particles measuring 10^{-15} centimetres (that is, the number one preceded by fifteen zeroes and a decimal point) and the so-called quasars that lie 10^{27} centimetres (the number one followed by twenty-seven zeroes) away from the earth. Prof. Salam is fascinated by the "black holes of gravity" in space occupied by celestial bodies that have collapsed under the weak but relentless force of gravity.

Theoretical physicists, such as those found at Trieste, try to explain the behaviour of elementary particles. But though he may use a computer, the physicist's main tool is his mind, and he must have contact with other minds if he is to keep it honed.

This problem, a major one for many scientists from the developing world, was explained to me by Dr. Paul Vitta who got his Ph. D. in the United States and is now teaching at the physics department of the university of Dar-es-Salaam, in Tanzania.

He has come to the centre to attend a two-month nuclear theory course that was just ending. "In Tanzania," he said, "I am the only nuclear physicist. I am in perfect isolation. With our teaching load, one very soon gives up all hope of research. So you pick up a textbook. It gets out of date, but you're stuck with it. I simply need to come to a centre like this."

Dr. Khaik Leang Lim, from the University of Malaya at Kuala Lumpur, is the only Malaysian theoretical nuclear physicist. "There might be some outside, but not in the country. If you're on your own, you can only read scientific journals. It's hard to keep up. Your interest fluctuates when you get tired reading. And there is no one to talk to."

Dr. Lim is an associate at the centre which means that he has the right to three three-month stays there over a period of five years. Trieste now has sixty such associates from more than twenty countries. It hopes to expand the list until it covers all the estimated 200 theoretical nuclear physicists in the developing world. The centre's activities are now being extended into mathematics as well.

He thinks that basic science is necessary to a developing country if only because over-specialized scientists have trouble adapting to change, but he certainly does not believe that Malaysia needs the whole gamut of theoretical physics. His own speciality, nuclear physics, requires fast computers that are not available at home. "Here, I must think of something to do that requires less computation. One cannot change from one field to another, but one can change within a field. In that respect, the centre helps the individual. He can meet people in the same or related fields, he can learn what's going on."

Dr. Lim would like to see a similar centre some day in South-east Asia, perhaps in Bangkok. He had to come all the way to Trieste to meet Dr. I. T. Cheon from Korea with whom he is now collaborating by correspondence.

It was once fashionable to remark that no Einstein can come out of the jungle, but the reply heard at Trieste to that is simply: "Why not?" A physicist can come from almost anywhere. Paul Vitta grew up on a farm in a village 600 miles from Dar-es-Salaam, the capital of his native Tanzania, where he went to boarding school. For the educational revolution is paying off.

Omar El Amin, a research worker at the University of Khartoum's radiation and isotope centre in the Sudan, is one of five brothers whose father was a crewman on a Nile river steamer. He reminded me that education in the Sudan is free. It enabled him to reach the point where he was able to go to the University of London for his M. Sc. in radiation physics. Of his brothers, one is a textile technician, another is also studying science, one is in the army and the fourth is working in electronics engineering at Kiev. Mr. El Amin, an experimental physicist, wanted to come to the Trieste centre to "see what theoreticians do with their long equations and their mathematics".

Scientists often say that the best way to look at a phenomenon is to study an extreme case. In that respect, Dr. Toshar Gujadhur certainly qualifies as the most isolated of the theoretical physicists in Trieste. His home is on the island of Mauritius and he was returning there after an absence of ten years that began when he went to Imperial College to earn his doctorate in mathematical physics.

He was about to take a post in a new teacher training institute on Mauritius. "I want to go back, my roots are there, but it will mean complete paralysis of the mind if I cannot get to Trieste every three years or so. I'm working in relativity and quantum mechanics. Learning to me is like food, I need it. It's a challenge; you do it in spurts. I'm here at least twelve hours a day, six days a week. I arrive around eight or nine in the morning, sometimes I go home on the last bus at 10:30 at night.

Some people prefer to work only at night so the place is open twenty-four hours a day."

To Dr. Gujadhur, the greatest advantage of the Trieste centre is its mere existence. "It is a meeting-ground, it offers post-doctoral training but, most of all, there is the fact that one can come back. I must be able to tell myself that I will always be able to come here for three months. Otherwise, one is just cut off."

That is a precarious situation for the physicist. Prof. Georges Ripka of the French Atomic Energy Commission at Saclay, who co-organised the nuclear theory course with Prof. Luciano Fonda of the University of Trieste, believes that the scientist in a developing country runs the risk of doing "perfectly good but irrelevant work". If he only reads scientific journals, he cannot keep up and there is the danger that he will do work already performed elsewhere. "In an hour's conversation with a physicist" said Prof. Ripka, "I learn more than in a day in the library."

The theoretician must also remain in touch with the experimentalist, as is the case at Trieste. "Research is not walking in a garden and waiting for a good idea," said Prof. Ripka. "Physics, unlike mathematics, is an approximate science. One must construct corrected theories when new experimental data come in. It's not that the old theories and experiments are bad, they are approximate. In physics, one is always guessing. Our guessing is never quite wrong, and never absolutely right."

A number of guest lecturers participated in the nuclear theory course as they do in all the training given at Trieste. The centre has become a sort of crossroad for physicists in Europe and it is not unusual for a man to come down from Germany or up from Yugoslavia to lecture for a day or two. The centre pays for his stay and his own institution pays his salary as a contribution by the scientific community to Trieste.

Participants in the course worked hard with three lectures a day followed by a seminar in which each one talked about what he was doing. Most of the lectures were in the forefront of the subject because, as Prof. Ripka remarked to me, it is just as easy to teach new material as it is to teach the old. Prof. Ripka was particularly pleased that many participants had arranged to collaborate by post. Such arrangements are valid, he thinks, if they are preceded by personal contacts, "In nuclear theory, contacts are invaluable. That's how I got started myself. I was invited to go from France to a conference in New York and I talked to a lecturer there. He got me going and he put me in touch with a student of his in Belgium. After speaking to him, I had the drive and I'm sure my case is typical. That is why I feel responsible to a fellow here. He must go home with all that he can."

Such courses are far from the centre's main function. In fact, every time one looks around, another function seems to appear. Besides courses, associateships, research workshops and occasional symposia, the centre has a system of federated institutions. They number twenty from sixteen countries and each has the right to send a scientist of its own choosing to Trieste for a period of up to 40 days a year. Such is the thirst for theoretical physics that some institutions send forty scientists for one day — and they stretch their subsistence allowance to cover nearly a week by staying in modest boarding houses or with friends.

Even a week is enough to become imbued with the atmosphere of the International Centre for Theoretical Physics. The long corridor that leads to Prof. Salam's office on the second floor is punctuated by portraits of the centre's spiritual fathers: Einstein, Niels Bohr, Oppenheimer, Werner Heisenberg, Wolfgang Pauli, Louis de Broglie, among others . . . and a humorous New Year's card from Lev Landau showing the fox who fished with his tail. The developing world flocks to Trieste and to what these names represent. To Prof. Salam, this is only the swing of history's pendulum.

He likes to tell how Michael the Scot left his native land in the 13th century to travel south to the Arab universities of Toledo and Cordova. Or how Mamum the Caliph of Baghdad in the 9th century, sent the Emperor of Byzantium a work concerning "a new path in mathematics called Algebra". Prof. Salam blames the Mongol invasions for the end of Islamic science. "The Mongols systematically destroyed libraries. Before printing, the destruction of a library meant the end of a tradition. When the libraries at Baghdad, Bukhara and Samarkand went up in flames, Islamic science was lost with them."

With men like Prof. Salam we are witnessing an exciting resurgence of Islamic science. Already his work has received recognition by the award of the "Atoms for Peace" prize — a fitting recompense for a man who name, Abdus Salam, means "Servant of Peace".

With such a name he was almost pre-destined to work for the United Nations. He now has another dream, a world university of which the Trieste centre would be but one campus.

Such a university could meet a number of needs. There is already strong movement for an institution devoted to the study of peace and disarmament which are at the heart of the problems that the United Nations must solve.

Secondly, institutes could be set up at the postgraduate level to conduct research in the basic sciences. Like his own at Trieste, Prof. Salam thinks that they should use the same built-in plug against brain drain by requiring their participants to spend most of their time in their own countries.

And thirdly, Prof. Salam foresees international campuses with truly international faculties for the applied sciences. "They could be anywhere: Kenya for health sciences, particularly tropical diseases, Iran for petroleum and petro-chemicals, Nigeria or Latin America or Pakistan for agriculture, and so forth." Gaps left in the network of UN centres would be filled by federated universities and research institutes. "I'm after fifty campuses, not five or six," said Prof. Salam. "This must be truly a world university. Don't worry, it will come, certainly not tomorrow but certainly within twenty years."

On that note, I left Abdus Salam. His prediction is just a dream, but this disconcerting man of faith and science must be one of the world's most realistic dreamers . . .

II.3
OTHER INITIATIVES

A World Federation of Institutes of Advanced Study

Reprinted from *Journal of the National Science Council of Sri Lanka,* 1, 7-17 (1973).

A number of groups have been working independently towards the project of setting up one or more world universities. That this is of importance in the context of the international future of mankind goes without saying. That at least one university did not come into existence at the same time as the United Nations organisation did in 1945 is something of which the world's academic community cannot feel proud. Recognising this, at its twenty-fourth session, in 1969, the General Assembly of the United Nations adopted the resolution 2573 (XXIV) inviting the Secretary-General to undertake a comprehensive expert study on the feasibility of an international university. In introducing this widely sponsored resolution, it was stated that "the establishment of an international university would satisfy the aspirations which were becoming apparent in all parts of the world and it would fulfill an obvious need".

There are at least four reasons for this universal interest in the setting up of one or more international world universities:

1. *The Idealistic Reason — International Understanding*

There is no instrument more potent in bringing an appreciation of different — at present national — points of view than the atmosphere of an international university.

2. *Global Studies*

Within the context of such a university there is the possibility of growth of international studies on global subjects — like international development, international economies, global environment, disarmament and the like.

3.   *Contacts of Scholars*

Human knowledge transcends national boundaries.   To a scholar interested even in his narrow speciality, there is nothing more valuable than the possibility of free contact with his peers from *all* countries. A well-constituted world university may resolve the present political difficulties in achieving such contacts.

4.   *Access to Specialised Knowledge for Scholars from Developing Countries*

In the past, when scholars and scientists have worried about international contacts, they have tended to feel concern about East and West contacts only.   One tends to forget the needs for contact of students and scholars from developing countries with their peers from developed countries.   Opportunities for such contacts do not exist — not for political reasons, but because of economic factors.   A world university, representing East, West *and the Third World*, is less likely to forget the needs of these students and scholars and more likely to afford them access to academic, scientific and technical areas at present the exclusive preserve of the richer countries.   The developing countries fully recognise that a truly international university — preferably under the UN auspices — is the one real guarantee for their scholars to receive their share — as of right — of the facilities and resources of the international institutions to be created.

In response to the General Assembly resolution a study has been carried out on behalf of the Secretary-General.   This study suggests the setting up of a set of postgraduate international institutions within the United Nations family — to be called UN International Universities — with two objectives:

(*i*) "To enable scholars from all parts of the world jointly to study research and reflect on the principles, moral imperatives, objectives, purposes, perspectives and needs of the UN system in the light of its fundamental laws and developing accords, declarations, resolutions and programmes".

(*ii*) "Secondly, to undertake a continuing and widely-based international scholarly effort of study and research, directed in consonance with Charter obligations towards social, economic and cultural progress through co-operation among nations and peoples. The universities would achieve these ends through emphasising . . . relevant international studies, largely interdisciplinary, of wide and generally global significance".

It is clear that the objectives of this particular response to the General Assembly resolution are limited to the special global studies related to global problems. This is not going to be a traditional university pursuing the traditional range of subjects, but a specialised institute or set of institutes.

Commendable as this response is, it falls short of the aspirations of at least two of the communities which have supported world university projects. By and large both these communities have had in mind the traditional range of academic disciplines, *in addition to global studies.* The two communities are:

1. Academic scholars and scientists in East and West who desire in their traditional disciplines more contacts with each other.

2. Developing countries who look upon the world university idea as the one way by which they can secure entry for their students and scholars into the privileged intellectual, scientific and technological club on terms of equality. Notwithstanding the fact that no stated bar operates against anyone from a developing country pursuing advanced studies and research at any of the world's great institutes, in practice the economic and other factors do operate in such a manner that at least the scientific and techno-logical gap between the poor and the rich countries grows ever wider. The developing countries look upon the world university project as a means to bridge this gap.

From this it would seem that nothing short of one or more full-fledged world universities in traditional discipline — *at least for post-graduate scientific and technological studies* — will satisfy these two groups.

Unfortunately, to develop full-fledged universities — and particularly under UN sponsorships — is not all that easy. One does not have to recount the difficulties which are likely to be met. Since the sums of money involved are large, it is out of the question that the United Nations

Organisation — even with the generous support of the World Bank — could finance such a venture. It is also unclear if one could get a number of the richer countries passionately interested in a project of this type and ready to back it. There are too many casualties among proposed international or regional institutions in the academic field already to give one great hopes of success, unless one proceeded in a gradual manner. Further, the choice of location of such a world university in one country in preference to another will always present difficulties. Even the choice of faculties to develop first is not going to be all that plain sailing.

One way to circumvent the difficulty of creating new institutes, and yet to achieve at least partly some of the objectives listed above, is to take advantage of existing centres of excellence and quality which would like to discharge international functions and to link such centres with the UN institutions for global studies proposed by the Secretary-General, the whole making up the beginning of a world university.

This note then is concerned with a world university idea emerging *gradually* from an amalgam of the UN institutes together with existing centres of advanced studies linked in a federation. *In the first instance the emphasis is on post-graduate research and training for research.* Later development of the ideas may envisage undergraduate studies and the corresponding institutions.

Let us consider the various stages of the *post-graduate* plan. The important point we wish to make is that every part of the plan has merits of its own, irrespective of whether the later stages follow or not. The first stage is the identification of such existing institutions which already operate substantial international programmes. There is around the world no dearth of institutions of quality which are to a lesser or larger degree international in character, even though their original charters do not specify this. The idea would be to make them even more consciously so. The hope is that a voluntary federation would help in this: at the least in defining norms and making it possible to share experience; at the best in raising new funds for the international operation. As a second part of the plan, and if this federation so chooses, a UN charter could be accepted and a formal link established with the UN Institution on Global Problems proposed by the Secretary-General. The centres constituting the federation and covering traditional disciplines together with the Secretary-General's UN University on Global Problems, would make a complementary whole — the beginning of a world university.

Such centres as should belong to the proposed federation must satisfy certain criteria. For example, such centres must possess the highest rating of quality; they must possess — to a lesser or greater degree — an international faculty of staff and research fellow; they must agree to spend a minimum proportion (to be fixed, perhaps between 15-25%) of their resources and their facilities towards furthering the work of high-grade scholars from developing countries.

To illustrate the working of one such centre, one may perhaps cite the example of the International Centre for Theoretical Physics at Trieste, Italy. This case is not typical because the institute is financed and run by two of the United Nations Agencies, but it does provide an example of the type of international academic faculty in actual operation. The centre was set up under the auspices of the International Atomic Energy Agency (IAEA) with the co-operation (and from 1970 equal participation) of the United Nations Educational, Scientific and Cultural Organisation (UNESCO). The centre is devoted to imparting *training for and conducting research in* all disciplines of theoretical physics at the highest level. It draws its scientific faculty (consisting mainly of visitors) and research fellows from (theoretically 100 but in practice) some 50 countries of the East, West and the Third World. *Some 50% of its facilities and junior and senior research positions are reserved for scientists from developing countries.* A unique feature is that the Centre offers dual appointments to active senior theoretical physicists from developing countries. Such appointments are held for periods of three to five years; the scholar spends the bulk of his time — about nine months of the year — in his own country, and the remaining three months of every year in Trieste. In addition, the Centre has built up federation links with some twenty research institutes in various countries — on a cost-sharing basis — which afford mobility of their staffs and research fellows. On the East-West co-operation side, as a UN-sponsored organisation, the centre plays an absolutely unique role: it is one of the few places in the world where physicists in subjects as sensitive as plasma research from the East and West meet regularly and for prolonged periods (quarters or years) and with no national pride or sensitivities inhibiting scientific concourse.

The proposed *World Federation of International Institutes of Advanced Study* would include centres with already a large international programme or desirous of starting one. The institutes which would join this federation may operate schemes of *dual appointments and federation* with corresponding centres both in developed and developing countries. From informal contacts one knows that a number of institutes in USA, USSR, Great Britain, France and other countries are extremely desirous of widening their faculties internationally to share staffs and visitors with others in the same disciplines and, through the strength given to their international programmes by the fact of belonging to such a federation, be obliged to throw their doors even more widely open to scholars from developing countries.

Why should a federation be created of institutes in diverse subjects? What advantages could come to the members of the proposed federation? Should it be *independent* institutes as well as institutes within national universities which should be invited to join? What about the financing of the international programmes? And the links to the UN family?

In answering these questions, one has to ask if the federation could be stronger in any way in carrying out the international aspects of its programmes than any one of its component units? Would, for example, the Trieste Centre get any benefit by being federated in a sort of loose link with the Institute for Advanced Study at Princeton, or the Salk Institute for Biological Studies?

In our opinion, the answer to the last question is an affirmative "yes". The fact that a federation exists is likely to have important repercussions:

1. To get the general idea of international staffs and international use of facilities of scientific institutes accepted in a more "official" manner by the governing bodies of the institutes.

2. To secure a mobility of high grade scientific personnel. Hopefully, there may emerge a UN Laissez Passer for academic personnel to travel freely, at least between the federating institutes, if the UN did get involved with the federation idea.

3. A commitment in respect of scholars from developing countries: a federation to which a fair number of reputable institutions belong would go much further in organising and getting accepted common standards. The committing of a certain *percentage* of resources to helping scholars from developing countries, and to scholars from countries with different political systems, is a new idea. Many institutes do set aside certain sums but there is no coherent policy about this. We are hoping that belonging to a federation would provide a visibility to these efforts and a better focus.

4. If we envisage that institutes from developing countries would also belong to such a federation, these institutes will in many cases have to raise their standards in order to qualify to join. This type of pressure would be an excellent tonic for them, and make the tasks of those running these institutes vis-a-vis their own governing councils — and their Governments — somewhat easier.

5. In respect of the question raised, whether it should be independent institutes which should federate or those located within universities, one should keep an open mind. In every case the permission of the governing bodies of the institutes would be needed. I believe this is easier for independent institutions. For the present we may envisage only such institutes being invited, but the matter should be dealt with pragmatically.

6. The question of financing international programmes is a difficult one. It is definitely envisaged that in the first instance the members of the federation would find funds from their own sources for this. Later, collective action may bring extra funding from outside — even from UN sources.

7. A first list of possible independent or semi-independent institutes which may consider forming initial memberships of the federation is suggested in the Appendix. It is suggested that a preliminary meeting of Directors of these institutes be held to gain acceptance of the ideas in this note.

Note added November 1972

This memorandum was circulated in 1970 in a mimeographed form. The late Professor Arne Tiselius, President of the Nobel Foundation took up the ideas and at two Serbelloni meetings held during 1971 and early 1972, the idea of an International Federation of Institutes of Advanced Study was hammered out.

This Federation, consisting at present of 24 Institutes, was inaugurated at a meeting at Trieste during October 1972. Its offices are located in the Nobel Foundation House, Stockholm. Its Chairman is Nils Stähle and its Secretary is Sam Nilsson.

The Federation may become the precursor of a World University.

Third World Higher Education and Italy

Speech delivered by Abdus Salam at the International Meeting on "La Cooperazione Universitaria: Bilancio e Prospettive delle Esperienze Europa – Paesi in Via di Sviluppo" held at Trieste, November 1985.

1. Italy is an incredible country. It is among Western Europe's BIG FOUR — together with the Federal Republic of Germany, France and the UK. It is a country of great science and great technology. It is the world's seventh ranking country in industrial production. It is also the seventh country in terms of its education system in so far as foreign students are concerned. What, however, is so incredible about Italy is that notwithstanding these superiorities, Italians have no colour complex, nor any of the attitudes to helping others often manifested by other developed nations. Personally, during the twenty-one years that I have worked in Trieste, I have received the warmest understanding for my ventures towards building up science and technology in the developing world. It is in this spirit of appreciation that I speak. If I appear to complain, it will be as a friend to a friend.

Before I go on to discuss problems of advanced scientific research and the role Italian universities and research institutions can play in this regard, I shall speak briefly of what Italy can do for foreign students in general. I shall speak principally of the university level but in some of the figures I shall give, I shall be concerned with the totality of foreign student population.

I wish to make three points:

(*i*) If we consider the numbers of foreign students in all categories versus the domestic student population in different European countries, Italy has the lowest percentage of any European country. (Table I)

Table I

Percentage of Students from Developing Countries
versus Total Number of Students

Host country	1974	1981
USA	13.4%	n.a.
France	6.2%	10.4%
UK	5%	4.5%
FRG	3%	2.7%
Canada	4.5%	2.8%
Italy	1%	1%
USSR	1%	n.a.

(*ii*) While there have been increases during the last 10 years of numbers of foreign students at the university level in other European countries, (figures have risen by an average of 50% during the last ten years) there has been no corresponding increase in Italian intake — in fact a slight decrease. (Table II)

(*iii*) On the positive side, Italy does not discriminate against foreign students in respect of fees and other scholarly expenditures. This happens elsewhere. Thus, the UK, which was one of the major European countries to which students at the university level went, began to discriminate since the mid 1960s, between foreign and domestic students. Fees charged from foreign students (except those from EEC countries) are now a factor of five or six higher than those charged from domestic students. This was not so when I was a student in the late 1940s. We all paid £70 a year for fees whether the student came from Pakistan or from the UK. Today, while a domestic UK student pays £800 as annual fees, a foreign student would pay £5,000. This has made the numbers from the poorer developing countries decline so far as the UK is concerned. (Figure 1).

No wonder the developing countries think of Italy more and more for making up for the lack of educational opportunity which this discrimination entails elsewhere.

Table II

Host country	Total no. of foreign students from developing countries (OPEC not included)		Increase in %	No. of foreign students (OPEC not included) in 1981			
	1973	1981		Africa	Latin America	Asia + Oceania	OPEC countries
USA	94,000	136,000	45%	13,500	32,500	90,000	73,500
France	38,500	74,000	92%	51,000	10,000	13,000	17,000
UK	17,000	28,000	65%	6,000	2,000	20,000	11,000
FRG	15,000	24,500	63%	3,500	3,000	18,000	12,000
Canada	21,000	23,500	12%	3,500	3,000	17,000	2,500
Italy	6,500	5,800	−10.7%	1,100	700	4,000	4,500

Figures are rounded up and based on 'UNESCO Statistical Yearbook', 1975 and 1984 editions.

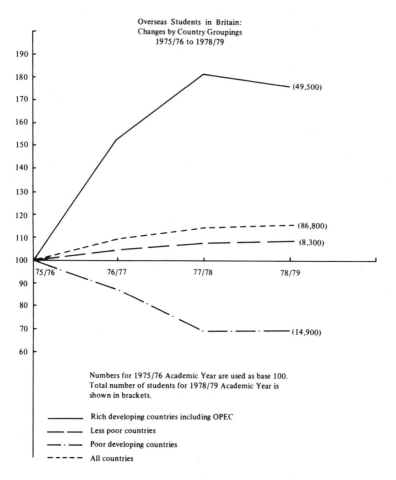

Overseas Students in Britain:
Changes by Country Groupings
1975/76 to 1978/79

Numbers for 1975/76 Academic Year are used as base 100.
Total number of students for 1978/79 Academic Year is
shown in brackets.

——————— Rich developing countries including OPEC
—— —— Less poor countries
—— · —— Poor developing countries
— — — — All countries

Based on: Philipps Alan, "British Aid for Overseas Students".

Figure 1

2. From the pre-graduate level, I now turn to the post-graduate level and then, finally, I shall speak of the post-doctoral situation.

So far as the post-graduate level is concerned, barring some exceptional institutions like the Institute for Advanced Studies (SISSA) at Trieste, there is no widespread awarding of Ph.D. degrees by Italian Universities. This has meant that few foreign students come to Italy for post-graduate, pre-doctoral work. This is one reform which could transform the situation if provision was made — like at SISSA — of encouraging a Ph.D. stream.

3. It is now universally recognised in the developing world that there is a need for institutions like the International Centre for Theoretical Physics here at Trieste which offers possibilities of post-doctoral work after the Ph.D. Fortunately, the Italian Government is conscious of this need for disciplines other than physics, as evidenced, for example, by the proposal to create an International Centre for Biotechnology at Trieste.

What I would like to speak about today — and this is my principal recommendation — is that even if no new centres are created, there could be a scheme of associateships, like the one which we run at the International Centre for Theoretical Physics, for scientists from developing countries. Such a scheme needs to be built up for sciences other than theoretical physics with the cooperation of Italian universities and other Italian research institutes.

Let me explain what I mean in detail. At the Trieste Centre, we recognised fairly early that one of the major obstacles for continuing scientific research in developing countries is the scientific isolation of the research workers. The science communities in our countries are small; journals, scientific equipment and apparatus are not available, and, much worse, there are no scientific peers with whom one may discuss one's scientific problems. At the International Centre for Theoretical Physics, we have tried to cure this — one of our modalities being an offer to leading scientists in developing countries of an appointment as an Associate of the Centre which holds for six years. During these six years, an Associate can pay three visits to the Centre at times of his choosing, each visit ranging from six weeks to three months. Our hope was that in this way a scientist, living and working in a developing country could plan his scientific life in such a way that, with periodical visits to the International Centre, he would be able to charge his intellectual batteries, work on research problems while here, and then return to his country carrying a new line of work, refreshed with new ideas and new interactions. This is the basic idea of the associateship. At the present time, we have 300 Associates who come regularly to the Centre, each appointed for six years. Their fares and per diems are paid by the Centre. There has never been a single case of anyone who was an Associate leaving his country and joining the brain drain. While we have had help for the Associateship Scheme from the Swedish Government, from the Government of Japan and others, our chief help has come from the Government of Italy and its Dipartimento per la Cooperazione allo Sviluppo.

This scheme has been running, so far as theoretical physics is concerned, for the last 21 years. With a special grant of one million dollars which the Dipartimento makes to us, we have also recently included in this scheme experimental physicists whom we select and send to Italian university laboratories. Almost all Italian university laboratories working on Biophysics, Climatology, Condensed Matter Physics, Geophysics, Laser Physics, Medical Physics, Microprocessors, Plasma Physics, have cooperated with us and accepted our Associates. We have also cooperated with LAMEL, Bologna, for Silicon Materials and Alternative Energy Sources; the EEC Research Centre, Ispra, for Alternative Energy Sources; IROE, Florence (a CNR Institute), the Institute of National Optics (INO), Florence, and CISE, Milano for Laser and Optical Fibres; MASPEC, Parma, for Condensed Matter Physics; the Foundazione Ugo Bordoni, Rome, for Radio Meteorology; CSELT, Torino, for Communication and Laser Physics as well as many others. Through these modalities, the community of physicists in the developing countries has kept up with modern ideas and modern techniques, and, although it is difficult to quantify, there is no question but that this scheme has transformed the status of physics in the Third World.

The question I would like to raise with you today is: can we do something similar for other disciplines, like Chemistry, Biology, Medicine, Geology and other sciences? There are no associateship schemes for any of these disciplines. What I would like to propose is that the Italian universities band together to receive such associates from developing countries — say four or five in each active department of Chemistry, Biology, Medicine and Geology. If funds of the order of $5 million were made available, say, to the Third World Academy of Sciences, which was founded two years ago here at Trieste, this Academy — in association with Italian Academies, like the Academia Nazionale dei Lincei or the Academia Nazionale delle Scienze (dei XL), or the Istituto Veneto di Scienze, Lettere ed Arti, and the Consiglio Nazionale della Ricerca (CNR), and the Istituto Nazionale di Fisica Nucleare (INFN), and the "Ettore Majorana" Centre for Scientific Culture at Erice — could devise and run such a scheme for the disciplines of Chemistry, Biology, Medicine, Geology and other sciences, to be tenable at Italian universities throughout the country. This would make for a real revolution in these subjects in the Third World, keeping the research studies of these subjects at a high level, as well as enrich Italian universities and Italian institutions.

In addition to the associateship scheme, we have federation links with 200 research institutes in physics around the world. These links provide for an agreement by which Federated Institutes can send scientists to the Centre for up to a total quantum of 120 or 180 or 240 days. We pay for the stay, while the cost for travel is shared. These linkages could be extended and strengthened with mutual benefit to both developing country institutions as well as to the universities here.

4. As I said, if a grant of $5 million[a] could be made by the Dipartimento per la Cooperazione allo Sviluppo for Associateships and for Federation linkages, the Third World Academy of Sciences (in association with Italian Academies and Italian universities), could undertake to run these schemes. We would then be well on the way of conquering the isolation of developing country science in all its areas, and not just for physics alone. This then is my concrete proposal.

5. I shall conclude by recounting to you a story which I told in my Nobel Lecture: Scientific thought and its creation is the common and shared heritage of mankind. In this respect, the history of science, like the history of all civilisation, has gone through cycles. Perhaps one can illustrate these cycles with an actual example.

Nearly seven hundred and seventy years ago, a young Scotsman left his native glens to travel south to Toledo, Bologna and Salerno. His name

[a]Note added in July 1986: Five million dollars over three years — have been promised to the Third World Academy of Sciences by the World Laboratory, set up by Antonino Zichichi, for awarding Fellowships tenable at Italian Universities to Third World scientists.

was Michael, his goal to live and work at these places and to master the newer developments in sciences, then available only at these Centres. Since Arabic was the language of science at that time, he had to learn Arabic.

Michael reached Toledo in 1217 AD, Bologna in 1220 and Salerno in 1224. Once he reached the South, Michael formed the ambitious project of introducing Aristotle to Europe, translating into Latin not from the original Greek, which he knew not, but from the Arabic translation of Aristotle then taught at Toledo. This was the first introduction of Aristotle's work into mediaeval Europe, together with the work of Arab-Muslim writers like Averros. At Salerno there flourished a great medical school chartered by Frederick in 1231. Here Michael met the Danish physician Henrik Harpestraeng — later to become Court Physician of Eric IV Waldemarsson. Henrick, the physician, had come to Salerno to compose his treatise on blood-letting and surgery. Henrik's sources were the medical canons of the great clinicians of Islam, Al-Razi and Avicenna, which Michael the Scot translated for Henrick from Arabic.

Toledo's, Bologna's and Salerno's schools, representing as they did the finest synthesis of Arabic, Hebrew, and Latin scholarship, were some of the most memorable of international assays in scientific collaboration in the Middle Ages. To these Centres came scholars not only from the rich countries of the East, like Syria, Turkey and Egypt, but also from developing lands of the West, like Scotland and Scandinavia. Then, as now, there were obstacles to this international scientific concourse, with an economic and intellectual disparity between different parts of the world. Men like Michael the Scot or Henrik Harpestraeng, the Dane, were singularities. They did not represent any flourishing schools of research in their own countries. With all the best will in the world their teachers at these universities doubted the wisdom and value of training them for advanced scientific research.

Today Arabic is no longer the language of science. In fact, the Third World produces little science at the highest level. We need the tolerant spirit of Emperor Frederick II once again and to prevail, and newer centres for sciences created where a candle can be lighted from a candle. I am certain the Italian universities will rise to this challenge today as much as their mediaeval counterparts did, and provide the world with the educational and research opportunities which it desperately needs.

I started by praising Italy. I should perhaps qualify my words a little by quoting Pavese, in *La Casa in Collina:*

"Professore', esclamo Nando a testa bassa,

'voi amate l'Italia?'

Di nuovo ebbi intorno le facce di tutti: Tono, la vecchia, le ragazze, Cate, Fonso sorrise.

'No', dissi adagio, 'non amo l'Italia. Gli italiani'."

("Professor', exclaimed Nando, lowering his head,

'do you love Italy?'

Once again all their faces were around me: Tono, the old woman, the girls, Cate. Fonso smiled.

'No', I said quietly, 'not Italy. The Italians'.")

Address to UNESCO

Address by Professor Abdus Salam to UNESCO Executive Board upon the award of the Einstein Medal by UNESCO and the Nobel Prize for Physics by the Swedish Academy of Sciences, 17 October 1979.

Mr. Chairman of the Executive Board, Mr. President of the General Conference, Mr. Director-General, I cannot describe to you how honoured I feel by your very kind invitation for me to speak here today, and for the award of the Einstein Medal. Over the years, UNESCO has become more and more a House of World Science, a crossroads of international scientific ideas, a place which scientists from both developed and developing countries consider as their own house. I render grateful thanks to Allah that this movement you have initiated will be helped by the celebration today of the honour the physics community and the Swedish Academy of Sciences have bestowed upon me.

The Holy Quran enjoins us to reflect on the verities of Allah's created laws of nature; however, that our generation has been privileged to glimpse a part of His design is a bounty and a grace for which I render thanks with a humble heart.

My first thought on this occasion is with the great European experimental laboratory at Geneva — CERN — in the founding of which UNESCO (through Professor P. Auger) played a major role. This laboratory in 1973 provided the first experimental evidence of Neutral Currents which are an essential part of the prediction of the theory. The apparatus used was the Gargamelle Bubble Chamber donated by the Government of our host country, France. My thoughts go equally to the Stanford Linear Accelerator Centre in the United States which last year in an epic experiment provided confirmation of the second aspect of the theory — its very heart — the unification of electromagnetic forces with the weak nuclear forces to one part in 4,000. An experiment at Novosibirsk by a group led by Professor Barkov further confirmed the findings of SLAC.

The Director-General in his remarks has spoken of the burgeoning growth of maturity in sciences in the developing countries. In this context I wish to recall that the history of sciences, like the history of all civilisation, has gone through cycles. Perhaps I can illustrate this with an actual example.

Seven hundred and fifty years ago, an impoverished Scotsman left his native glens to travel South to Toledo in Spain. His name was Michael, his quest to live and work at the Arab Universities of Toledo and Cordova, where the greatest of Jewish scholars, Moses Bin Maimoun, had taught a generation before.

Michael reached Toledo in 1217 AD. Once in Toledo, Michael formed the ambitious project introducing Aristotle to Latin Europe, translating, not from the original Greek, which he knew not, but from the Arabic translation then taught in Spain.

Toledo's school, representing as it did the finest synthesis of Arabic, Greek, Latin and Hebrew scholarship, was one of the most memorable of international assays in scientific collaboration. To Toledo and Cordova came scholars not only from the rich countries of the East, like Syria, Egypt, Iran and Afghanistan — but also from developing lands of the West like Scotland. Then as now, there were obstacles to this international scientific concourse, with an economic and intellectual disparity between different parts of the world. Men like Michael the Scot and his contemporary Alfred the Englishman were singularities. They did not represent any flourishing schools of research in their own countries. With all the best will in the world, their teachers at Toledo doubted the wisdom and value of training them for advanced scientific research. At least one of his masters counselled young Michael to go back to clipping sheep and to the weaving of woollen cloths.

In respect of this cycle of scientific disparity, perhaps I can be more quantitative. George Sarton, in his monumental five-volume History of Science, chose to divide his story of achievement in sciences into ages, each age lasting half a century. With each half century he associated one central figure. Thus 450 BC–400 BC Sarton calls the Age of Plato; this is followed by half centuries of Aristotle, of Euclid, of Archimedes and so on. From 600 AD to 650 AD is the Chinese half centuries of Hsuan Tsang, from 650 AD to 700 AD that of I-Ching, and then from

750 AD to 1100 AD — 350 years continuously — it is the unbroken succession of the Ages of Jabir, Khwarizmi, Razi, Masudi, Wafa, Biruni and Omar Khayam; — Arabs, Turks, Afghans and Persians — men belonging to the culture of Islam. After 1100 appear the first Western names; Gerard of Cremona, Jacob Anatoli, Roger Bacon — but the honours are still shared with the names of Ibn-Rushd (Averroes), Moses Bin Maimoun, Tusi and Ifn-Nafis — the man who anticipated Harvey's theory of circulation of blood. After 1350, however, the developing world loses out except for the occasional flash of scientific work, like that of Ulugh Beg — the grandson of Tamurlane — in Samarkand in 1400 AD, or of Maharaja Jai Singh of Jaipur in 1720 — some 40 years after the setting up of Greenwich — who corrected the serious errors of the then Western tables of eclipses of the sun and the moon by as much as six minutes of arc. As it was, Jai Singh's techniques were surpassed soon after with the development of the telescope in Europe. As a contemporary chronicler wrote: "With him on the funeral pyre, expired also all Science in the East". And this brings us to this century when the cycle begun by Michael the Scot turns full circle, and it is we in the developing world who turn to the West for Science. As Alkindi wrote 1100 years ago: "It is fitting then for us not to be ashamed to acknowledge truth and to assimilate it from whatever source it comes to us, even if it is brought to us by foreign peoples. For him who scales the truth there is nothing of higher value than truth itself; it never cheapens or abases him."

During this period, starting with this century, in the world of physics the first name is that of C. V. Raman — the Nobel Laureate of 1930, then of Yukawa, Tomonaga and Esaki; in between Lee, Yang and Ting. And lastly today we heard the news of the Nobel award to the great Jamaican economist, Sir Arthur Lewis, in which I am sure all of us rejoice.

In this context the question we must ponder is this; are we today firmly on the road to a renaissance in sciences — as the West was in the 12th Century at the time of Michael the Scot? Unfortunately the answer is no.

There are two prerequisites to this renaissance: one, availability of places like Toledo for international concourse, where one can light a candle from a candle. Second, the interest in our own developing societies to give the topmost priority to the acquisition of knowledge, as for example was given by the Japanese Constitution after the Meiji Revolution.

Regarding the first point, regretfully the opportunities for international scientific concourse are fast shrinking, with greater and greater restrictions in the traditional countries like UK and USA on acceptance of overseas scholars, including those from developing countries. It is becoming increasingly clear that the developing world will need internationally run — UN and UNESCO run — institutions — Universities of science — not just for research, but for the high level teaching of traditional technology and sciences, both pure and applied.

The second prerequisite for development of science and technology is a passionate, consuming desire on the part of the developing countries and the removal of all internal barriers in its acquiring. Unfortunately the prognosis in this respect is not very bright.

Some of you may recall that on 5 May this year UNESCO celebrated Einstein's birthday and the Director-General graciously invited me to speak. I recalled then how Einstein may have been lost to physics, but for a series of accidents — such was the measure of financial and economic and other frustrations which he faced, even in a country like Switzerland. Unfortunately the same applies so far as developing countries are concerned in a measure still greater. Perhaps I shall illustrate this with my own case.

The fact that I became and remained a research physicist is due to three accidents. First, the Second World War; as soon as I showed some competence in sciences, my well wishers, my parents, all those around me, destined me for a career in the then prestigious Indian Civil Service. As it happened, with the war the Civil Service Examination was suspended for the duration. But for this I would be a Civil Service functionary today. The second accident which sent me to Cambridge for research was again connected with the war. The then Prime Minister of my home state — the Panjab — collected some funds for the "War Effort". The war ended; the funds were left unutilised. He decided to institute "Small Farmers Son's Scholarships" for study abroad. A number were offered; I was one of those fortunate to be selected and sail the same year — 1946 — to Cambridge. Several other scholarships were awarded; unhappily the other scholars were promised admission for subsequent years. In between, the subcontinent was partitioned and with it the scholarships disappeared. The entire exercise of the then Prime Minister succeeded in one thing only — in sending me for research at St. John's College, Cambridge, where Professor Dirac (whom you honoured on 5 May) lived and worked. You can understand why I started this lecture with humble praise to Allah for his Grace and Beneficence.

The third accident happened when I returned to Pakistan from Cambridge to teach and to try to found a school of research. In no uncertain terms, it was made plain for me that this was impossible. I must either leave physics research or my country. With anguish in my heart, I made myself an exile — and it was this anguish which led me to propose the creation of an International Centre for Theoretical Physics, with the most active sponsorship of the Government of Pakistan and other developing countries. The ideas was to award what we call "Associateships" of the Centre, so that a deserving young man may spend his period of vacation in an invigorating environment in close touch with his peers in research, to charge his batteries with new ideas, while still spending nine months of his academic year in his own country, working at his university.

I do not have to tell the executive Board of UNESCO how the idea of the Centre fared. With UNESCO's active help, and with very generous assistance from the Government of Italy and the Town of Trieste, the Centre was created by the IAEA in Trieste in 1964. UNESCO joined as equal partners with IAEA in 1970. Over the 15 years that the Centre has existed now, it has veered from emphasis on fundamental and basic physics towards subjects on the interface of pure and applied physics — subjects like Physics of Materials, Physics of Energy, Physics of Fusion, Physics of Reactors, Physics of Solar and other unconventional sources, Geophysics, Physics of Oceans, and Deserts, Systems Analysis — this, in addition to High Energy Physics, Quantum Gravity, Cosmology, Atomic and Nuclear Physics and Mathematics. This shift from pure to applied physics was not made because we thought that pure physics is less important for developing countries. It was simply that there was not and still is not any other international institute responsive to needs of technological hunger involving the discipline of physics. Every year around 1,500 physicists — half from developing countries — spend of the order of six weeks or more at the Centre attending extended symposia or research workshops. The Centre has brought credit to UNESCO in the Comity of International Scientific Scholarship — pure, applied and technological — besides strengthening physics in the developed and developing world.

But over the 15 years that I have directed the Centre, I have felt more and more strangled, and never more so than now. I used to pride myself on spending half a day every day in research, half a day in administration.

Progressively over the last five years this has become impossible. This is not because the task of administration has become more arduous; it is simply because the uncertainty of the Centre's standing in the ecology of international institutes has increased, despite its success, despite its demonstrated need. Its very existence is uncertain from year to year.

Briefly, half the budget of the Centre comes from the Italian Government; the other half is shared between IAEA and UNESCO. I am fully aware of UNESCO's limitations in shouldering the responsibilities for such an enterprise; UNESCO defined its mandate 24 years ago as a catalyst of new institutions and not as their long-term sustainer. The realities of the situation now demand a revision of this mandate and newer stable funds to achieve this. As I said earlier, the world needs today international institutions with requisite stability, e.g. on the applied side, institutes like the Wheat and Rice Research Institutes, and on educational and physics side, institutes like the International Centre for Theoretical Physics. To my distinguished colleagues on the Executive Board of UNESCO I have come today to say: if you cannot find ways and means of keeping alive an initiative you took in 1970 and now universally acclaimed and recognised to be essential to the health of physics in the developing world, then no one else can and no one else will. Institutes like the International Centre for Theoretical Physics must become parts of the normal, continuing, stable United Nations scene; otherwise the science and technology gap of the North and the South will never, never be bridged.

In sciences, as in other spheres, this world of ours is divided between the rich and the poor. The richer half — the industrial North and the centrally managed part of humanity — with an income of 5 trillion dollars, spends 2% of this — some 100 billion dollars — on nonmilitary science and development research. The remaining half of mankind — the poorer South, with one fifth of this income (1 trillion dollars) — spends no more than 2 billions on science and technology. On the percentage norms of the richer countries, they should be spending ten times more — some 20 billions. At the United Nations run Vienna Conference on Science and Technology six weeks ago, the poorer nations pleaded for international funds to increase two billions to four billions. They obtained promise, not of two billions, not of one billion, but only one-seventh of this. UNESCO's programmes will be the poorer; and with them the International Centre at Trieste.

I would like to conclude with three appeals. The first to the International Community — both of Governments and of the Scientists. A world so divided between the haves and have-nots of science and technology cannot endure; at present an International Centre for Theoretical Physics (with a budget of 1.5 million) is all that is internationally available for physics in 100 developing countries. Compare this with European joint projects involving physics alone, of half billion dollars annually. Somehow, somewhere, a break must come.

My second appeal is to the developing countries. In the end, science and technology among them is their own responsibility. Speaking as one of them, let me say this: your men of science are a precious asset. Prize them, give them opportunities, responsibilities for scientific and technological development of their own countries. The goal must remain to increase their numbers tenfold, to increase the two billions spent on science and technology to twenty billions. Science is not cheap; but expenditure on it will pay tenfold.

And then finally, and in all humility, I wish to make a particular appeal to my brothers in the Islamic countries. To some of you Allah has given a bounty — an income of the order to sixty billion dollars. On the international norms these countries should be spending one billion dollars annually on science and technology. It is their forebears who were the torchbearers of international scientific research in the 8th, 9th, 10th and 11th centuries. It was these forebears who funded Bait-ul-Hikmas — Advanced Institutes of Sciences — where concourses of scholars from Arabia, Iran, India, Turkey and the Byzantium congregated. Be generous once again. Spend the billion dollars on international science, even if others do not. Create a Talent Fund — available to all Islamic, Arab and developing countries, so that no potential high level talented scientist is wasted. My humble personal contribution to this Fund will be all I possess — the $60,000 the Swedish Academy has so generously awarded me. Rabbana Taqqabal Minna.

A Proposal for the Creation of an International Centre for Science

1. The Third World as a whole is slowly waking up to the realisation that Science and Technology are what distinguish the South from the North. On Science and Technology depend the standards of living of a nation. The widening gap in Economics and in influence between the nations of the South and the North is basically the Science gap.

2. The creation of the International Centre for Theoretical Physics (ICTP) in 1964, has been an important influence in correcting this imbalance. There are three aspects of scientific research: (1) *basic* scientific research; (2) *applied* scientific research; and (3) research and development in *science-based technology*. The ICTP was initially created to help developing countries — together with those from the North — to make contributions to the basic aspects of High Energy, Nuclear and Condensed Matter Physics, as well as Physics of the Earth and the relevant Mathematics. Since no new international centres with a similar mandate have been created for Applied Physics, or for physics-based technology, the ICTP has added over the years — within its programme — related areas of Applied Physics (Soil Physics, Cloud Physics, Physics of the Oceans and the Atmosphere, Seismology), as well as High Technology — Microprocessors, Fibre Optics, Communication Physics and Physics of Materials for Non-conventional Energy Sources and has set up training laboratories for some of these.

3. The Centre, in addition, has pioneered new modalities — whose efficacy has come to be universally recognised for imparting the latest in sciences

— in particular, sciences supporting high technology.

Besides its research functions, the Centre organises extended courses to which some 2,500 physicists come annually both from developing and developed countries, including some from Italy, the host country. As an anti-brain-drain device, it receives of the order of 150 associates a year (these are first-class men and women from developing countries who are given six-year appointments and who come to the Centre (at its expense) three times, for periods of up to three months, during these six years). There are additionally 264 institutions in the developing world which are federated to the Centre and are empowered to send their members to the Centre.

4. For experimental physicists there is the modality, financed by the Government of Italy, through which 100 experimenters among the Centre's members, are accredited to experimental laboratories located in Italy, for a one year period. This experience enriches these laboratories as well as the countries of origin of the physicists. In addition, through a generous grant (again from the Italian Government), the Centre has embarked on external liaison which permits help to Centres created in developing countries in the image of the Trieste Centre, and also help with the research and training activities in physics held outside Italy.

All in all, the Centre has been extraordinarily fortunate, both in its donors as well as in the service which it has been able to render to the physics community the world over and, in particular, to the building up of physics communities in the developing countries.

5. It seems the time has come when the concept and modalities of the ICTP should be extended further to the other basic Sciences, like Biology, Chemistry and Geology. Regarding Biology, there is the proposed UNIDO Centre of Biotechnology (fashioned after the ICTP) in two components — one in Trieste, sponsored by the Italian Government and the other sponspored by the Indian Government in Delhi. On the high technology (and Basic Physics, Basic Chemistry and Basic Biology) side, there is the proposal to create an Italian Synchrotron Radiation Laboratory in the Research Area of Trieste.

6. The proposal which we are making is that an *International Centre for Science* should be created, which should have as its units (1) the

International Centre for Theoretical Physics (to take care of Physics and Mathematics); (2) the International Centre for Genetic Engineering and Biotechnology (which should be extended to take care of fundamental advances in Biology), (3) an International Centre for High Technology and Material Sciences (of Microprocessors, Microelectronics, Lasers, Fibre Optics, Communication Physics, High Temperature Superconductors, Computational Sciences and Aspects of Space Sciences — this Centre may also have help, as one of its supporting laboratories, from the Synchrotron Radiation Laboratory in Trieste); (4) an International Centre for Chemistry, both Pure and Applied (for fine pharmaceutical as well as bulk chemicals); and finally, (5) an International Centre for Earth Sciences (which would take care of research in and of imparting the recent advances in geology, prospecting, soils as well as the environmental aspects of Earth Sciences).

7. Such an *International Centre for Sciences,* although envisaged as a single entity, will function as a loose federation of existing and new centres in Trieste. (1) It may be constituted as an independent organisation under the Italian Government auspices; (2) Its units may be affiliated to the IAEA, UNIDO, FAO or WNO and UNEP; or (3) It may become affiliated to UNESCO. The Third World Academy of Sciences, through its Third World Fellowship, would play a crucial role in bringing this new Centre into existence.

8. If the last suggestion regarding UNESCO involvement were accepted, UNESCO Science would be divided into two parts: firstly, the ICSU related global aspects of Science, which would be headquartered in Paris as is the case now (with its programmes on man and the biosphere plus classical engineering technologies). The second would be a Trieste type of operation set up in Italy which would concentrate on the *International Centre for Science* and have federation links with similar centres in the developing countries. This is the type of operation which UNESCO has not undertaken in a big way just now and would represent an additional programme area as far as UNESCO is concerned. (The International Centre for Sciences may become a University institution in its own right — capable eventually of awarding doctoral degrees).

9. The need for such a Centre to take care of the developing countries' sciences, both basic and applied, as well as science-based high technology

is clear. The modalities which the International Centre for Theoretical Physics has pioneered have become known all over the world, particularly in the Third World and their extension to other sciences would be deeply appreciated by the Third World countries. Such a Centre would also make a contribution to Italian science in that it will bring to Italy some of the best and most prestigious names and techniques from all over the world. The utility of such a Centre for the European science should also be clear when one recognises that already the International Centre for Theoretical Physics has played an important role, particularly in enhancing and consolidating Solid State Physics within Europe. Through the momentum which such a Centre will provide, Europe wil benefit and so will world science as a whole.

10. The timeliness of the Centre's creation is also clear. Two of its units already exist, that is to say, the International Centre for Theoretical Physics and the International Centre for Genetic Engineering and Biotechnology. Of the other three units, the International Centre for High Technology and Material Sciences would be the most costly; it should have a budget of around 15 − 20 million dollars a year when fully operational. The Centres for Earth Sciences and for Chemistry need not cost that amount. They may be budgeted at the peak to cost of the order of 5 − 10 million dollars. A start could be made for all these three centres with funds of the order of about ten million dollars a year, if buildings could be contributed by the Region of Trieste. These buildings should ideally be near the ICTP.

11. The practical creation of an International Centre for Science may follow a three-phased gradual approach.

(*i*) *Definition Phase*, lasting for approximately one year with roughly an expenditure of US$1.5 million. During this phase, the project for setting up the three new Centres will be defined. The Third World Academy, through its Fellows from the developing countries, would play a major role in providing the necessary scientific inputs while the relevant UN bodies will be contacted in order to obtain the broadest possible scope of cooperation in synergy with their present activities (each new Centre may receive a different mix of support from the competent UN bodies). At the same time, a basic joint managerial structure may be set up with a well-defined directive: three scientific committees to define programmes and scientific structures for each

Centre as well as contact with possible scientists who may lead the three Centres.

(*ii*) *Execution Phase*, lasting for approximately two to three years with an average expenditure of about US$10 million each year. Each of the three Centres could start independently as soon as properly defined and would be financed through a Trust Fund. Interim programmes could start, if necessary, in provisional facilities: such activities would help in building up a good scientific staff and would constitute a tangible attraction for potential donors. During this execution phase, funds should come mostly from the Italian Foreign Ministry as seed money and, to the extend possible, from other concerned UN organisations.

(*iii*) *Consolidation Phase*, lasting for a further few years during which the three new Centres should reach their full operational level. In parallel to this but without any interference with their harmonious development and scientific activities, a new legal framework could be defined to coordinate the activities of the ICTP, ICGEB, and the three new Centres. In this exercise, the "Central Centre" should be conceived in such a broad way that it would be able to attract the participation of all competent and interested international organisations as well as the goodwill of future potential donors such as governments, non-governmental organisations, private foundations, industry and others.

Finally, the International Centre for Science should operate an interdisciplinary Institute for Science and Technology Policy and Management, bringing together scientists, technologists and planners. These activities could grow together with those described above starting early in Phase II.

Inaugural Address to the Third World Academy of Sciences by Secretary-General of the United Nations, Mr. Javier Perez de Cuellar

Trieste, Italy – 5 July 1985

Professor Salam, Distinguished Academicians.

Ladies and Gentlemen,

It is a great pleasure for me to open this meeting of the Third World Academy of Sciences. Standing before this distinguished gathering, I feel a sense of contact with the scientific leadership of the developing countries. This is an exhilarating experience for me.

No one who is familiar with the situation of underdevelopment in Asia, Africa and Latin America and aware of its roots can question the proposition that what the Third World needs is not only economic and social transformation but also an intellectual renovation. The key to it lies in the cultivation of the sciences, of course, but the pursuit will remain uncertain if there is not a re-growth of the scientific spirit combined with a sense of human values. I am, therefore, deeply appreciative of the opportunity you have given me to address this meeting. What I am going to say will be in general terms. Advisedly: I have too much regard for your independence as scientists to want to impinge on it by imposing a bias on your discussion.

Even though we are gathered here in a Third World context, I remain conscious of the fact that science as such has no national affiliation. The supposition that only a specific cultural tradition is conducive to science finds no support in historical evidence. The history of science, indeed, involves the history of diverse civilisations. From the time of the Babylonians through Greece, China, India and the Arab-Islamic civilisation to the modern age, the scientific enterprise has shown no dependence on any particular cosmology or social system. In choosing your vocation, therefore, you have become citizens of the world. Cosmopolitan in outlook but, mentally and spiritually, you have not migrated from your own societies.

As far as science is concerned, the important distinctive feature of modern times is not to my mind the emergence of something that did not exist before. It is, rather, the systematic organisation of the scientific activity. As a result, what was formerly episodic has become continuous and cumulative. This seems to have a dual effect on the practitioner of science; it both circumscribes and enriches him. Within the world of science, the scientist does not enjoy the kind or degree of autonomy that disciplines like philosophy or poetry confer upon their practitioners. The poet or the philosopher is much less subject to pressures: he can carve out his own domain; and the greater his mastery of the craft, the more he will be recognised as sovereign over it.

The scientist, on the contrary, finds his place in a constantly evolving network of research and discovery; he can turn his back on the relevant findings and activities of others only at peril to his pursuit. Furthermore, it is the sheer intellectual challenge of a problem which is the main spur of his activity; while, if he is lucky, the solution is his. The problem at least is suggested by others. Insofar as it is an operative organisation, science maintains its own momentum and most scientists are caught in it.

This may be one of the reasons why human priorities sometimes tend to be forgotten in the scientific pursuit. The scientist does not always enjoy the freedom and the mental leisure to stand back and reflect on the question whether or not the solution of the problem which pre-occupies him will significantly lessen human suffering and enhance human dignity. In saying this, I do not have the slightest intention of denigrating the scientist's sense of social responsibility. Leaders of science themselves bewail the misdirection of science and the misuse of the scientific talent that is represented by the on-going quest for ever more destructive weapons. With the fear and the insecurity it has caused and with the senseless waste of material assets that it inevitably entails, the arms race is a chastening demonstration of the necessity of re-orienting the scientific and technological outlook towards the goal of human betterment.

It is no doubt a historical fact that the invention of newer weapons of war has absorbed technicians throughout the ages and in all societies. Our age, however, has demonstrated that, when carried too far, it reaches the absurd limit of threatening the very existence of the human species and perhaps even the continued habitability of the planet.

I believe that there are three phenomena of the current human situation which should serve as eye-openers about the direction of the scientist's mission. The first is what I just mentioned: the continuous refinement of the technology of destruction. The second is the degradation of the environment and the tension between human society and its natural surroundings. The third is the continued prevalence of poverty, ignorance and avoidable disease in large parts of the world at a time when humanity disposes of enough resources and skills to redress the situation. The last may superficially appear to be an economic and administrative problem but I need hardly labour the point that it does involve a challenge to the scientific community, particularly of the developing countries.

Though these three phenomena are regrettable, they can, if fully appreciated, have the effect of stimulating scientific activity in or for the Third World in the right direction. As we are gathered here, the thought must be uppermost in our minds that the present is a most propitious historical moment for human society to encourage a more purposeful and less indiscriminate use of science and technology. I feel impelled to repeat that the addition of new dimensions to the arms race, the hazards of industrialisation without proper care, the pollution of air and water, the depletion of natural resources, the world's food supply lagging behind the growth of population and the persistence of widespread disease in developing countries — all these bring home the truth to us that we, scientists as well as non-scientists, have to chart a course different from what was taken at a time of exuberance over scientific discovery and invention.

In the world of the imaginative arts, the theory of art for art's sake was discredited long ago. Perhaps in the world of science too, the theory of science for the sake of science should not be allowed to take root.

You, the scientists of developing countries, can play a very influential role in this re-orientation. The very fact that you are confronted with illiteracy, poverty and disease in your countries lends its own significance to your activity. I realise that you do not command the facilities that industrialised societies offer to your colleagues. However, you also have certain privileges. The cultural traditions which have originated in the third world, whatever their strengths and weaknesses, did inculcate respect for learning and scholarship. Because of this and also because of the fact that your smaller number gives you a position of prominence in your own societies often greater than what your peers from developed nations enjoy in theirs, you can wield a powerful influence over the cultural development of your countries. The educative effect of the cultivation of science even on non-scientists can be felt more in countries which have recently embarked on the long process of development. The victories you win in your own chosen spheres can constitute a demonstration of the triumph of rationality and empiricism. It is not altogether naive to hope that the search for objectivity, the avoidance of exaggeration, the discipline of testing hypotheses through experiment and observation required by the scientific pursuit will have a spill-over effect and help lend a balance to the mental outlook and culture of your societies. It may even be a factor in the building of institutions which lead to political maturity and a stable democratic order. These are your privileges and these are your responsibilities as well. Both, of course, imply your full engagement in the great tasks of development. Never did science have to respond to such compelling human necessities and never did it have a more beneficent mission to perform as in developing countries today.

At this meeting you will devote your discussions to the subject of the collaboration of north and south for the development of sciences in the Third World. This is a very important aspect of the wider question of economic and social development and of narrowing the gap between the rich and the poor on which the United Nations has been striving its utmost to initiate a purposeful north-south dialogue. It was within the framework of these effects that the United Nations Conference on Science and Technology, held six years ago, adopted the Vienna Programme of Action. Many of you were central participants in that conference. You are no doubt aware of the manifold difficulties we face. Primarily, these arise from opposed perceptions. I feel, however, that in your own defined sphere the process might be much smoother. This optimistic expectation is based on the fact that scientists all over the world constitute a fraternity. They speak the same language, share the same thought-patterns, and I believe, are equally interested in the development of science and in ensuring the widest access to its fruits. I need hardly assure you that I will be deeply interested in your conclusions.

It seems to me that one way of stimulating greater scientific co-operation between the north and the south would be for scientists and those engaged in preparing national development plans in the Third World jointly to identify with precision the problems whose solution requires the application of the latest results of scientific research without creating severe problems of social adjustment. The development of agriculture and the conquest of avoidable disease are broadly known to be the priorities for most of the developing countries. However, it is only scientists who can define with clarity the relevance of the latest research and the gaps that still need to be filled. The judgments of social planners about the feasibility of utilising modern discoveries will be sounder if they are arrived at after full discussions with scientists.

The very exercise will involve much greater exchange than exists at present not only among the scientists of the developing countries but also between them and economic planners. I believe it will also arouse the deep and sympathetic interest of the scientists of the north who share the human concern to alleviate suffering in the world.

I will conclude by saying that the present situation of mankind calls for a re-orientation of science towards what are really human priorities. In extending to you my warmest wishes for a productive conference, I am voicing the hope that your discussions will lead towards that most desirable goal.

The Founding of the Third World Academy of Sciences (TWAS) and the Third World Network of Scientific Organizations (TWNSO)

Nine hundred years ago, a great physician of Islam, Al Asuli, living in Bokhara, wrote a medical pharmacopeia which he divided into two parts, "Diseases of the Rich" and "Diseases of the Poor". If Al Asuli were alive and writing today about the afflictions wrought upon itself by mankind, I am sure he would divide his pharmacopeia into the same two parts. One part of his book would speak of the threat of the nuclear annihilation inflicted on humanity by its richer half. The second part of his book would speak of the major affliction which poor humanity suffers — underdevelopment, undernourishment and famine. He would add that both these diseases spring from a common cause — excess of science and technology in the case of the rich, and the lack of science and technology in the case of the poor. He might also add that the persistence of the second affliction — underdevelopment — was the harder to understand, considering that resources — scientific and material — are indeed available, to eradicate poverty, disease and early death for the whole of mankind, in this age of scientific knowledge and scientific miracles.

What obstacles prevent mankind from deploying its scientific resources for the eradication of want? First, there is a general lack of political will to use science and technology for this purpose. Second, there is the uneven distribution of scientific resources — research and capability for research, between the rich and the poor. Among these precious resources is also the supply of distinguished men of sciences, particularly those from the Third World.

Such men and women are undoubtedly there, but hitherto they had not united themselves in a world-wide endeavour with a forum of their own.

The Third World Academy of Sciences, with membership of 200 Fellows from 42 developing countries of the Third World today, is an expression of such banding together. Of these Fellows, 10 are Nobel Laureates in Sciences and Economics. Fifty of them are also members, in their own right, of nine of the world's most prestigious Academies of Science — the Academia Nazionale dei Lincei of Italy, the Pontifical Academy of Sciences of the Vatican, the Royal Society of the United Kingdom, the Academie des Sciences of France, the Soviet Academy of Sciences of the USSR, the Royal Swedish Academy of Sciences, the American Academy of Arts and Sciences, the Accademia Nazionale delle Scienze detta dei XL of Italy and the National Academy of Sciences of the US.

So far as the adventure of science is concerned, on all criteria, ours — the twentieth — has been one of the greatest centuries in the history of civilisation. Quantitatively, there has been an explosion of discovery; while, in the deeper understanding of God's design, a number of principles of great synthesis have been established: for example, in Genetics — the double helix principle, in Astrophysics — the standard Big Bang Model, in Geology — the principle of plate tectonics, in Physics — the principles of relativity, quantum theory, and now in my own subject — the unification of fundamental forces. All mankind has contributed to this international enterprise of science and to the pursuit of scientific truth. But, in recent times, in his adventure of discovery on the frontier, unlike as in the past ages, the South has not been able to play a commensurate role principally because of lack of opportunity. This, however, is not a situation which young men and women from the Third World will accept. They enviously, and deservedly, long to participate in this exciting adventure of scientific creation on equal terms. How to make this possible within the resources of their own societies, and how the scientific community at large can help in an organised manner in strengthening science in the Third World, both in teaching and research, will be one of the concerns of our Academy.

The second theme of our Academy is science as the instrument of change — science in application — both globally and within our countries. There is no question but that our present world is a creation of modern science in application. We tend to forget that it was the science of physics which brought about the modern communications revolution and gave a real meaning to the concept of One World and its mutual interdependence. We tend to forget that it was the science of medicine which brought about the penicillin revolution, leading to the present level of population. We

tend to forget that it was the sciences of chemistry and genetics in application which brought about the fertilisers and the green revolutions to
feed the world's population. And we tend to forget that it is to these
same sciences — the wealth-producing sciences of physics and geophysics,
and the survival sciences of medicine, molecular biology, cell culture and
chemistry — to which the Third World must turn for resolution of some of
its present problems.

With these ideas in mind, the inaugural meeting of the Third World
Academy was held in Trieste on 5 July 1985, with an opening address
by the Secretary General of the United Nations, Mr. Javier Perez de Cuellar.
In addition to the Fellows of the Third World Academy, 250 delegates
representing 50 of the world's academies and research councils from the
South and North as well as representatives of international organisations
were also present at the meeting. This meeting decided to set up a
Consultative Committee of the presidents of all academies and research
councils of the South. At the same time, an African Academy of Sciences
was founded, a new international scientific project was started to study
the problem of drought, desertification and food deficit in Africa, in
collaboration with the National Academy of Sciences of the USA, the
World Bank, the MacArthur Foundation and the Dipartimento per la
Cooperazione allo Sviluppo of the Government of Italy, and it was decided
to create African Institutions of Scientific Research to help with the
resolution of the problems identified.

What have we achieved since the founding of the Academy in July 1985?
First and foremost was the problem of setting our own house in order.
We created three offices for the Academy in Asia, Africa and Latin America.
The Academy was fortunate in having had the blessings of the major
countries in the Third World with messages from the presidents of ten
developing countries. A magnificent annual grant of US$1.5 million, which
On. Giulio Andreotti made available to us on behalf of the Government of
Italy and its Dipartimento per la Cooperazione allo Sviluppo, enabled us
to start our operations. A sum of Cdn$450,000 was given by the Canadian
International Development Agency (CIDA). A grant of US$50,000 was
given by the Kuwait Foundation for the Advancement of Sciences for our
publications. The Government of Jordan also gave US$4,000, while the
Government of Sri Lanka continues to contribute US$1,000 annually.
Special grants were provided by the United Nations University (UNU),
the world Meteorological Organisation (WMO) and the United Nations

Educational, Scientific and Cultural Organisation (UNESCO). In addition, the Governments of India, China and Brazil were requested to provide 50 Fellowships each for use by scientists from other developing countries who might work in the institutions of India, China and Brazil. The Third World Academy will provide travel grants.

In respect of the operational programmes which we have supported so far, these fall into three major categories: programmes designed to help individual scientists, programmes designed to build up scientific institutional infrastructure in developing countries and programmes designed to highlight the relevance of science and scientific literacy in the Third World.

Regarding the first set of programmes designed to help individual scientists, we have instituted:

(*i*) research grants to young scientists in the developing countries. These grants — up to US$10,000 for three years — cover cost of equipment, expendable supplies, scientific literature and field studies in areas of pure and applied mathematics, experimental physics, molecular biology and biological chemistry. Some 300 of these have been made available so far (99 in Biochemistry/Molecular Biology, 138 in Physics and 42 in Mathematics);

(*ii*) South-South Exchange Fellowships. So far, 123 scientists from some 40 countries have been awarded Fellowships to visit institutions in Third World countries. In this context, the Academy is happy that governments and scientific organisations in Argentina, Brazil, Chile, China, Colombia, Ghana, India, Iran, Kenya, Madagascar, Mexico, Vietnam and Zaire have agreed to provide local hospitality for a total of 257 annual visits.

(*iii*) grants to research workers working and living in developing countries to work for a period of a year or so at research laboratories in those developed countries which have provided us with funds for this purpose. These include Italy and Canada. This programme will cover biological, medical and chemical sciences. So far as Italy is concerned, 14 Italian laboratories have already agreed to receive scientists, from the Third World. Sixty-eight scientists (27 biologists, 34 chemists, 7 geologists) from 26 Third World countries are currently visiting laboratories in Italy. Senator S. Agnelli announced in October 1988 on behalf of the Italian Government that $1.6 million will be allotted by the Italian Government for this programme.

So much for the programmes which are designed to enhance the work of individual scientists. Regarding the programmes to enhance scientific institutions in developing countries and South-South links, we donate to 250 libraries in the developing world international scientific journals and scientific books. We have organized South-North Roundtables which bring together leaders of research from the South and the North with the objective of forging future links between individual laboratories and research institutes in the North and South. A number of such Roundtables have been convened: the first was held in Khartoum in November 1985 on "The role of a laboratory in the control of soil erosion and sand movement" and the second in Trieste on "Synchrotron radiation and its uses in the developing countries". Subsequent Roundtables have focused attention on Mossbauer Spectroscopy, Haemoglobinopatrics, Plant Breeding, Chemistry of Solid Materials, International Centre for Computers and Informatics, South-South Cooperation 1986, and International Centre for Applied Physics. In October 1988, we organized a conference on the Role of Women in the Development of Science and Technology in the Third World.

In the context of increasing the awareness of scientific achievement and scientific literacy in the Third World, the Academy has instituted prestigious prizes for Third World scientists. Four Prizes, each of US$10,000 are awarded every year in the fields of biology, chemistry, mathematics and physics. We are planning to institute awards for outstanding researchers in medicine and agriculture from 1989 onwards.

The Academy has also instituted a History of Science Prize to bring to light the scientific achievements of Third World scientists prior to the 20th Century, whose work hitherto has not been clearly recognised. The first History of Science Prize was awarded in 1988 to Dr. David A. King (U.K.) for his essay on Shams Al-Di Al-Khalili and the culmination of the Islamic science of astronomical timekeeping.

We are also helping individual academies in the third world to institute prizes of their own with our assistance so that young scientists can be encouraged. Around US$30,000 will be spent on such endeavours to build up sciences in the developing countries themselves. Likewise, we are supporting a number of local language journals devoted to science education and scientific literacy in developing countries and grants have been made for this.

In addition, we have decided to institute a scheme of Third World lecturers — prestigious men of science who will go out to the countries of the third world and lecture there.

But in all these endeavours to help science communities in developing countries, we are handicapped, by and large, by the lack of interest within the countries themselves in such endeavours. There are many aspects of this. The first is the lack of utilisation of scientists in nation-building activities. The second, so far as newer entrants to the subject goes, is the unfortunate fact that not enough new, younger people are entering the profession in developing countries. This has various reasons. It may be due to indifferent science teaching at the tertiary level; it may be due to the fact that the career structures for scientists are poor, it may be due to science not figuring as a valid profession, at least so far as development efforts in our countries are concerned.

How to ensure the utilisation of scientists by the planners in our countries, by our own societies and by our governments, is the hardest task of all. The Third World Network of Scientific Organisations was formed in October 1988 with precisely this objective in view. The creation of the Network linking ministries of Science and Technology and Higher Education, plus Academies, plus Research Councils in the Third World, was first proposed by the Third World Academy of Sciences' President at the opening of the TWAS Second General Conference in Beijing, China, in September 1987.

The proposal was subsequently discussed and endorsed by the participants of the Conference. At the meeting in Beijing, Sig. Balboni of the *Direzione Generale per la Cooperazione allo Sviluppo* of the Government of Italy pledged finanical support for the Network of a quarter of a million dollars annually.

An ad-hoc Committee, chaired by Professor J. Aminu (Nigeria), was set up to institute the Network. Since then, 99 Academies, Research Councils and Ministries from 59 Third World countries have joined the Network.

The purpose of the Network is to increase the effectiveness of Science in the South, through collaboration and communication among Ministries of Science and Technology and Higher Scientific Education, Research Councils and Academies. What collective action can mean need hardly be stressed, since Science has become more and more global. The South can collectively make substantial input in global science programmes and frontiers of science research (examples of these are the ICSU/UNESCO

global programmes, the biological studies of the human genome, thermo-
nuclear fusion, space research, marine science and so on). This will
eventually enhance the effectiveness of South-North collaboration in
these programmes.

The Network would also focus its efforts on alert systems for the
South for natural man-made hazards, such as desertification, pollution,
the greenhouse effect and the dumping of toxic and nuclear waste, the
global and frontier problems of Science and Technology and governmental
involvement in these as well as the whole problem of political action needed
on Science and Technology for development.

Forty-four members of the Network, including 15 Ministers, attended
the first meeting in October 1988 which may be remembered by future
generations as the Group of 44 meeting for Science and Technology.

We hope to have the second meeting of the Network in Bogota,
Colombia, during the period 16−20 October 1989 in conjunction with
the Third General Conference of the Academy which will be hosted by
the Government of Colombia and the third meeting in Kuwait in conjunction with the Fourth General Conference of the Academy which will
be hosted by the Government of Kuwait in 1991.

III

ISLAM AND SCIENCE

Scientific Thinking: Between Secularisation and the Transcendent

(Dedicated to my father who taught me Islam)

This paper combines two talks delivered by Abdus Salam: "Islam and Science" which was given at the UNESCO Headquarters, Paris, in April 1984, and "Scientific Thinking: Between Secularisation and the Transcendent: An Islamic View-Point" which was delivered at the Conference organised by the Giovanni Agnelli Foundation, Torino, Italy, from 21 to 23 June 1988.

1. Introduction

I would like to begin by offering my profound appreciation to the Giovanni Agnelli Foundation for contributing $3,000 to the Fund for Physics. This is a Fund which I created for the developing countries' scientists, from the proceeds of the Prize which was given to me in 1979. The Holy Prophet of Islam taught us to "thank men and women whenever they do good for you" — for "whosoever does not thank people, does not thank Allah." In keeping with this teaching, I express my gratitude to the Giovanni Agnelli Foundation. Since the Fund for Physics represents a charitable purpose, I unashamedly say that I would appreciate it if others could join in keeping the Fund alive.

I have divided my remarks today into six parts. First, I wish to consider an issue raised by my esteemed colleagues — some of these are Nobel Laureates — who have opined that "Science is the creation of the Western, democratic, Judeo-Christian tradition". I disagree with them and I shall explain why. Secondly, I would like to reflect — in the context of the three Abrahamic religions (Judaism, Christianity and Islam) — on the topic which we have been asked to speak on, i.e. Scientific Thinking (as a bridge) between Secularisation and the Transcendent. (I shall not speak about the high traditions of the three non-Abrahamic world-creeds — Hinduism, Buddhism and Confucianism. This is simply because of ignorance.)

Thirdly, I shall address myself to the question — why I am a believer. In this context, let me say right away that I am both a believer as well as a practising Muslim. I am a Muslim as well as a scientist, because I believe in the <u>spiritual</u> message of the Holy Quran (Note 1). In the subsequent three parts, I shall state the present picture of science in the Islamic countries, suggest steps needed to bring about change, and finally make my concluding remarks.

Part I

2.   Openness and Democracy Within the Scientific Community

First I want to take up the rise of modern Science and its relation to the Western democratic tradition. Clearly the Western tradition cannot be the whole story, since this, for example, would exclude the rise of excellent science created in Japan in the present century.[a] My own view has always been that Science is the shared creation and joint heritage of all mankind and that as long as a society encourages it, Science will continue to flourish within that society.

I am sure it is painful for some of you to hear, and painful for me to say it, but the truth of the matter is that excellence in Sciences is dependent on the freedom and openness within the scientific community (assuming that such a community is large enough) and **not** necessarily upon the openness or democracy within the society at large. In Sciences, if there is a considerable body of persons who appreciate what one is engaged on, if there is a body of persons who are permitted to discuss freely and openly their doubts, express their reservations of each other's work as well as speak out their own visions, if there is a body of persons who can discard the beliefs they cherished, if empirical evidence goes against these beliefs, then Science at the highest level will continue to be created among such people. In the USSR, for example, there is a fairly large community consisting of a quarter of a million scientists who work for the Academy of Sciences and similar institutions, and who have their own reward systems and value judgements. Thus till such time as openness and scientific democracy continues to prevail among this community, science in the USSR shall flourish. Of course it will flourish even more if the whole society accepts democracy as Mr. Gorbachev would like it to.

Second is the question of whether modern Science and Technology is a cultural phenomenon. I am prompted to ask this question because

---

[a]Lest the Japanese be credited with preoccupation with technology only, it is good to remember that the finest Encyclopaedia of Mathematics in the English language is the Japanese, translated into English by the Massachusetts Institute of Technology Press.

recently my illustrious colleague Professor Ilya Prigogine posed this question about China: In spite of the well-known Chinese pre-eminence in Technology till the 15th – 16th Century and their numerous discoveries of Science, in application, why did modern science not emerge from China? He goes on to say that among the many reasons put forward is that of the "absence of a sovereign, law-giving God, a concept deeply entrenched in European thought at the end of the Middle Ages. . . . What today we call classical science was born in a culture dominated by the notion of an alliance between man, situated at the interface between the divine order and the natural order, and a law-giving, apprehensible god, a sovereign architect conceived in our own image."

According to Prigogine, "the correspondence between the German mathematican and philosopher Gottfried Leibniz and the English philosopher Samuel Clarke, acting as Isaac Newton's mouthpiece, is very revealing in this respect. The correspondence followed a criticism made by Leibniz in which he accused Newton of having a very poor opinion of God, since, according to Newton, his handiwork was less perfect than that of a good clockmaker. Newton had, in fact, spoken of the constant intervention of God, the creator of a world whose activity he unceasingly nourished. In reply, Newton and Clarke accused Leibniz of reducing God to the status of a Deus otiosus, an idler king, who, after a once and for all act of creation, retires from the stage. In classical science, dominated by the notion of the possibility of an omniscient power indifferent to the passage of time, Leibniz's view prevailed."

This is an interesting thought. If this is true, then, in all justice, there ought to be a recognition of the Islamic, in addition to the Greco-Judeo-Christian legacy. Since this is an important point for the scientific future of one quarter of mankind, to which I belong spiritually and culturally, I shall take time to explain why I believe this to be the case in the next few sections (Secs. 3 – 5) of this paper and in the Appendices.

3. The Holy Quran and Science

Let me affirm once again that so far as a Muslim is concerned, the Quran speaks to him directly and emphasises for him the necessity for reflection on the Laws of Nature, with examples drawn from cosmology, physics and biology. Thus

> *"Can they not look up to the clouds, how they are created;*
> *And to the Heaven how it is upraised;*
> *And the mountains how they are rooted,*
> *And to the earth how it is outspread?"* (88:17)

and again,

> *"Verily in the creation of the Heavens and of the earth,*
> *And in the alternation of the night and of the day,*
> *Are there signs for men of understanding."* (3:189–190)

Seven hundred and fifty verses of the Quran (almost one eighth of the Holy Book) — exhort believers "to study Nature, to reflect, to make the best use of reason in their search for the ultimate and to make the acquiring of knowledge and scientific comprehension part of the community's life." The Holy Prophet of Islam (Peace be on Him) emphasised that the "quest for knowledge (and sciences) is obligatory upon every Muslim, Man and Woman".

4.   Classical Science, A Greco-Islamic Legacy?

How seriously did the early Muslims take these injunctions in the Holy Quran and of the Holy Prophet?

Barely a hundred years after the Prophet's death, the Muslims had made it their task to master the then-known sciences. Founding institutes of advanced study (*Bait-ul-Hikmas*), they acquired an ascendancy in the sciences and technology that lasted up to around 1450 AD when Constantinople fell to the technologically superior Turkish cannonade.

An aspect of reverence of the sciences in Islam was the patronage they enjoyed in the Islamic Commonwealth. H.A.R. Gibb wrote the following in the context of literature: "To a greater extent than elsewhere, the flowering of the sciences in Islam was conditional . . . on the liberality and patronage of those in high positions. So long as, in one capital or another, princes and ministers found pleasure, profit or reputation in patronising the sciences, the torch was kept burning."

The Golden Age of Science in Islam was doubtless the Age around the year 1000 AD, the Age of Ibn-i-Sina (Avicenna), the last of the medievalists — and of his contemporaries, the first of the moderns, Ibn-al-Haitham and Al Biruni.

Ibn-al-Haitham (Alhazen, 965–1039 AD) was one of the greatest physicists of all time. He made experimental contributions of the highest order in optics. He "enunciated that a ray of light, in passing through a medium, takes the path which is the easier and 'quicker'". In this he was anticipating Fermat's Principle of Least Time by many centuries. He enunciated the law of inertia — later, and independently — to become part of Galileo's and Newton's law of motion. He was the first man to conceive of the Aswan Dam though he was unable to build it because the technology of the time could not keep up with his ideas. (He had to feign madness in order to escape the wrath of the Fatimid Caliph, Al Hakim of Egypt, for having proposed the idea of the dam and **not** actually building it.)

Al Biruni (973 – 1048 AD), Ibn-i-Sina's second illustrious contemporary, worked in today's Afghanistan. He was an empirical scientist like Ibn-al-Haitham; as modern and as unmediaeval in outlook as Galileo, six centuries later.

However, it is commonly alleged that Islamic Science was a derived science; that Muslim scientists followed the ("un-Islamic") Greek theoretical tradition blindly.

This statement is false. Listen to this assessment of Aristotle by Al Biruni:

"The trouble with most people is their extravagance in respect of Aristotle's opinions, they believe that there is no possibility of mistakes in his views, though they know that he was only theorizing to the best of his capacity".

Or Al-Biruni on mediaeval superstition:

"People say that on the 6th [of January] there is an hour during which all salt water of the earth gets sweet. Since all the qualities occurring in the water depend exclusively upon the nature of the soil . . . these qualities are of a stable nature . . . . Therefore this statement . . . is entirely unfounded. Continual and leisurely experimentation will show to anyone the futility of this assertion".

And finally, Al-Biruni on geology, with this insistence on observation:

". . . But if you see the soil of India with your own eyes and meditate on its nature, if you consider the rounded stones found in earth however deeply you dig, stones that are huge near the mountains and where the rivers have a violent current: stones that are of smaller size at a greater distance from the mountains and where the streams flow more slowly: stones that appear pulverized in the shape of sand where

*the streams begin to stagnate near their mounths and near the sea —
if you consider all this, you can scarcely help thinking that India was
once a sea, which by degrees has been filled up by the alluvium of
the streams".*

In Briffault's words: "The Greeks systematised, generalised, and
theorised, but the patient ways of detailed and prolonged observation and
experimental inquiry were altogether alien to the Greek temperament . . .
What we call science arose as a result of new methods of experiment,
observation, and measurement, which were introduced into Europe by
the Arabs . . . (Modern) science is the most momentous contribution
of the Islamic civilisation . . .".

These thoughts are echoed by George Sarton, the great historian of
Science: "The main, as well as the least obvious, achievement of the
middle Ages was the creation of the experimental spirit and this was
primarily due to the Muslims down to the 12th century". Clearly it is
not enough to call Modern Science, with its insistence on experiment
and observation, a Grecian legacy. If anything, it is truly a Greco-Judeo-
Christian-Islamic legacy.

5.   The Decline of Sciences in Islam

Why did creative Science die out in Islamic civilisation?

No one knows for certain. There were indeed external causes, like the
devastation caused by the Mongol invasion. In my view however, the
demise of living science within the Islamic commonwealth had started
earlier. It was due much more to internal causes — firstly the inward-
turning isolation of our scientific enterprise, secondly — and in the main,
of active discouragement to innovation (*taqlid*) by the fanatical attitudes
of the religious establishment. The later parts of the eleventh and early
twelfth centuries in Islam (when this decline began) were periods of
intense politically-motivated, sectarian, and religious strife. Even though
a man like Imam Ghazali, writing around 1100 AD, could say "A grievous
crime indeed against religion has been committed by a man who imagines
that Islam is defended by the denial of the mathematical sciences, seeing
that there is nothing in these sciences opposed to the truth of religion",
the temper of the age had turned away from creative science, either to
Sufism with its other-worldliness or, more importantly, to a rigid orthodoxy
with a lack of tolerance (*taqlid*) for innovation (*ijtihad*), in all fields of
learning, including the sciences (Note 2 and Appendix IV).

Part II — Reflections on Transcendence and Secularisation

6.   Science as Anti-Religion

It is generally stated that Science is anti-religion and that Science and Religion battle against each other for the minds of men. Is this correct? (Note 3)

Now if there is one hallmark of true science, if there is one perception that scientific knowledge heightens, it is the spirit of wonder; the deeper that one goes, the more profound one's insight, the more is one's sense of wonder increased. This sentiment was expressed in eloquent verse by Faiz Ahmad Faiz:

کئی بار اس کی خاطر ذرّے ذرّے کا جگر چیرا

مگر یہ چشمِ حیراں جس کی حیـــــرانی نہیں جاتی

"Moved by the mystery it evokes, many a time have I dissected the heart of the smallest particle. But this eye of wonder; its wonder-sense is never assuaged!"

In this context, Einstein, the most famous scientist of our century, has written: "The most beautiful experience we can have is of the mysterious. It is the fundamental emotion which stands at the cradel of . . . true science. Whoever does not know it and can no longer wonder, no longer marvel, is as good as dead, and his eyes are dimmed. It was the experience of mystery — even if mixed with fear — that engendered religion. A know-ledge of the existence of something we cannot penetrate, our perceptions of the profoundest reason and the most radiant beauty, which only in their most primitive forms are accessible to our minds — it is this knowledge and this emotion that constitute true religiosity; in this sense, and in this alone, I am a deeply religious man".

Einstein was born into an Abrahamic faith; in his own view, he was deeply religious.

Now this sense of wonder leads most scientists to a Superior Being — der Alte, the Old One — as Einstein affectionately called the Deity — a Superior Intelligence, the Lord of all Creation and Natural Law. But then the differences start, and let us discuss these.

The Abrahamic religions claim to provide a meaning to the mystery of life and death. These religions speak of a Lord who not only created (1) Natural Law and the Universe in His Glory, his own Holiness and His Majesty; but also created *us*, the human beings in His own image, endowing us not only with speech, but also with spiritual life and spiritual longings. This is one aspect of transcendence. (2) The second aspect is of the Lord who answers prayers when one turns to him in distress. (3) The third is of the Lord who, in the eyes of the Mystic and the Sufi, personifies eternal beauty and is to be adored for this. These transcendent aspects of religion as a rule lead to a heightening of one's obligation towards living beings. (4) The fourth is of the Lord who endows some humans — the prophets and His chosen saints — with divinely inspired knowledge through revelation.

Regarding what may (in the present context) be called the "societal", "secularist" thinking, Abrahamic religions speak of (1) the Lord who is also the Guardian of the Moral law — the precept which states that "Like one does, one shall be done by"; (2) the Lord who gives a meaning to the history of mankind — the rise and fall of nations for disobedience to His commandments; (3) the Lord who specifies what should be human belief as well as ideal human conduct of affairs (Note 4); (4) and finally, the Lord who rewards one's good deeds and punishes wrong doing (like a Father), in this world, (or in the life hereafter).

While many scientists in varying degrees do subscribe to the first three aspects of transcendentalism, not many subscribe to the "societal" aspects of religiosity (Note 5). Scientists have their own dilemmas in this respect.

7.   The Three View Points of Science

Let us start with Natural Law, which governs the Universe. There are scientists who would take issue with Einstein's view, that there is a sublime beauty about the laws of nature and that the deepest (religious) feelings of man spring from the sense of wonder evoked by this beauty. These scientists would instead like to deduce the laws of nature from a self-consistency and "naturalness" principle, which made the Universe come into being spontaneously. This should be something like the doctrine of spontaneous creation of life and its Darwinian evolution — only now carried to the realm of all laws of nature and the whole Universe. If successful, this, in their view, would lead to an irrelevance of a deity (Note 6). Man's spiritual dimension, so called, would be nothing but a particular manifestation of physiological processes occurring inside the human brain (not fully understood at present), but their hope would be that a molecular basis would one day be discovered for this (Note 7). Contrasting with this is the view of the anthropic scientist who believes that the Universe was created purposefully with such attributes and in such a manner that sentient beings could arise. These then are the three viewpoints — first the (religious and transcendental) attitude of an Einstein, second the anthropic view (which in a way supports the first), and third the viewpoint of the self-consistent scientist in whose scheme of things, the concept of a Lord is simply irrelevant.

Regarding what I have called the secularist sentiments in general, Einstein has this to say: "I am satisfied with the mystery of the eternity of life and with the awareness and a glimpse of the marvellous structure of the existing world, together with the devoted striving to comprehend a portion, be it ever so tiny of the Reason that manifests itself in nature. ... (But) I cannot conceive of a God who rewards and punishes his creatures, or has a will of the kind that we experience in ourselves. ... The existence and validity of human rights are not written in the stars." Instead his belief was that "the ideals concerning the conduct of men toward each other and the desirable structure of the community have been conceived and taught by enlightened individuals in the course of history".

Apart from the subjective character of the opinion, note Einstein's silence about the spiritual dimension of Religion.

8. Modern Science and Faith

In the formation of such attitudes towards religion, could it be that the mediaeval Church was partly responsible through its opposition to Science? Could it be that these attitudes are a legacy of the battles of yesterday when the so-called "rational philosophers", with their irrational and dogmatic faith in the cosmological doctrines they had inherited from Aristotle, found difficulties in reconciling these with their faith?

One must remind oneself that the battle of Faith and Science was fiercely waged among the schoolmen of the Middle Ages. The problems which concerned the schoolmen were mainly problems of cosmology; "Is the world located in an immobile place; does anything lie beyond it; Does God move the primum mobile directly and actively as an efficient cause, or only as a final or ultimate cause? Are all the heavens moved by one mover or several? Do celestial movers experience exhaustion or fatigue? What was the nature of celestial matter? Was it like terrestrial matter in processing inherent qualities such as being hot, cold, moist and dry?" When Galileo tried, first, to classify those among the problems which legitimately belonged to the domain of Physics, and then to find answers, only to them, through physical experimentation, he was persecuted. Restitution for this is, however, being made now, three hundred and fifty years later.

At a special ceremony in the Vatican on 9 May 1983, His Holiness the Pope declared: "The Church's experience, during the Galileo affair and after it, has led to a more mature attitude . . . The Church herself learns by experience and reflection and she now understands better the meaning that must be given to freedom of research . . . one of the most noble attributes of man. ... It is through research that man attains to Truth. ... This is why the Church is convinced that there can be no real contradiction between science and faith. ... (However), it is only through humble and assiduous study that (the Church) learns to dissociate the essential of the faith from the scientific systems of a given age, specially when a culturally influenced reading of the Bible seemed to be linked to an obligatory cosmology".

9. The Limitations of Science

In the remarks I have quoted, His Holiness the Pope stressed the maturity which the Church had reached in dealing with science; he could equally have emphasised the converse — the recognition by the scientists from Galileo's times onwards, of the limitations of their disciplines — the recognition that there are questions which are beyond the ken of present (or even future) Sciences and that "Science has achieved its success by restricting itself to a certain type of inquiry".

We may speculate about some of them, but there may be no way to verify empirically our metaphysical speculations. And it is this empirical verification that is the essence of modern science. We are humbler today than, for example, Ibn Rushd (Averroes) was. Ibn Rushd was a physician of great originality with major contributions to the study of fevers and of the retina; this is one of his claims to scientific immortality. However in a different scientific discipline — cosmology — he accepted the speculations of Aristotle, without recognising that these were speculations, and that the future may prove Aristotle wrong. The scientist of today knows when and where he is speculating; he would claim no finality even for the associated modes of thought. And about accepted facts, we recognise that newer facts may be discovered which, without falsifying the earlier discoveries, may lead to generalisations; in turn, necessitating revolutionary changes in our "concepts" and our "world-view". In Physics, this happened twice in the beginning of this century; first with the discovery of relativity of time and space, and secondly with quantum theory. It could happen again; with our present constructs appearing as limiting cases of the newer concepts — still more comprehensive, still more embracing.

Permit me to elaborate on this.

I have mentioned the revolution in the physicists' concepts of the relativity of time. It appears incredible that the length of a time interval depends on one's speed — that the faster we move the longer we appear to live to someone who is not moving with us. And this is not a figment of one's fancy. Come to the particle physics laboratories of CERN at Geneva which produce short-lived particles like muons, and make a record of the intervals of time which elapse before muons of different speeds decay into electrons and neutrinos. The faster muons take longer to die, the slower ones die and decay early, precisely in accord with the quantitative law of relativity of time first enunciated by Einstein in 1905.

Einstein's ideas on time and space brought about a revolution in the physicist's thinking. We had to abandon our earlier modes of thought in physics. In this context, it always surprises me that the professional philosopher and the mystic — who up to the nineteenth century used to consider space and time as their special preserve — have somehow failed to erect any philosophical or mystical systems based on Einstein's notions!

The second and potentially the more explosive revolution in thought came in 1926 with Heisenberg's Uncertainty Principle. This Principle concerns the existence of a conceptual limitation on our knowledge. It affirms, for example, that no physical measurements can tell you simultaneously that there is an electron on the table — here — and also that it is lying still. Experiments can be made to discover precisely where the electron is; these same experiments will then destroy any possibility of finding at the same time whether the electron is moving and if so at what speed. There is an inherent limitation on our knowledge, which appears to have been "decreed in the nature of things" (Note 8). I shudder to think what might have happened to Heisenberg if he was born in the Middle Ages — just what theological battles might have raged on whether there was a like limitation on the knowledge possessed by God.

As it was, battles were fought, but within the twentieth century physics community. Heisenberg's revolutionary thinking — supported by all known experiments — has not been accepted by all physicists. The most illustrious physicist of all times, Einstein, spent the best part of his life trying to find flaws in Heisenberg's arguments. He could not gainsay the experimental evidence, but he hoped that such evidence may perhaps be explained within a different "classical" theoretical framework. Such framework has not been found so far. Will it ever be discovered?

10.   Faith and Science

But is the Science of today really on a collision course with metaphysical thinking? The problem, if any, is not peculiar to the faith of Islam — the problem is one of Science and Faith in general, at least so far as the Abrahamic religions are concerned. Can Science and Faith at the least, live together in "harmonious complementarity"? Let us consider some relevant examples of modern scientific thinking.

My first example concerns the "metaphysical" doctrine of creation from nothing. Today, a growing number of cosmologists believe that the most likely value for the density of matter and energy in the Universe is such that the mass of the Universe adds up to zero, precisely. The mass of the Universe is defined as the sum of the masses of matter — the electrons, the protons and neutrinos, which constitute the Universe as we believe we know it — plus their mutual gravitational energies (converted into mass. The gravitational energy of an attractive force is negative in sign.) If the mass of the Universe is indeed zero — and this is an empirically determinable quantity — then the Universe shares with the vacuum state the property of masslessness. A bold extrapolation made around 1980 then treats the Universe as a quantum fluctuation of the vacuum — of the state of nothingness.

Attractive idea, but at the present time, measurements do not appear to sustain it. This has led to an ongoing search for a new type of matter — the so-called "dark matter" — which is not luminous to us, but would show itself to us only through its gravity.

We shall soon know empirically whether such matter exists or not. If it does not, we shall discard the whole notion of the Universe arising as a quantum fluctuation. This may be a pity, but this points to a crucial difference between Physics and Metaphysics — experimental verification is the final arbiter of even the most seductive ideas in Physics.

11. Anthropic Universe

My second example is the Principle of the anthropic Universe — the assertion by some cosmologists, that one way to understand the processes of cosmology, geology, biochemistry and biology is to assume that our Universe was conceived in a potential condition and with physical laws, which possess all the necessary ingredients for the emergence of life and intelligent beings. "Basically this potentiality relies on a complex relationship between the expansion and the cooling of the Universe after the Big Bang, on the behaviour of the free energy of matter, and on the intervention of chance at various (biological) levels", as well as on a number of "coincidences" which, for example, have permitted the Universe to survive the necessary few billion years.

Stephen Hawking, the successor of Newton in the Lucasian Chair at Cambridge, in his recent book, *A Brief History of Time: from the Big Bang to Black Holes* (Bantam Press, 1988), has stated the anthropic principle most succinctly:

"There are two versions of the anthropic principle, the weak and the strong. The weak anthropic principle states that in a universe that is large or infinite in space and/or time, the conditions necessary for the development of intelligent life will be met only in certain regions that are limited in space and time. The intelligent beings in these regions should therefore not be surprised if they observe that their locality in the universe satisfies the conditions that are necessary for their existence. It is a bit like a rich person living in a wealthy neighbourhood not seeing any poverty.

"One example of the use of the weak anthropic principle is to "explain" why the Big Bang occurred about ten thousand million years ago — it takes about that long for intelligent beings to evolve. ... An early generation of stars first had to form. These stars converted some of the original hydrogen and helium into elements like carbon and oxygen, out of which we are made. The stars then exploded as supernovas, and their debris went to form other stars and planets, among them those of our solar system, which is about five thousand million years old. The first one or two thousand million years of the earth's existence were too hot for the development of anything complicated. The remaining three thousand million years or so have been taken by the slow process of biological evolution, which has led from the simplest organisms to beings who are capable of measuring time to the big bang.

"Few people would quarrel with the validity or utility of the weak anthropic principle. Some, however, go much further and propose a strong version of the principle. According to this theory, there are either many different universes or many different regions of a single universe, each with its own initial configuration and, perhaps, with its own set of laws of science. In most of these universes, the conditions would not be right for the development of complicated organisms; only in the few universes that are like ours would intelligent beings develop and ask the question: 'Why is the universe the way we see it?' The answer is then simple: if it had been different, we would not be here!

"The laws of science, as we know them at present, contain many fundamental numbers, like the size of the electric charge of the electron and the ratio of the masses of the proton and the electron. We cannot, at the moment at least, predict the values of these numbers from theory — we have to find them by observation. It may be that one day we shall discover a complete unified theory that predicts them all, but it is also possible that some or all of them vary from universe to universe or within a single universe. The remarkable fact is that the values of these numbers seem to have been very finely adjusted to make possible the development of life. For example, if the electric charge of the electron had been only slightly different, stars either would have been unable to burn hydrogen and helium, or else they would not have exploded. Of course, there might be other forms of intelligent life, not dreamed of even by writers of science fiction, that did not require the light of a star like the sun or the heavier chemical elements that are made in stars and are flung back into space when the stars explode. Nevertheless, it seems clear that there are relatively few ranges of values for the numbers that would allow the development of any form of intelligent life. Most sets of values would give rise to universes that, although they might be very beautiful, would contain no one able to wonder at that beauty. One can take this either as evidence of a divine purpose in Creation and the choice of the laws of science or as support for the string anthropic principle."

Another example of anthropic principle at work is provided by the recently discovered "electroweak" force. It is interesting to ask why Nature has decided to unify the electromagnetic and weak nuclear forces into one electroweak force. (The electroweak together with the strong nuclear and the gravitational forces constitute the three fundamental forces that we know about in Nature.) One recent answer to this question seems to be that this unification provides one way to understand why in the biological regime one finds amino acids which are only left-handed, and sugars which are only right-handed. (Left and right-handedness refers to the polarisation of light after its scattering from the relevant molecules.) In the laboratory, both types of molecules, left-handed as well as right-handed, are produced in equal numbers. Apparently over biological times, one type of handed molecule decayed into the other type.

According to some scientists, the handedness of naturally occurring molecules is predicated by the fact that electromagnetism (the force of chemistry) is unified with the weak nuclear force — a force which is well-known to be handed (e.g. the weak neutral force exists only between left-handed neutrons and left-handed electrons). This fact plus the long biological times available for life to emerge, apparently was responsible for the observed handedness of biological molecules in neutrons.

But where does anthropic principle come into this? One indication of this could be as follows. As stated earlier, penicillin molecules produced in the laboratory are both right and left-handed. The right-handed ones among them successfully interfere and destroy the coatings of the naturally-occurring right-handed molecules on the bacterial skins which they attack. The penicillin miracle would thus be impossible except for the unification of electromagnetism and weak forces.

12. Self-Consistency Principle

Finally, there is the third category of scientists, who use "self-consistency" and "naturalness" to motivate the architecture of the Universe. To illustrate self-consistency as applied in physics, I shall take a recent example.

As an extension of the recent excitement in physics — that is, of our success with the electroweak force, our success in unifying and establishing the identity of two of the fundamental forces of Nature, the electric force and the weak nuclear forces — some of us are now seriously considering the possibility that space-time may have 10 dimensions. Within this context, we hope to unify the electroweak force with the remaining two basic forces, the force of gravity and the strong nuclear force. This is being done nowadays (1988) as part of "supersymmetric string theories" in ten dimensions. The attempt, if successful, will present us with a unified "Theory of Everything — T.O.E."

Of the ten dimensions, four are the familar dimensions of space and time. The other six dimensions are supposed to correspond to a hidden internal manifold — hidden because these six dimensions are assumed to have curled in upon themselves to fantastically tiny dimensions of the order of 10^{-33} cms. We live in a 6-dimensional manifold in the 10-dimensional space-time: our major source of sensory apprehension of these extra dimensions is the existence of familiar charges — electric and nuclear, which in their turn produce the familiar electric and the nuclear forces.

Exciting idea, which may or may not work out quantitatively. But one question already arises: why the difference between the four familiar space-time dimensions and the six internal ones?

So far our major "success" has been in the understanding of why ten dimensions in the first place (and not a wholesome number of dimensions like thirteen or nineteen). This apparently has to do with the "quantum anomalies" which plague the theory (and produce unwanted infinities) in any but ten dimensions. The next question which will arise is this: were all the ten dimensions on par with each other at the beginning of time? Why have the six curled in upon themselves, while the other four have not?

The unification implied by the existence of these extra dimensions curling in upon themselves is one of the mysteries of our subject. At present, we would like to make this plausible by postulating a "self-consistency and naturalness" principle. (This has not yet been accomplished.) But even if we are successful, there will be a price to pay — there will arise subtle physical consequences of such self-consistency — for example, possibly remnants, just like the three-degree radiation which we believe was a remnant of the recombination era following on the Big Bang. We shall search for such remnants. If we do not find them, we shall abandon the idea.

Creation from nothing, extra and hidden dimensions — strange topics for late twentieth century physics — which appear no different from the metaphysical preoccupations of earlier times; however they are all driven by a self-consistency principle. So far as Physics is concerned, mark however the insistence on empirical verification at each stage.

Part III

13. The Essence of Belief

For the agnostic, self-consistency (if successful) may connote irrelevance of a deity; for the believer, it provides no more than an unravelling of a small part of the Lord's design — its profundity, in the areas it illuminates, only enhanced his reverence for the beauty of the design itself. (Note 6)

But belief has to come first. No one can ordain it: no one can argue it into existence. According to the Mystic, "It is a part of the Grace from Allah". Belief may come from one's early life — as it did in my case from my father's teachings and from his precept — or it may come later, through some experience one may undergo.

But where does one's Science come into this?

Heinz Pagels recounts the following story about Feynman, one of the great physicists of our times — perhaps the greatest — who died earlier this year. "He was in a sensory-deprivation tank and had an exosomatic experience — he felt that he came out of his body and saw the body lying before him. To test the reality of his experience he tried moving his arm, and indeed he saw his arm on his body move. As he described this, he said he then became concerned that he might remain out of his body and decided to return to it. After he concluded his story, I asked him what he made of his unusual experience. Feynman replied with the observational precision of a true scientist: 'I didn't see no laws of physics getting violated'." (Note 9)

I have myself never seen any dichotomy between my faith and my science — since faith was predicated for me by the timeless spiritual message of Islam, on matters on which *physics is silent and will remain so.* It was given meaning to by the very first verse of the Holy Quran after the Opening:

> *"This is the Book*
> *Wherein there is no doubt,*
> *A guidance to the God-fearing,*
> *Who believe in the Unseen."*.

"The Unseen"; "Beyond the reach of human ken" — "The Unknowable"; the original Arabic words are

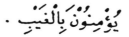

There are other good reasons why I am a believer. Maybe Einstein was oblivious to such a need, but personally I do have faith in the efficacy of prayer at times of distress. I could elaborate on this intensely personal thought but I shall forbear to do this. I am also a believer in what Islam

teaches me about doing good for mankind. I believe in the Moral Lord — that there is a metaphorical balance; on the one side are placed one's good deeds, on the other one's transgressions against humanity. One is rewarded — already here, on this earth — if one side outweighs the other. My greatest desire before I die is that Allah in His Bounty may grant me the mystical vision — so that I too can partake first-hand of what was vouched to the Seers in the past.

Part IV

14.   The Present Picture of Sciences in the Islamic Countries

What is the picture of science and technology in the Islamic Commonwealth? For purpose of identification, the Islamic peoples fall into six geographical regions. First and foremost are the nine countries of the Arabian Peninsula and the Gulf. The second region consists of the Arab northern tier; Syria, Jordan, Lebanon, the Palestinian West Bank and Gaza. The third region comprises Turkey, Muslim Central Asia, Iran, Afghanistan and Pakistan. The fourth (most populous) region consists of Bangladesh, Malaysia, Indonesia, (plus the large Muslim minorities in India and China). In the fifth region are the Arab countries of North Africa, while the sixth region would comprise the non-Arab African countries. (Tables I, II)

If we consider the present enrolment in scientific and technological education in the 18–23 year age group at the universities as an index of high scientific potential, the Islamic countries average 2% of the relevant age group compared to the norms of around 12% for the developed countries (Table III). A similar ratio of 1:6 prevails also in respect of GNP expenditures on scientific and technological research and development. No detailed statistics of numbers of those engaged in scientific research are available. However, in the Background Paper submitted to the first meeting of the Science Commission of the Organisation of Islamic Conference, held in Islamabad during 10–13 May 1983, a figure of around 68,000 research and development scientists and engineers was given for the entire Islamic world, compared to one and a half million in the USSR and four hundred thousand in Japan.

According to A.B. Zahlan, who taught at the American University of Beirut, an analysis of these and similar figures reveals that at least as far as physics is concerned, the Islamic community is around one-tenth in size and one-hundredth in scientific creativity in research publication,

TABLE I

Participation of scientists from Arab-Islamic Countries to the activities of
the International Centre for Theoretical Physics (ICTP)
and of the Third World Academy of Sciences (TWAS), Trieste

Member States	No. of scientific visitors 1970–1988 (30 June)	ICTP External Activities 1986–1988	ICTP Visits to Italian Labs 1983–1988	Federation Agreements 1988	TWAS Visits to Italian Labs 1986–1988	TWAS South-South Fellowships 1986–1988	Research Grants 1986–1988	GNP[1] (US$ billions) 1985	Population[1] (millions) 1985
GROUP I									
Bahrain	3	—	—	—	—	—	—	4	0.42
Kuwait	110	—	—	2	—	—	—	24.8	1.7
Oman	—	—	—	—	—	—	—	8.4	1.2
Qatar	16	—	—	1	—	—	—	5.1	3.2
Saudi Arabia	69	—	—	2	—	—	—	102.1	11.5
United Arab Emirates	1	—	—	—	—	—	—	26.1	1.4
Yemen AR	44	1	—	1	—	—	—	4.1	8
Yemen PDR	4	—	—	1	—	—	—	1.1	2.1
GROUP II									
Iraq	137	1	3	4	—	—	—	n.a.	15.7
Jordan	140	3	7	4	2	—	—	4	3.5
Lebanon	124	—	—	2	—	—	—	n.a.	2.6 (1983)
Syria	113	2	2	6	2	—	2	17.1	10.5
West Bank	2	1	—	4	—	—	—	n.a.	n.a.
GROUP III									
Afghanistan	11	—	—	—	—	—	—	n.a.	17.02 (1983)
Iran	299	2	13	21	—	1	2	n.a.	45.2
Maldives	—	—	—	—	—	—	—	0.05	0.17
Pakistan	598	8	13	9	8	3	19	36.2	95
Turkey	585	4	19	20	2	—	3	56	49.4
GROUP IV									
Bangladesh	319	5	7	3	1	6	4	14.8	100.6
Indonesia	167	1	1	—	2	—	—	86.6	162.2
Malaysia	179	7	—	2	1	—	2	31.9	15.6

TABLE I (Cont'd.)

Member States	No. of scientific visitors 1970–1988 (30 June)	External Activities 1986–1988	ICTP Visits to Italian Labs 1983–1988	Federation Agreements 1988	Visits to Italian Labs 1986–1988	TWAS South-South Fellowships 1986–1988	Research Grants 1986–1988	GNP[1] (US$ billions) 1985	Population[1] (millions) 1985
GROUP V									
Algeria	169	1	6	4	—	—	1	55.2	21.8
Djibouti	1	—	—	—	—	—	—	n.a.	0.4
Egypt	962	5	10	26	2	2	2	32.2	47.1
Libya	123	—	2	4	1	—	—	27	3.6
Morocco	159	—	3	8	2	2	3	13.4	21.9
Sudan	241	3	2	3	2	5	1	7.4	22
Tunisia	85	1	—	6	—	—	—	8.7	7.1
GROUP VI									
Burkina Faso	15	—	—	—	—	—	—	1.1	7.8
Cameroon	32	—	—	1	—	1	—	8.3	10.2
Central African Rep.	3	—	—	—	—	—	—	6.2	2.58
Chad	2	—	—	—	—	—	—	n.a.	5
Comoros	2	—	—	—	—	—	—	0.1	0.4
Cote d'Ivoire	29	5	1	3	2	1	—	6.2	10
Ethiopia	96	1	—	2	—	—	—	4.6	47.27
Gabon	5	—	—	1	—	—	—	3.3	1
Gambia	4	—	—	—	—	—	—	0.17	0.7
Guinea	38	—	2	3	—	3	—	1.9	6
Guinea-Bissau	—	—	—	—	—	—	—	0.15	0.9
Mali	51	1	4	3	—	—	—	1.1	7.5
Mauritania	15	—	—	1	—	—	—	0.7	1.7
Niger	11	—	—	—	—	—	—	1.3	6.4
Nigeria	533	8	32	20	8	4	14	75.9	99.7

TABLE I (Cont'd.)

Member States	No. of scientific visitors 1970–1988 (30 June)	ICTP External Activities 1986–1988	ICTP Visits to Italian Labs 1983–1988	Federation Agreements 1988	Visits to Italian Labs 1986–1988	TWAS South-South Fellowships 1986–1988	Research Grants 1986–1988	GNP[1] (US$ billions) 1985	Population[1] (millions) 1985
Senegal	63	2	1	1	—	—	—	2.4	6.6
Sierra Leone	70	1	2	1	—	3	3	1.4	3.7
Somalia	21	—	5	1	—	—	—	1.5	5.4
Togo	34	—	—	—	1	—	—	0.75	3
Uganda	58	—	—	—	1	—	—	n.a.	15.4
Grand total	5743	63	135	170	34	31	56		

[1] From *World Bank Atlas*, 1987

From 1986 to 1988 TWAS has provided scientific books and journals, as well as spare parts for scientific equipment to 27 countries and financially supported 55 scientific meetings in 14 countries.

TABLE II

R & D Manpower in Islamic Countries

Country	Population in 1985[1] (millions)	R & D Scientists and Engineers[2]
Afghanistan	17.02 (1983)	330[3] (1966)
Algeria	21.80	242[3] (1972)
Bahrain	0.42	–
Bangladesh	100.6	–
Burkina Faso	7.8	–
Cameroon	10.2	–
Chad	5	85[3] (1971)
Comoros	0.4	–
Djibouti	0.4	–
Egypt	47.1	19,941 (1982)
Gabon	1	8[3] (1970)
Gambia	0.7	–
Guinea	6	1,282 (1984)
Guinea Bissau	0.9	–
Indonesia	162.2	24,895 (1984)
Iran	45.2	3,104 (1985)
Iraq	15.7	1,486[3] (1972)
Jordan	3.5	1,241 (1982)
Kuwait	1.7	1,511 (1984)
Lebanon	2.6 (1983)	180 (1980)
Libya	3.6	1,100 (1980)
Malaysia	15.6	–
Maldives	0.17	–
Mali	7.5	–
Mauritania	1.7	–
Morocco	21.9	–
Niger	6.4	93 (1976)
Nigeria	99.7	2,200 (1977)
Oman	1.2	–
Pakistan	95	9,325 (1986)
Qatar	3.2	229 (1986)
Saudi Arabia	11.5	–
Senegal	6.6	522 (1972)
Sierra Leone	3.7	–
Somalia	5.4	–
Sudan	22	3,806 (1986)
Syria	10.5	–

TABLE II (Cont'd.)

Country	Population in 1985[1] (millions)	R & D Scientists and Engineers[2]
Tunisia	7.1	–
Turkey	49.4	7,747 (1983)
Uganda	15.4	–
United Arab Em.	1.4	–
Yemen A.R.	8	60 (1983)
Yemen P.D.R.	2.1	–
Total		79,387

[1] From: *World Bank Atlas*, 1987.
[2] From: *UNESCO Statistical Yearbook*, 1987.
[3] From: "Islamic Conference on Science and Technology", *Secretariat Report*, May 1983.

compared to the international norms. Pakistan, which is one of the most scientifically advanced of Islamic countries, had in 1983 nineteen universities, but only 13 professors of physics, and a total of 42 physics Ph.D. teachers and researchers in all its universities — this for a population of 90 millions. To compare the corresponding numbers at one college at one university in the United Kingdom — the Imperial College of Science and Technology — there are 12 professors and more than 125 researchers.

To give an outside observer's assessment, writing in the prestigious scientific journal, *Nature,* of 24 March 1983, Francis Giles raises the question "What is wrong with Muslim Science?" This is what he says: "At its peak about one thousand years ago, the Muslim world made a remarkable contribution to science, notably mathematics and medicine. Baghdad in its heyday and southern Spain built universities to which thousands flocked: rulers surrounded themselves with scientists and artists. A spirit of freedom allowed Jews, Christians and Muslims to work side by side. Today all this is but a memory.

"Expenditure on science and technology may have increased in recent years though that increase has been, perforce, limited to oil-rich countries . . . Trade structures are dominated by imported technology and most countries have economic and scientific systems geared to imitation rather than originality.

"Even the recent wealth provided by all exports makes relatively little difference . . . science policy and politics, much to the displeasure of many scientists, are closely linked in the Middle East. The region is dominated by dictatorships, benevolent or otherwise ... further complicating any attempt to allow science to take root indigenously. Not surprisingly the brain drain to industrialised countries continues to debilitate intellectual life throughout the Middle East".

Harsh criticism, but much of it factual and deserved.

The same issue of *Nature* contains another article on research man-power in Israel from which I quote: "The need for a substantial increase in the number of academically trained people to work in research and development is widely accepted. The National Council for Research and Development has urged that their country will need 86,700 such people in 1995, compared with 34,800 in 1974 — an increase of 150 per cent". Compare the Israeli figure of 34,800 with around 68,000 researchers in all Islamic countries (the population ratio is 1 : 200).

The article continues: "In the 1960s Professor Derek de Solla Price of Yale University developed a method for measuring scientific manpower in various countries based on the total of researchers who had papers published in major professional journals and concluded that in this country there are five times as many scientists as would be expected for its population and gross national product. Price insists that 'the situation is no different today; the country still possesses an enormous reservoir of trained people, something for which she has every reason to be grateful because her scientists and technicians more than compensate for the lack of oil and minerals' ".

15.   Renaissance of Sciences in Islam

Can we turn the pages of history back and once again lead in sciences? I would humbly like to submit that we can — provided society as a whole — and our youth in particular — come to accept this as a cherished goal, in keeping with our ideological beliefs, and in keeping with our own experience of early centuries of Islam. We must, however, remember that there are no short cuts to this Renaissance. In the conditions of today a nation's youth have to be fired and the nation commit itself with a passionate commitment to this goal; it must impart *hard* scientific training to more than half of its manpower, it must pursue basic and applied sciences with $1-2\%$ of its GNP spent on research and development; at least one quarter to one third of this on pure sciences alone.

This was done in Japan with the Meiji revolution when the Emperor took an oath that knowledge will be acquired from wherever it can be found from the far corners of the earth. This was done in the Soviet Union sixty years ago when the Soviet Academy of Sciences, created by Peter the Great, was asked to expand its numbers and was set the ambition of excelling in all sciences. Today it numbers a self-governing community of half a million scientists working in its institutes, with priorities and privileges accorded to them in the Soviet system that others envy. According to Academician Malcev, this principally came about in 1945, at a time when Soviet economy lay shattered by the war. Stalin decided at that time to increase the emphasis on sciences. Without consulting anyone else, he apparently decided to increase the emoluments of all scientists and technicians connected with the Soviet Academy, by a factor of three hundred per cent. He wanted bright young men and bright young women to enter massively the profession of scientific research.

A similar emphasis on sciences is now being placed in a planned manner and at a frantic speed by the People's Republic of China, with a defined target of catching up and surpassing the United Kingdom in space sciences, in genetics, in microelectronics, in high energy physics, in fusion physics and in the control of thermo-nuclear energy by the end of this century. The Chinese have recognised that all basic science is relevant science; that the frontier of today is tomorrow's application and that they must remain at the frontier. In this context one may recall that the GNP of the Islamic nations exceeds that of China, while the human resources are not significantly smaller. And China has a lead of no more than a few decades over us in sciences. Shall we set ourselves the goal, at the least, of emulating the Chinese?

The societies I have mentioned are not seduced by diversionary slogans of "Japanese" or "Chinese" or "Indian" science. They recognise that though the emphasis in the choice of disciplines on which to research may differ from society to society, the laws, the traditions and the modalities of science are universal. They do not feel that the acquiring of "western" science and technology will destroy their own cultural traditions: they do not insult their own traditions by believing that these are so weak.

I have spoken earlier of patronage for sciences. One aspect of this is the sense of security and continuity that a scientist-scholar must be

accorded for his work. Today, an Arab or a Muslim scientist and technologist — and on Zahlan's count, there are more than thirty thousand of
them — can be sure of a life-long welcome in the United States or in
the United Kingdom if he possesses the requisite quality. He will have
security, respect and equality of opportunity for his work and advancement. We must ask ourselves if this is true within our societies. We must
ask ourselves if we discriminate against, or even at times terminate the
services of scientists because they happen to have originated in a country
with which our government may temporarily have differed.

There is no question but that the United States of America built up
its present ascendance in sciences in a telescoped period of time, by
welcoming the community of scientists who had to flee from Europe
of the inter-war years. But this welcome was not superficial; these men
were accorded rights of citizenship; there was no expectation that they
would return to their countries of origin after "their tour of duty" was
finished. These scientists learned English, settled and raised families
in the United States. There is the well-known story of Enrico Fermi,
who went to USA, just after the ceremony in Stockholm, paying his
own and his family's fare from the Nobel Prize he was awarded in
December 1938. In the USA he was commissioned to build the first
atomic reactor, while still waiting for official clearance for his immigration:
the higher authorities dared not accelerate these procedures for fear of
alerting the Axis intelligence. But all security rules were bent, for the US
had faith in Fermi. The question is — are our countries making a similar
bid for at least the highest level among the scientists they have imported?
Do we accord such men security and personal peace: do we welcome
them with open arms, so that they can build up schools of research for
us with the fullest involvement?

In my view, there is need of a Commonwealth of Science for the
Islamic countries, even if there may be no political commonwealth yet
in sight. Such a Commonwealth of Science was a true reality in the
great days of Islamic Science, when Central Asians like Ibn-Sina and
Al-Biruni would naturally write in Arabic, or their contemporary and my
brother in physics, Ibn-ul-Haitham could migrate from his native Basra
in the dominions of the Abbasi Caliph to the Court of his rival, the Fatmi
Caliph, sure of receiving respect and homage, notwithstanding the political
and sectarian differences, which were no less acute then than they are
now. A new Islamic Commonwealth of Science needs conscious articula-

tion, and recognition once again, both by us, the scientists, as well as our governments. Today we, the scientists from the Islamic countries, constitute a very small community — one-hundredth to one-tenth in size, in scientific resources, and in scientific creativity compared to international norms. At the least, we need to band together, to pool our resources, to feel and work as a community, at science centres which run for all Islamic countries. To foster this growth, could we possibly envisage from our governments a moratorium, a compact, conferring of immunity, for say the next twenty five years, during which the scientists from within this Commonwealth of Science, this Ummat-ul-Ilm, could be treated as a special sub-community with a protected status, so far as internal political and sectarian differences are concerned, just as was the case in the Islamic Commonwealth of Sciences in the past?

And finally, there is the isolation of our scientific effort from international science. It is amazing to find that with the exception of Egypt, which is a member of sixteen unions, no other Islamic country uniformly subscribes to more than five International Scientific Unions in the diverse subjects of science. No international centres of scientific research have been created or are located within our confines; few international scientific conferences are organised there; very few of us, if living and working in our own countries, can travel to scientific institutions and meetings outside; such travel, as a rule, is considered wasteful luxury.

It was this isolation which prompted me to propose the creation of the International Centre for Theoretical Physics so that physicists from developing countries do not make exiles of themselves in order to keep themselves abreast in newer developments in their subject. This Centre belongs to two United Nations Agencies — IAEA and UNESCO; some one hundred and seventy five Arab and Muslim physicists (out of around 1,000 from developing countries as a whole) are supported at the Centre every year. Of these, fifteen are supported by the Kuwait Foundation for Science and Kuwait and Qatar Universities; the rest come with benefactions I may secure for them from Italy or Sweden.

And it is not just the physical isolation of the individual scientist that we suffer from. There is also the isolation from the norms of international science, the gulf between the way we run the scientific enterprise in our countries and the self-governing manner in which it is run in the West or within the community of scientists in the USSR Academy. We seem to have no developed system of professional organisations, no internal review committees, no independent studies of state of art or quality, no science foundations administered by the scientists, no independent sources of grants.

To summarise, the renaissance of sciences within an Islamic Commonwealth is contingent upon five cardinal preconditions: passionate commitment, generous patronage, provision of security, self-governance and internationalisation of our scientific enterprise.

$$\text{إِنَّ اللهَ لَا يُغَيِّرُ مَا بِقَوْمٍ حَتَّى يُغَيِّرُوا مَا بِأَنْفُسِهِمْ}\ .$$

That such an orientation towards sciences will be resisted by some should not be doubted. The tragedy is that such people wrongly claim to speak in the name of the Islamic theological tradition. Even today there are those whose views on science are represented by the following quotes from a widely circulated Islamic monthly, published from London.

"Was the science of the Middle Ages really 'Islamic' science? . . .

"The story of famous Muslim scientists of the Middle Ages such as Al Kindi, Al Farabi, Ibn-al-Haitham and Ibn-Sina shows that, aside from being Muslims, there seems to have been nothing Islamic about them or their achievements. On the contrary, their lives were distinctly unIslamic. Their achievements in medicine, chemistry, physics, mathematics and philosophy were a natural and logical extension of Greek thought . . .

"Al-kindi held Mutazalite beliefs . . . Ibn-al-Haitham was another Aristotelian. In the words of one scientific historian, De Boer, "Al Haitham considered the various doctrines and came to recognise in almost all of them more or less successful attempts to approximate the truth." Truth to him was only that which was presented as material for the sense perception. No wonder that he was generally regarded as a heretic, and has been almost totally forgotten in the Muslim world . . ."

There is no question about it, we do not speak the same language. After this incredible outburst against men in whose work most Muslims take pride, the writer goes on to advocate a policy of the same type of isolationism that destroyed our scientific tradition in the past:

"Countries that have escaped the political dominance of the West have done so by unilaterally imposing an eclectic isolation on themselves. This is the case with both Russia and China, and would also have been the case with Japan had not Commander Perry made the opening of trade ties between Japan and America a precondition for postponing the conquest of Japan . . .

". . . Muslim countries must develop a science policy that makes them capable of dispensing with the need to import both western science and western technology."

. اَلْحِكْمَةُ ضَالَةُ الْمُؤْمِنِ

I could not agree more, so far as technology is concerned. But for science, one is reminded of the story of Al-Biruni, who was accused by a contemporary divine of heresy when he used the Byzantine (solar) calender for an instrument he had invented for determining the times of the prayers. Al Biruni retorted by saying, "The Byzantines also partake of bread. Will you now promulgate a religious sanction against bread?"

Part V

16. Steps Needed for Building up Sciences in the Islamic Countries

I shall outline in brief the concrete steps needed for building up sciences in Islam. One must realise at the outset that the pursuit of science is not cheap and we are trying to redress the neglect of centuries.

16.1 *Science Education*

The Holy Book places strong emphasis on Al-Taffakur (reflection on and discovery of laws of nature) and Al-Taskheer (acquiring a mastery over nature through technology). Taking this and the realities of modern living into account, one of the first requisites of the Ummah in Islam is to encourage scientific and technological education from the secondary and the tertiary, through the university stages.

The present level of net enrolment as percentage of population receiving education in the Islamic countries is illustrated in the accompanying Table III. This table (issued by the World Bank in April 1980) unfortunately makes no distinction between scientific and technical or other categories. It, however, illustrates the stark fact that many of the Islamic countries have a long leeway to make to reach even the average level achieved by the developing countries in general — let alone the averages achieved by the developed countries.

TABLE III

Net Enrolment in Islamic Countries and Countries with Large Muslim Minorities 1985*

	Primary (6 to 11 years of age) %	Secondary (12 to 19 years of age) %	Tertiary (20 to 24 years of age) %	ICTP visitors 1970–88 (30 June)
Low-income economies (GNP/Cap less than 400 US$)	99 w	34 w	—	
1 Gambia	75**	20**	—	4
2 Burkina Faso	32	5	1	15
3 Bangladesh	60	18	5	319
4 Mali	23	7	1	51
5 Tanzania	72	3	—	118
6 Somalia	25	17	—	21
7 Central African Rep.	73	13	1	3
8 India	92	35	—	1988
9 Sierra Leone	—	—	—	70
10 Sudan	49	19	2	241
11 Pakistan	47	17	5	598
12 Mauritania	—	—	—	15
13 Senegal	55	13	2	63
14 Afghanistan	—	—	—	11
15 Chad	38	6	—	2
16 Guinea	30	12	2	38
Total				3557

TABLE III (Cont'd.)

	Primary (6 to 11 years of age) %	Secondary (12 to 19 years of age) %	Tertiary (20 to 24 years of age) %	ICTP visitors 1970–88 (30 June)
Middle-income economies Lower middle-income (GNP/Cap between 400 and 1600 US$)	104 w 104 w	49 w 42 w	14 w 13 w	
17 Yemen PDR	66	19	–	4
18 Indonesia	118	39	7	167
19 Yemen Arab Rep.	67	10	–	44
20 Philippines	106	65	38	107
21 Morocco	81	31	9	159
22 Nigeria	92	29	3	533
23 Egypt Arab Rep.	85	62	23	962
24 Cameroon	107	23	2	32
25 Turkey	116	42	9	585
26 Tunisia	118	39	6	85
27 Jordan	99	79	37	140
28 Syrian Arab Rep.	108	61	17	113
29 Lebanon	–	–	–	124
Total				3055

TABLE III (Cont'd.)

	Primary (6 to 11 years of age) %	Secondary (12 to 19 years of age) %	Tertiary (20 to 24 years of age) %	ICTP visitors 1970–88 (30 June)
Upper middle-income (GNP/Cap between 1600 and 4000 US$)	105 w	57 w	16 w	
30 Malaysia	99	53	6	179
31 Algeria	94	51	6	169
32 Oman	89	32	1	–
33 Iran, Islamic Rep.	112	46	5	299
34 Iraq	100	55	10	137
Total				784

TABLE III (Cont'd.)

	Primary (6 to 11 years of age) %	Secondary (12 to 19 years of age) %	Tertiary (20 to 24 years of age) %	ICTP visitors 1970–88 (30 June)
Developing Countries	101 w	39 w	8 w	
Oil Exporters	107 w	44 w	10 w	
Exporters of manufactures	109 w	40 w	–	
Highly indebted countries	104 w	47 w	16 w	
Sub-Saharan Africa	75 w	23 w	2 w	
High Income oil exporters	86 w	56 w	11 w	
35 Saudi Arabia	69	42	11	69
36 Kuwait	101	83	16	110
37 Bahrain	111**	86**	10*	3
39 United Arab Emirates	99	58	8	1
39 Libya	127	87	11	123
Total				306
Grand Total				7702

Without availability of reliable figures for science versus non-science student-enrolment for the Islamic countries, one cannot make firm statements. It is, however, my impression that comparatively, the situation for science enrolment is much worse; we reach on average a proportion of science enrolment ranging between 1/4 and 1/3 compared with the norms prevailing in developed countries, with a much lower level in comparative quality. In the latest report of the United Kingdom University Grants Commission (issued in 1984) the figure of 52 : 48 is cited for populations of scientists and technologists versus art students. And at the secondary level, whereas in China, or Japan, all science subjects are compulsory — in the USSR, even the future musicians or footballers or seamstresses must study physics, chemistry, mathematics and biology till they are sixteen — there is no such compulsion, for example, in Pakistan's educational system.

Thus, proportionately, too few Muslims are learning sciences. We simply must encourage more of our students to study scientific and technical subjects at the school and the university.

To ensure this, there is need for provision of science teaching at schools, there is need of qualified teachers and of science equipment. And perhaps even more important, there is need for inducements to be provided to the brighter ones among the young students, to remain in science and not drop out. Such dropout takes place in many cases, for reasons of financial stringency. The parents cannot afford to give their charges the long years of education needed for careers in sciences.

Treating the Ummah as a whole, there is thus a need for a Talent Fund for Sciences which should encourage young Muslims to pursue scientific and technological studies say from the age of 14 upwards. In a recent visit to India, at a meeting of Muslim educationalists, it was estimated that for twenty of the larger cities in Northern India alone, there would be need of scholarships for this purpose, amounting modestly to around five million dollars a year, if Indian Muslims are at all to come up to the level attained by the other Indian communities. This would mean the setting up of a fund capitalised at around 50 million dollars to guarantee five millions annually of talent scholarships for sciences. Unfortunately, the Indian Muslim community is too depressed financially to afford this. Such a fund must be created for them and for other indigent parts of the Ummah.

To cater for the whole world of Islam one would need an Islamic Science Talent Fund, available to all Islamic countries, of around fifty million dollars a year. Since the creation of such a fund on a cooperative Islamic basis is not an easy project, at least the Muslim OPEC countries may take a lead and set up *their own Science Talent Funds* on a liberal scale. Such funds may then be thrown open to other Muslim countries, with specialisation to geographical areas.

16.2 *Science Foundations in Islam*

In 1973, the Pakistan Government, on my suggestion, requested the Islamic Summit in Lahore to sanction at least one Foundation for science for Islam, equal in size to the Ford Foundation, with a capital of one billion dollars. Eight years later, in 1981, such a Foundation was created but with just 50 million dollars promised instead of the one billion requested. It may have been more charitable not to have deceived ourselves by this creation. At any rate, what I wrote then (1973) in my memorandum, I will reproduce below.

ISLAMIC SCIENCE FOUNDATION

(*i*) This is a proposal for the creation of a Foundation, by Islamic countries, with the objective of promotion of science and technology at an advanced level. The Foundation (working in conjunction with the Islamic Conference) would be sponsored by the Muslim countries, and operate within these, with an endowment fund of $1,000 million and a projected annual income of around $60−70 million. The Foundation will be *non-political, purely scientific, and run by eminent men of science and technology from the Muslim world.*

(*ii*) *Need.* No Muslim country, in the Middle East, in the Far East or Africa possesses high-level scientific and technological competence attaining to any international levels in quality. The major reason is the persistent neglect by Governments and society in recent times in acquiring of such competence. In relation to international norms (around .3% of economically active manpower engaged in higher scientific, medical and technological pursuits, with around 1% to 2% of GNP spent on these) the norms reached in the Islamic world are one-tenth of what one should expect for a modern society.

(*iii*) *Objectives of the Foundation.* It is suggested that a well-endowed Islamic Science Foundation be created with two objectives; *building up of high level scientific personnel and building up of scientific institutions.* In pursuit of these objectives:

(*a*) The Foundation will create *new communities* of scientists in disciplines where none exist. It will strengthen those communities which do exist. This will be done in a systematic manner, with the urgency of a crash programme.

(*b*) The Foundation will help in building up and in strengthening *institutions* for advanced scientific research at international level, both in pure and applied fields, relevant to the needs of Muslim countries and their development.

The emphasis of the Foundation's work would lie in building up sciences to *international standards of quality and attainment.* Of the two objectives listed above, the building up of high level scientific personnel will receive the higher priority in the first stages of the Foundation's work.

(*iv*) *Programme.* In pursuance of its twin objectives (*a*) of building up high-level scientific manpower in a systematic manner, and (*b*) of employing this manpower for advanced work for the betterment and strength of Islamic societies the Foundation will pursue the following programme:

(*a*) *Building up of scientific communities*

(*i*) Scholars will be sponsored by the Foundation to acquire knowledge of advanced sciences, wherever available, in areas where gaps exist and where there are no existing leaders of sciences. After their return to their countries, the Foundation will help them to continue with their work. Funds of the order of $10 million would support some 4,000 scholars annually while they are receiving advanced training, and support around 1,000 scholars and the needed facilities on their return.

(*ii*) Programmes will be organised around existing scientific leaders in order to increase high-level scientific manpower. For this purpose, contracts will be awarded to university departments to strengthen their work in selected fields. *Quality* of the university faculties will be the criterion for the award of these contracts. Funds to the total of around $15 million may be spent annually for these contracts.

(*iii*) *Contact of scholars from the Islamic world with the world scientific community.* Existing science in Muslim countries is weak because of its isolation. There are no contacts between scholars in Muslim countries and the world scientific community, principally on account of distance. Science thrives on the interchange of ideas and on continuous criticism. In countries with no international scientific contacts, science ossifies and dies. The Foundation will endeavour to change this. This will entail frequent two-way visits to fellows and scholars, and holding of international symposia and conferences. Funds of the order of around $5 million will subsidise some 3,000 visits a year of around two month's duration. This, spread over around ten sciences and over 15 countries, is about 20 visits a year from any one country in any one science.

(*b*) *Sponsoring of relevant applied research*

The Foundation may spend around $25 million for the strengthening of existing, and the creation on new research institutes on problems of development in the Middle East and the Islamic world. These new institutes of international level and standing would be devoted to research in problems of health, technology (including petroleum technology), agricultural techniques and water resources. These institutes may also become units of the United Nations University system in order to attain international standards of quality and achievement through contact with the international community. (A successful institute like the International Rice Research Institute in the Philippines costs about $5 – $6 million to create, and about the same amount to run annually at an *international* level).

(*c*) The Foundation may spend around $5 million in making the general population of Islamic countries technologically and scientifically minded. This will be achieved through instruction using mass media, through scientific museums, libraries and exhibitions, and through the award of prizes for discoveries and inventions. An appreciation of science and technology by the peoples is crucial if there is to be a real impact of science and technology.

(*d*) The Foundation will help with the task of modernising syllabi for science and technology at the high school as well as university levels.

(*v*) *Functioning of the Foundation*

(*a*) The Foundation will be opened to sponsorship by all Islamic countries which are members of the Islamic Conference.

(*b*) The Foundation will have its headquarters office at the seat of the Islamic Conference. In order to retain active and continuous contact with the research centres and projects it endows, it may set up subsidiary offices as well as employ scientific representatives, resident or at large.

(*c*) The Board of Trustees of the Foundation, which will be responsible for liaison with the Governments, will consist of representatives of the Governments, preferably scientists. The endowment fund of the Foundation will be vested in the name of the Board of Trustees.

(*d*) There will be an Executive Council of the Foundation which will consist of scientists of eminence from the Muslim countries. The first Council and its Chairman (who will also be the Chief Executive of the Foundation) will be appointed by the Board of Trustees for a five-year term. This Council will decide on the Foundation's scientific policies, the expenditure of the funds, their disbursement and their administration. The work of the Foundation and the Executive Council will be free from political interference. The Board of Trustees, through the statutes, will be charged with the responsibility of ensuring this.

(*e*) The Foundation will have the legal status of a registered non-profit-making body and would have a tax-free status both in respect of its endowments as well as emoluments of its staff.

(*f*) The Foundation will build up links with the United Nations, UNESCO and the United Nations University system, with the status of a non-Governmental organisation (NGO).

(*vi*) *Financing of the Foundation*

(*a*) It is envisaged that the sponsoring countries would pledge themselves to provide the endowment fund of $1,000 million in four yearly instalments.

(*b*) The proportion of this endowment fund to be contributed by each sponsoring country will be a fixed fraction of the export earnings of the country. The 1972 schedule of export earnings for the Muslim countries is appended. In future years these earnings are expected to increase. However, even at the 1972 level of 25

billion dollars per year, a contribution of less than one per cent
per country per year would suffice to build up the initial endowment capital of one billion dollars over four years.

2 July 1973"

This memorandum was written in the economic climate of 1973. If I
were writing today, I would not be content with one Ford-size Foundation. On standard norms, the Islamic world needs and deserves fifty
independent foundations for science, technology and science education.
This is because in the fifteen intervening years, the GNP of Islamic countries as a whole has risen many fold — it is in excess of 400 billion dollars
now. There are five giants among our countries (Saudi Arabia, Iran,
Turkey, Indonesia and Nigeria) each with GNP in excess of 50 billions,
with another eight countries (Iraq, Pakistan, Malaysia, Algeria, Libya,
Kuwait, Egypt and UAE) with GNP in excess of 20 billions each.

Regarding our collective responsibility towards the Ummah, on the
Day of Judgement, *nations* as well as *individuals* — those designated as
مُتْرَفِيهَا in the Holy Book — will surely be questioned on the uses
they made of what Allah had bestowed on them.

رَبَّنَا لَا تُؤَاخِذْنَا إِنْ نَسِينَا أَوْ أَخْطَأْنَا رَبَّنَا
وَلَا تَحْمِلْ عَلَيْنَا إِصْرًا كَمَا حَمَلْتَهُ عَلَى الَّذِينَ مِنْ قَبْلِنَا

16.3 *Technology in Our Countries*

And this brings me to technology. The Holy Book throughout places
equal emphasis on Taskheer and Taffakur — on acquiring mastery of
nature, through scientific knowledge as much as on the creation of knowledge. The Holy Book holds forth for us the examples of David and
Solomon, with their mastery of the technologies of their day.

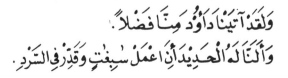

وَلَقَدْ آتَيْنَا دَاوُدَ مِنَّا فَضْلًا .
وَأَلَنَّا لَهُ الْحَدِيدَ أَنِ اعْمَلْ سَابِغَاتٍ وَقَدِّرْ فِي السَّرْدِ .

"And we made iron soft for him . . ." *"We subjected the winds for him" and "under his command he had jinns . . .",*

$$وَلِسُلَيْمَٰنَ الرِّيحَ غُدُوُّهَا شَهْرٌ وَرَوَاحُهَا شَهْرٌ .$$
$$وَمِنَ الْجِنِّ مَنْ يَعْمَلُ بَيْنَ يَدَيْهِ بِإِذْنِ رَبِّهِ .$$

that is, in my humble interpretation, controlled powers of technology of his day, which were used to fashion building blocks, palaces, dams and reservoirs. And then we are reminded of Dhul-qarnain, building defences with blocks of iron and molten copper. Thus are the technologies of metallurgy, heavy construction, technology of wind-power and of communications emphasised. As every Muslim knows, the Holy Book does not relate, except as an exhortation for the future and as an example to be followed by the community.

$$تِلْكَ الْأَمْثَالُ نَضْرِبُهَا لِلنَّاسِ لَعَلَّهُمْ يَتَفَكَّرُونَ .$$

An example of this was set by our Prophet, who was avid to acquire the latest technology of defense. Witness his use of the "khandaq" for the first time in Arabia. Or his ordering of the setting up of Byzantine "manjaniqs" to reduce "Khaibar" — though as it was, the fort fell before these — for the Arabs, novel — instruments of war could be erected.

What are the obstacles in our societies to our acquiring the highest proficiency in technology — and in particular with newer science-based high technologies? After all, never before in human history has so much effort and such magnitude of funds gone into creating technical facilities in such a short duration of time as in the Arab Muslim lands during the last decade. Thus, according to Zahlan, even by 1978 more than 400 billion dollars had been spent on major technological contracts between these countries and foreign suppliers. These projects ranged over hydrocarbons and petrochemicals (160 billions), civil works including transport (80 billions), industrial plants including iron and steel, pharmaceuticals, fertiliser plants (40 billions).

Unfortunately, most of these projects were executed in the technology-free turnkey mode; their execution had no association, no employment of the incipient (research and development) community of Arab men of technology and engineering. And one of the reasons was the fragmentation of the projects. Thus, according to Zahlan, the 584 projects executed by 1976 in the field of petrochemicals, were designed by 83 international firms. The projects included 16 for urea plants; of these Algeria had one, Egypt one, Iraq two, Kuwait four, Libya one, Qatar two, Saudi Arabia one, Sudan one, Syria one, UAE one. No Arab country or combination of countries in the entire Arab Commonwealth had — or now has, after the experience — the technical base to provide the design and construction services for these projects, nor the competence to upgrade these and modify them if need arises.

As contrasted with this, consider Japan with a population nearly equal in size to the Arab nations, and which entered the field of petrochemicals machinery twenty years ago. Right from the outset the Japanese had made up their minds to export such machinery; thus, during the last twenty years every third Japanese plant had been exported. The Japanese had the will as well as the competent men. If competent men in the technology concerned did not exist, they were easy to produce for the basic scientific knowledge within the nation was there. And such men, respected for their high scientific attainments, knew precisely what to go for, when it came to transplant technology.

To emphasise how we have lagged behind in these matters, with neglect which dates back over centuries, one may recall that in the year 1800 CE, William Eton, the then British Ambassador in Instanbul, wrote of his impressions of the Sublime Porte as follows: "No one has the least idea of navigation and the use of the magnet ... Travelling, that great source of expansion and improvement to the mind is entirely checked by arrogant spirit of their religion and by the jealousy with which intercourse with foreigners ... is viewed in a person not invested with an official character ... Thus the man of general science ... is unknown: anyone, but a mere artificer who should concern himself with the founding of cannons, the building of ships or the like, would be esteemed little better than a mad man." He concludes with the remark, with an ominous modern ring: *"They like to trade with those who bring to them useful and valuable articles, without the labour of manufacturing".*

What has been the reason today for this lack of attention to the concept of attaining self-sufficiency in manufacture? Except for a few Islamic countries, like Indonesia, the answer is uniformly the same: the decision-maker is as a rule a non-technical person; our countries, at best, are the paradise of the planner, the administrator; by and large, the technologist has no part in decision-making. In Pakistan, for example, the Planning Commission did not have a science and technology cell until three years ago. Even worse, inheriting a tradition from the British-Indian Civil Service, in Pakistan at least, it is still assumed that a technologist is incapable of making any decisions; his is not the broad vision. We seem not to have noticed that in Japan, in China, in Korea, in Sweden, in France — in all the countries with successful records of self-reliant growth — the most complete accord, participation, involvement and trust exists between the scientist, the technologist, and those who run the development machinery of the state and of the industry.

Besides industrial and science-based technology, there is the whole area of science in agriculture, in public health, in biotechnology, in energy systems, in science-based communications and in defense. The story in all these spheres, and particularly defense, is unfortunately the same — defense purchases, yes; defense production and technology, no. One despairs of whether we shall ever wake up. In the forceful words of Ibn Khaldun:[4] "What sets some above others is their seeking of higher qualities ... When parsimony (in respect of these) ... becomes rampant in a city or a nation, then will Allah's decree ... come into force and this is the meaning of His words in the Holy Quran:

$$وَإِذَا أَرَدْنَا أَنْ نُهْلِكَ قَرْيَةً أَمَرْنَا مُتْرَفِيهَا$$

$$فَفَسَقُوا فِيهَا فَحَقَّ عَلَيْهَا الْقَوْلُ فَدَمَّرْنَاهَا$$

$$تَدْمِيراً.$$

Part VI

17. Concluding Remarks

Why am I so passionately advocating our engaging in this enterprise of creating scientific knowledge? This is not just because Allah has endowed us with the urge to know, this is not just because in the conditions of today this knowledge is power and science in application the major instrument of material progress; it is also that as members of the international world community, one feels that lash of contempt for us – unspoken, but still there – of those who create this knowledge.

I can still recall a Nobel Prize winner in physics some years ago from a European country say this to me: "Salam, do you really think we have an obligation to succour, aid, feed and keep alive those nations who have never created or added an iota to man's stock of knowledge?" And even if he had not said this, my self-respect suffers a shattering hurt whenever I enter a hospital and reflect that almost every potent life-saving medicament of today, from penicillin upwards, has been created without our share of inputs from any of us from Islam. I am sure our men of religion feel exactly the same way; for didn't Imam Ghazzali in the first chapter of his great Ihaya ulum-ud-din ("The Revival of Religious Learning") lay stress upon the acquiring and creating of at least those sciences that are necessary for the development of Islamic society, specifically mentioning medical sciences? He designated active cultivation and advancing of such sciences as Farz-e-Kefaya – an obligation for the whole community, but one which can be discharged on its behalf, by a certain number of its members – otherwise the entire community would consist of transgressors.

I have addressed in this paper three categories of "those with the word" among us: these are the affluent whom Allah has endowed with substance, our ministers and princes responsible for our science policies, and our men of religion.

As I have repeatedly emphasised, science is important because of the underlying understanding it provides of the world around us, of the immutable laws and of Allah's design; it is important because of the material benefits and strength in defense its discoveries can give us; and finally it is important because of its universality. It could be a vehicle of co-operation of all mankind and in particular among the Islamic nations. We owe a debt to international science, which, in all self-respect, we must discharge.

I am now living and working in a small and not particularly rich city of one-quarter of one million inhabitants. In this city there is a bank – Cassa di Risparmio – that donated 1.5 million dollars in 1963 for the building in which the International Centre for Theoretical Physics (which

I had suggested the creation of) is housed. This city has now pledged from its regional resources 40 million dollars for the proposed UNIDO Centre for Biotechnology. I feel amazed at their love of science and their perceptiveness. Shall our cities and banks not rival this example? Just a few days ago I learned with envy that the Keck Foundation — founded by a (relatively obscure) US oil family — has given the California Institute of Technology a sum of 70 million dollars to build the largest telescope on earth — ten meters in diameter. And this in a discipline we — in the past — used to take pride in cultivating — Astronomy.

The international norms of one to two per cent of GNP I have spoken about would mean expenditures of five to ten billion dollars annually for the Islamic world on research and development, one-quarter to one-third of this spent on basic sciences. In the past centuries we had rich traditions in this respect. Imam Ghazzali, you may recall, paid a tribute in the 11th century, to the lands of Iraq and Iran when he said: "There are no countries in which it is easier for a scholar to make a provision for his children." This was at the time when he was planning to become a recluse and to cut himself off from the world. Today, we need not one but many such science foundations, run by the scientists themselves; we need international centres of higher learning within and without our universities, providing generous and tolerant continuity, for our men and their ideas. Let no future Gibb record that in the fifteenth century of the Hijra, the scientists were there in Islam, but there was a dearth of merchants, ministers and princes to provide for the facilities needed for their work.

رَبَّنَا وَآتِنَا مَا وَعَدْتَنَا عَلَى رُسُلِكَ وَلَا تُخْزِنَا

يَوْمَ الْقِيَامَةِ إِنَّكَ لَا تُخْلِفُ الْمِيعَادُ .

And let me finally repeat, for those who are worried about the impact of modern science on Islam, that to know the limitations of science, one must be part of living science; otherwise one will continue fighting yesterday's philosophical battles today. Believe me, there are high creators of science among us — and potentially among our youth. They have the strongest urge to join in the adventure of knowing. Trust them; their Islam is as deeply founded, their appreciation of the spiritual values of the Holy Book as profound, as anyone else's. Provide them with facilities to create science in its standard norms of inquiry. We owe it to Islam. Let them know science and its limitations from the inside. There truly is no disconsonance between Islam and modern science.

Let me conclude with two thoughts. One is regarding the urge to know. As I said before, the Holy Quran and the teachings of the Holy Prophet emphasis the creating and the acquiring of knowledge as bounden duties of a Muslim, "from cradle to the grave". I spoke of Al Biruni who flourished at Ghazna one thousand years ago. The story is told of his death by a contemporary who says: I heard, Al Biruni was dying. I hurried to his house for a last look; one could see that he would not survive long. When they told him of my coming, he opened his eyes and said: Are you so and so? I said: Yes. He said: I am told you know the resolution of a knotty problem in the laws of inheritance of Islam. And he alluded to a well-known puzzle. I said: Abu Raihan, at this time? And Al Biruni replied: Don't you think it is better that I should die knowing, rather than ignorant? With sorrow in my heart, I told him what I knew. Taking my leave I had not yet crossed the portals of his house when the cry arose from inside: Al Biruni is dead.

As my last thought, I would like to quote again from the Holy Book — a Book, the very sounds of which, in the words of Marmaduke Pickthall "move men to tears and ecstacy". More than anything else I know of, it speaks of the eternal wonder I have personally experienced in my own Science:

> *"Though all the Trees on Earth were Pens*
> *and the Sea was Ink*
> *Seven seas, after, to replenish it,*
> *Yet would the Words of Lord be never spent,*
> *Thy Lord is Mighty and All Wise."* (31 − 27)

REFERENCES

1. A.J. Arberry, *Revelation and Reason in Islam* (George Allen and Unwin, London, 1957) p. 19.
2. H.J.J. Winter, *Eastern Science* (John Murray, London, 1952) p. 72 − 73.
3. Briffault, *Making of Humanity,* pp. 190 − 202 quote taken from Muhammad Iqbal. *The Reconstruction of Religious Thought in Islam* (M. Ashraf, Lahore, 1981) pp. 129 − 130.
4. Ibn Khaldun, *Al-Muqaddimah* Part IV, Chapter 18.

NOTES

Note 1

A.J. Arberry, who was Professor of Arabic at the University of Cambridge, and whose translation of the Holy Book is probably the most poetical, had this to say:

"The Quran was the prime inspiration of a religious movement which gave rise to a civilisation of wide extent, vast power and profound vitality. The literature and fine arts of all Muslim peoples spring from this fountain head ... No man seeking to live in the same world as Islam can afford to regard lightly, or to judge ignorantly, the Book that is called The Quran. It is among the greatest monuments of mankind."

Note 2

As examples of the Muslim religious establishment's resistance to Science and Technology in the later centuries, one may cite the following:

1. The opposition to printing. This opposition succeeded in retarding the introduction of printing (and literacy) into the Islamic World till the Napoleonic occupation of Egypt in 1798 – full three hundred and fifty years after Gutenberg's first printing and dissemination of the Bible. In Turkey (apart from a short period between 1729 and 1745) printing could not be introduced till 1839 for secular books and not till 1874 for the Holy Quran on account of the opposition of the religious establishment (for details see D.J. Boorstin, *The Discoverers,* (Vintage Books, New York, 1983)). I myself was astonished to see a text of the Holy Quran printed in Venice towards the end of the sixteenth century, preserved in the University Library of the Aligarh Muslim University in India.

2. The blasting of the last observatory in Islam – the Instanbul Observatory in 1580 (recounted in Appendix IV of this paper).

3. As an example of obscurantist faith today, note an astonishing publication from a religious scholar (no less than the President of the Council for Guidance and Religious Sanctions of a leading Islamic country) who declared recently in 1982 that any (Muslim) who believes (presumably with Galileo) that the sun is stationary and that the earth is moving must be excommunicated and declared a heretic, that he should be hunted to death, and his property confiscated. (To be fair, one must

also register the protest at this unseemly statement, made by another religious scholar, which is noted in the same publication.)

By religious establishment, I do not mean the great Imams — like Imam Abu Hanifa — who accepted no public funds for his teachings, and whose stand against the Abbasi caliphs is legendary, or Imam Ghazali, whom I quoted earlier, or Imam Jafar Sadiq, who strongly stressed the value of scientific knowledge for the Islamic community. I have more in mind a class of people — particularly in my own country of Pakistan — who have arrogated to themselves the mantle of the Holy Prophet, without possessing a knowledge of even the rudiments of their own great and *tolerant* religion. (See my paper on "Religious Liberty in Islam", delivered at the "International Religious Liberty Congress" in Rome, 4 September 1984.)

I am also not speaking here of the recent movement on "Liberation Theology" directed against repression, however much I may approve of this, nor of the (justifiable) prohibitions which a religious tradition imposes on its followers in order to forge a community out of them. Nor will I discuss the (cultural) dimension which endows some of the best specimens of a religious tradition with highly desirable human qualities — like "courage in the face of adversity" plus a "complete lack of colour discrimination" in Islam, or like the "passion for learning" in Judaism, or like "compassion" in Christianity. I forbear to do this, for this is a lecture on Science, and not on Politics or Culture.

Note 3

One must recognise at the outset that religion is one of the strongest "urges of mankind", which can make men and women sacrifice their all, including their lives, for its sake.

Note 4

A Jew like Einstein was Jewish because he subscribed to the ostensibly "cultural aspects" of the Jewish faith, rather than any "fundamentalist" belief in the teachings regarding "ideal human conduct" in the Old Testament. Freud expressed himself, in a similar vein, in his preface to the Hebrew translation of *Totem and Taboo.* Referring to the emotional position of an author (himself) who is ignorant of the language of holy writ and estranged from the religion of his fathers, he says that if the question were put to him, "Since you have abandoned all these common characteristics . . . , what is there left to you that is Jewish?" he would reply, "A very great deal, and probably its very essence." He said he "could not now express that essence clearly in words which someday, no doubt, would become accessible to the scientific mind".

Note 5

The specification of "Ideal Belief and Conduct" unfortunately has almost always led to intolerance, ex-communication, fanaticism and repression, particularly of minorities. There is *devisiveness in the very concept of the chosen people.* In its worst manifestations, this devisiveness may sanction murder for religious disagreements — often making a mockery of a religion's own tolerant teachings. In this respect, the late Professor Sir Peter Medawar, Nobel Laureate in Biology and Medicine, in his book *The Limits of Science* (Oxford University Press, 1986) had this to say: "Religious belief gives a spurious spiritual dimension to tribal enmities . . . . . . The only certain way to cause a religious belief to be held by everyone is to liquidate nonbelievers. . . . The price in blood and tears that mankind generally has had to pay for the comfort and spiritual refreshment that religion has brought to a few has been too great to justify our entrusting moral accountancy to religious belief. By 'moral accountancy' I mean the judgement that such and such an action is right or wrong, or such a man good and such another evil."

Note 6

I find the creationist creed especially insulting in that while we ascribe subtlety to ourselves in devising these self-consistency modalities, the only subtlety we are willing to ascribe to the Lord is that of the potter's art — kneading clay and fashioning it into man.

Note 7

Since the twentieth century has been called "The Century of Science", I wish I could somehow convey the depth of the miracle of Modern Science, both basic and applied. The 20th century has been a century of great synthesis in science — the syntheses represented by quantum theory, relativity and unification theories in physics, by the Big Bang idea in cosmology, by the genetic code in biology and by ideas of plate techtonics in geology. Likewise in technology, the conquest of space and the harnessing of atomic power. Just as in the 16th century when the European man discovered new continents and occupied them, the frontiers of science are being conquered one after another. I have always felt passionately that our men and women in Arab-Islamic lands should also be in the vanguard of making these conquests today as they were before the year 1500.

Note 8

One of the most difficult questions which the self-consistent scientist has to answer is — "Why this decree?"

Note 9

According to Maurice Bucaille in his essay on "The Bible, the Quran and Science", there is not a single verse in the Quran where natural phenomena are described and which contradicts what we know for certain from our discoveries in the sciences.

APPENDICES I–V

After the monumental work done by Joseph Needham on Chinese Science, the first important study of Islamic Technology has been made by A.Y. Al-Hassan and D.R. Hill, in their book *Islamic Technology,* published by Cambridge University Press in 1986. The appendices taken from this study, I hope, will bear out my contention regarding modern Science and particularly modern Technology being a continuation from the Islamic tradition. (The debt owed to China by scientists in Islam is not discussed here.)

APPENDIX I. Transfer of Technology from the Islamic World to the West

"The traditional view of Western historians is that European culture is the direct descendant of the classical civilisations of Greece and Rome. According to this theory, the works of classical authors — mostly in Latin, but some in Greek — were preserved by the Church during the centuries that followed the fall of the Roman Empire, to re-emerge as a potent source of inspiration in the later Middle Ages and the Renaissance. Few would deny the strong influence of classical literature on European thought. ...

"In science, however, the situation is very different. During the twelfth century AD, the writings of such scholars as al-Farabi, al-Ghazali, al-Farghani, Ibn Sina and Ibn Rushd were translated into Latin, and became known and esteemed in the West. The works of Aristotle, soon to become a predominating influence on European thought, were translated from the Arabic together with the commentaries of Ibn Sina and Ibn Rushd (Avicenna and Averroes to the medieval Europeans). These commentaries were as important as the works of Aristotle himself in forming European scientific and philosophical thought. Many other scientific works, which had originally been translated from Greek into Arabic centuries earlier, were now translated into Latin. ...

"Charles Singer, in the Epilogue to the second volume of *A History of Technology*, discussed some of these points ... Referring to the Eurocentrism of Western historians, Singer wrote: "Europe, however, is but a small peninsula extending from the great land masses of Afrasia. This is indeed its geographical status and this, until at least the thirteenth century AD, was generally also its technological status." In skill and inventiveness during most of the period AD 500 to 1500, Singer continues "the Near East was superior to the West ... For nearly all branches of technology the best products available to the West were those of the Near East ... Technologically, the West had little to bring to the East. The technological movement was in the other direction."

"We shall now indicate how this technology transfer occurred, and give some examples of the transfer of ideas and techniques from Islam to the West. The adoption by Europe of Islamic techniques is reflected by the many words of Arabic derivation that have passed into the vocabularies of European languages. (In English these words have often ... entered the language from Italian or Spanish.) To cite but a few examples: in textiles — muslin, sarsanet, damask, taffeta, tabby; in naval matters — arsenal, admiral, in chemical technology — alembic, alcohol, alkali; in paper — ream; in foodstuffs — alfalfa, sugar, syrup, sherbet; in dyestuffs — saffron, kermes, in leather-working — Cordovan and Morocco. As one would expect, Spanish is particularly rich in words of Arabic origin, especially in connection with agriculture and irrigation. We have, for example, *tahona* for a mill, *acena* for a mill or water-wheel, *acequia* for an irrigation canal.

"Many Arabic works on scientific subjects were translated into Latin in the later Middle Ages, the translation bureau in Toledo in the twelfth century AD, where hundreds of such works were rendered into Latin, is a notable example of this activity. ... About AD 1277, in the court of Alfonso X of Castile, a work in Spanish entitled *Libros del Saber de Astronomia* was compiled under the direction of the King, with the declared objective of making Arabic knowledge available to Spanish readers. It includes a section on timekeeping, which contains a weight-driven clock with a mercury escapement. We know from other sources that such clocks were constructed by Muslims in Spain in the eleventh century, hence about 250 years before the weight-driven clock appeared in northern Europe. ... About AD 1277 the secrets of Syrian glass-making were communicated to Venice under the terms of a treaty made between Bohemond VII, titular prince of Antioch, and the Doge. ...

"... Relations between the Islamic World and Christian Europe were not always hostile. Muslim rulers were ... enlightened men and tolerant towards their Christian subjects, an attitude enjoined upon them by the precepts of the Holy Qur'an. Furthermore commercial considerations led to the establishment of communities of European merchants in Muslim cities, while groups of Muslim merchants settled in Byzantium where they made contact with Swedish traders travelling down the Dnieper. There were particularly close commerical ties between Fatimid Egypt and the Italian town of Amalfi in the tenth and eleventh centuries AD. (The ogival arch, an essential element of Gothic architecture, entered Europe through Amalfi — the first church to incorporate such arches being built at Monte Casino in AD 1071.) ...

"The most fruitful exchanges took place in the Iberian peninsula, where over many centuries the generally tolerant rule of the Umayyad Caliphs and their successors permitted friendly relationships between Muslims and Christians. Muslim operations in agriculture, irrigation, hydraulic engineering, and manufacture were an integral part of everyday life in the southern half of the peninsula, and ... Muslim ideas in these ... fields, and in others, passed from Spain into Italy and northern Europe.

"These transmissions were not checked by the Reconquistra. Indeed, they were probably accelerated, since the Christians took over the Muslim installations and maintained them in running order in the ensuing centuries. The Muslim irrigation systems with their associated hydraulic works and water-rising machines remained as the basis for Spanish agriculture and were in due course transferred to the New World. Other installations passed into Christian hands. Industrial plants, such as the paper mill at Jativa near Valencia, were taken over. . . . These few examples serve only as indicators of the passage of Muslim ideas into Europe."

APPENDIX II. Incendiary Weapons, Gunpowder and Cannon

Incendiary Weapons

"From the start of Islamic history, incendiary weapons were used on various occasions, both in local conflicts and against the Byzantines. The Muslim armies of the Abbasids (eighth to thirteenth centuries AD) formed special incendiary troops (*naffatun*) who wore fireproof clothing and threw incendiary materials. Military fires had been used in the Near East from ancient times, due it seems to the existence of natural seepages of *naft* (petroleum). . . . The incendiary materials used in warfare in pre-Islamic times included: (a) liquid petroleum, which was available in Iraq, Iran and from the shores of the Caspian Sea; (b) liquid pitch; (c) mixtures of pitch, resin and sulphur; (d) mixtures of quicklime and sulphur, which ignited on contact with water; and (e) mixtures of quicklime and sulphur with other flammable materials such as bitumen, resin, *naft*, etc. that also ignited on contact with water. All these materials continued in use with advent of Islam but after the eight century AD important developments took place.

"About the year AD 673 a Syrian architect from Baalbek called Callinicus defected to Byzantium; this was forty years after the establishment of the Arab government in Syria and just before Constantinople (Istanbul) was besieged by the Muslims . . . It seems that Callinicus had brought with him the secret of a new fire which helped the Byzantine Empire defend its capital for centuries against Muslim attacks, as well as those of the West Europeans and the Slavs; not until AD 1453 was the city conquered by the Muslim Ottomans. (This new fire was called 'Greek fire' by the Crusaders, though it was never called 'Greek' by the Byzantines.)

"Evidence of this is to be found in a number of instances. For example, Jabir ibn Hayyan describing the preparation of what we now call nitric acid used the term 'flowers of nitre', which may well refer to a crystalline form of salpetre, while in a later Arabic manuscript written in Syriac characters and probably composed in the tenth or eleventh centuries AD saltpetre is listed among the 'seven salts'. Its descriptions cover *buraq al-sagha* (borax — the hydrated sodium borate used by jewellers) which, it says, is white and resembles *al-shiha* (i.e. saltpetre), and lists it immediately afterwards, describing it as 'the salt that is found at the feet of walls'. Saltpetre was also known under other names: flower of assius *(shawraj)*'; wall salt *(milh al-ha'it)*; snow of China *(thalj al-Sin)*; salt of China *(milh al-Sin)*; *al-shiha; ashush;* tanner's salt; *barud*." . . .

Gunpowder in Military Pyrotechnics

"With the rise of Saladin in AD 1139 a new era of military pyrotechnics began. The Muslims used incendiary weapons in every battle, and the story of the technician from Damascus who prepared *naft* pots which destroyed the siege towers of the Crusaders is well known. According to many historians incendiaries containing gunpowder were the deciding factor at the Battle of al-Mansura in AD 1249 when King Louis IX of France was taken prisoner. Both sides relied heavily on the skills of their engineers, but the mastery of the Muslims in the use of incendiary weapons gave them a decided superiority. Indeed it has been said that these weapons 'were real artillery and the bombardment had a terrifying effect on the Crusaders'. The famous chronicler Jean, Sire de Joinville (AD 1224 – 1319), who was one of the fighting officers at the battle, recorded that when the French commander saw the 'Saracens' preparing to discharge the fire 'he announced in panic that they were irretrievably lost'. The fire was discharged, it is said, from a large ballista (giant crossbow) in 'pots of Iraq' *(qidr'Iraqi)*. . . . De Joinville wrote, 'It was like a big cask and had a tail the length of a large spear: the noise it made resembled thunder and it appeared like a great fiery dragon flying through the air, giving such a light that we could see in our camp as clearly as in broad daylight'. When it fell it burst and liquid was ejected, spreading a trail of flame. James Partington, the celebrated historian of chemistry, remarked that 'the Crusaders believed that anyone struck by it was lost . . . and it seems to have been regarded as a kind of old-fashioned atomic bomb'. Though incendiary weapons had been used since the time of the first Crusade some hundred and fifty years before, the effects had never been so terrifying; the cause was the secret ingredient, gunpowder.

"From the time gunpowder was introduced, military engineers began to play an increasing role. Corps of engineers were formed, and artisans such as smiths, carpenters, bronze founders, *naft* craftsmen and others came under the command of military engineers who were responsible direct to an army Amir. With their siege machinery and gunpowder, they were the effective elements when in AD 1291 Acre was besieged and fell, putting a final end to the Crusades. During this siege, it was said that the Sultan's trebuchets and catapults flung their 'pottery containers filled with an *explosive mixture* at the walls of the city or over them into the town', and that the Sultan had 'a thousand engineers against each tower.' . . .

Early Cannon

"Ibn Khaldun, who wrote his history about AD 1377, described what are clearly cannons in his account of the siege of Sijilmasa in the Maghrib by Sultan Abu Yusuf a century earlier. He claims that the Sultan

> *installed siege engines . . . and gunpowder engines (hindam al-naft), which project small balls (hasa) or iron. These balls are ejected from a chamber (khazna) placed in front of a kindling fire of gunpowder; this happens by a strange property which attributed all actions to the power of the Creator.*

"But cannon did not appear suddenly in the Maghrib. . . . *Barud* was already a familiar product there, and gun-powder had been used in fire-pots during the Crusades. . . . Jose Conde, the Spanish historian, claimed that in AD 1204 cannons were used by the Almohad Caliph al-Nasir in his siege of al-Mahdiyya in North Africa, and Peter, Bishop of Leon, reported the use of cannon in Seville in AD 1248. In addition, cannon appeared in the Islamic East in the second half of the thirteenth century AD . . . . Indeed, we would advance the view that in the Maghrib, where petroleum was not available, cannon may have developed into a siege engine somewhat earlier than in the Islamic East, and that the appearance of cannon at Sijilmasa as described by Ibn Khaldun was a natural development . . . .

"In the Mamluk kingdom cannon of a lighter type appeared in battles against the Mongols, who suffered several setbacks after their destruction of Baghdad in AD 1258. The first defeat of the Mongol Hulagu's army by a Mamluk force, led in this instance by Sultan Qutuz, took place in AD 1260 at 'Ayn Jalut in Galilee and was one of the decisive battles of history. The last battle occurred at Marj al-Suffar, south of Damascus in AD 1303 when the Mamluks defeated the Mongol army of Ghazan in another decisive battle that brought the menace of the Mongols to an end. The new cannons were not yet truly effective weapons for they were still in an early stage of development. Nevertheless there are several military manuscripts in Leningrad, Paris, Istanbul and Cairo which not only report the use of light cannon in battles with the Mongols but also say that the 'Egyptians' had a cavalry force specially equipped with such cannon (*midfa'*) and with crackers (*sawarikh*) which used to frighten the enemy and cause confusion in their ranks. In some of the manuscripts reference is made to 'the defeat of Halawun at Jalud', and one also speaks of their use against Ghazan. We have then two further dates for battles in which early cannon were used by the Mamluks, one in AD 1260 and the other in AD 1303. But since the cannon had developed into a siege engine in Maghrib by AD 1274, it seems reasonable to assume that it was invented before the battle of 'Ayn Jalut in AD 1260.

"It was during this period that cannons as siege engines became common in the Mamluk kingdom. Shihab al-Din b. Fadl al-Allah al-'Umari (AD 1301 – 1349), historian, encyclopaedist, high government official, and an expert in state affairs, wrote a number of books, among them *al-Ta'rif bi'l mustalah al-sharif (The noble [book] of definitions of established custom)*, which was a guide for senior government officers (AD 1340). Al-'Umari devotes a chapter to siege engines in action during the reign of Sultan al-Nasir, which ran from AD 1309 – 1340. Six kinds were in common use: the *manjaniq* (trebuchet), the *ziyarat* (mechanical crossbow), the *sata'ir* (protective coverings), *khita'i* (arrows), the *makahil al-barud* (gunpowder cannon) and *qwarir al-naft* (pots of *naft*). The last three were gunpowder weapons and al-'Umari gives a literary description of each. His remarks about cannon refer to gunpowder and to the destructive power of red-hot balls which break arches and damage structures.

"In a series of sieges and battles in Spain between 1340 and 1343 AD both Western and Arabic sources state that cannon took part. In AD 1340 cannon were used by the Muslims at the battle of Tarif, and also in AD 1342 at the Siege of Aljazira. The iron balls were fired from cannon charged with gunpowder and it is reported that the Earls of Derby and Salisbury, who were both at the battle, were the persons responsible for bringing back a knowledge of firearms to England. It is also believed by some Western historians that the Spaniards derived their knowledge of artillery from the Arabs of Granada, who had early on become acquainted with the use of gunpowder. Thus as with other things, the transfer to Europe of knowledge about gunpowder and cannon took place by way of Spain.

"In the year of the Aljazira siege, the Syrian and Egyptian Amirs decided to overthrow the newly appointed Sultan al-Nasir Ahmad because they found him unfit for his high office. They therefore sent a force to al-Karak and besieged him there, but it is reported that he 'installed on top of the citadel five *manjaniqs* (trebuchets) and many *madafi'* (cannon)' and so defended himself.

"The first battle in the West in which cannon are reported to have been used was in 1346 at Crecy. English cannon were also deployed to block the entrance to Calais harbour during the same campaign.

". . . After AD 1342 the development of the weapon was continued by the Mamluks who used them extensively in siege operations, though gradually their manufacture spread to other Islamic countries, and the Ottomans soon adopted them on a large scale. The most spectacular operation involving cannon was the capture of Constantinople (Istanbul) by the Ottomans in AD 1453. One bronze gun in the siege had a bore of over 88 centimetres and the ball it fired weighed over 270 kilograms. A ball which struck a Venetian ship cut it in two . . . ; in this case the ball weighed no less than 400 kilograms and was fired from a distance of 2.4 kilometres. Each pair of such large guns required, it is said, 70 oxen and up to 1000 men to move them."

In this context it may be interesting to note that Sultan Selim III did introduce studies of algebra, trigonometry, mechanics, ballistics and metallurgy in Turkey as long ago as 1799, creating special schools for these disciplines with French and Swedish teachers. His purpose was to modernise the army and rival European advances in gun-foundaries. Since there was no corresponding emphasis on research in these subjects and since the scholarly establishment in the medresas who called themselves scientists (*alims*) had nothing but contempt for these new technological schools (*funun*), Turkey did not succeed. In the long run, in the conditions of today, technology unsupported by science simply cannot flourish.

APPENDIX III. Paper-Making

"The introduction and spread of the paper-making industry in the Near East and Western Mediterranean was one of the main technological achievements of Islamic civilisation. It was a milestone in the history of mankind.

"The early history of paper is now becoming clear. While it is possible that during the third century BC some types of paper were made in Asia, a more reliable date is that of AD 105, when paper was being produced in China from mulberry bark. It has, however, also been claimed that the Chinese of that time eliminated all non-Chinese centres for paper-making in Asia and so monopolised its manufacture. Whatever the truth of this, Arabic sources report that their paper industry started at Samarqand in the middle of the eighth century AD when some Chinese prisoners of war were taken there after the Battle of the Talas River in AD 751. Al-Qazwini, quoting another source, said 'Prisoners of war were brought from China. Among these was someone who knew [about] the manufacture of paper and so he practised it. Then it spread until it became a main product for the people of Samarqand, from whence it was exported to all countries.

"With the introduction of paper-making to the Muslim world and its spread during the eighth to ninth centuries AD a revolution took place in the industry. Writing material was freed from monopoly and paper became a very inexpensive product.

"The oldest extant Egyptian paper bearing Arabic script dates from the period AD 796–815, though the earliest dated paper only goes back to AD 874. However, we also know that factories for paper-making were established in Baghdad (eighth century AD), and according to the historian of Science, Robert Forbes, there were floating paper mills on the Tigris by the tenth century AD. Paper mills then spread to Syria and were established at Damascus, Tiberias, Tripoli, and other places. Egypt followed, with a mill at Cairo, and then North Africa, where Fez became a noted paper-making centre. Finally, the manufacture of paper reached Muslim Sicily and Spain, Jativa in the Valencia region becoming famous for its products. Only later did paper-making spread to Europe and then rather slowly, the first paper mill being established at Fabriano in Italy in 1276 AD; it took another century and more before a mill was established at Nuremberg in Germany in 1390.

"The manufacture of paper created a cultural revolution. It facilitated the production of books on an unprecedented scale, and in less than a century hundreds of thousands of manuscripts spread throughout the Islamic countries. Books became available everywhere and the profession of bookseller (*warraq*) flourished. ..."

APPENDIX IV

"While Islamic religion was the main impulse behind the renaissance of science at the zenith of Muslim Arab civilisation, it was partly the post-sixteenth century rise of a clerical faction which froze this same science and withered its progress. Western Christendom had similar religious set-backs, apostatic movements which tried to hinder the scientific revolution in the West. But the triumph of religious fanaticism over science in Muslim lands would not have succeeded had there been sufficient economic prosperity to generate a demand for science and technology. For Islam, as we have mentioned, was the driving force behind the Muslim scientific revolution when the Muslim state had reached its peak. In the ages of decadence, however, the movement of religious fanaticism against science was no other than an outstanding symptom of political and economic disintegration. The tragedy of the demolition of the last observatory in Islam, established in Constantinople by Taqi al-Dinn in 1580, exemplifies this victory of the clerical faction over science. And it is deplorable to note the inherent irony of the fact that the first observatory in the West was built around the same period, by Tycho Brahe. ..."

APPENDIX V. Iqbal on Ijtihad

Excerpts from "The Principle of Movement in the Structure of Islam" (taken from The Reconstruction of Religious Thought in Islam by Dr. Muhammad Iqbal, poet-philosopher (Reprinted by Ashraf Press, Lahore, 1971)).

"... It is only natural that Islam should have flashed across the con-sciousness of a simple people untouched by any of the ancient cultures,

and occupying a geographical position where three continents meet
together. ... Islam, as a polity, is only a practical means of making this
principle a living factor in the intellectual and emotional life of mankind.
It demands loyalty to God, not to thrones. And since God is the ultimate
spiritual basis of all life, loyalty to God virtually amounts to man's loyalty
to his own ideal nature. ... What then is the principle of movement in
the structure of Islam? This is known as *Ijtihad*.

"... The idea, I believe, has its origin in a well-known verse of the
Quran — 'And to those who exert We show Our path'. We find it more
definitely adumbrated in a tradition of the Holy Prophet. When Ma'ad
was appointed ruler of Yemen, the Prophet is reported to have asked him
as to how he would decide matters coming up before him. 'I will judge
matters according to the Book of God,' said Ma'ad. 'But if the Book of
God contains nothing to guide you?' 'Then I will act on the precedents
of the Prophet of God.' 'But if the precedents fail?' 'Then I will exert to
form my own judgement.' ... In this paper I am concerned with the
first degree of *Ijtihad* only, i.e. complete authority in legislation. The
theoretical possibility of this degree of *Ijtihad* is admitted by the Sunnis,
but in practice it has always been denied ever since the establishment of
the schools. ... Such an attitude seems exceedingly strange in a system
of law based mainly on the groundwork provided by the Quran which
embodies an essentially dynamic outlook on life. It is therefore, necessary,
before we proceed further, to discover the causes of this intellectual
attitude which has reduced the Law of Islam practically to a state of
immobility. ... The real causes are, in my opinion, as follows:

"1. We are all familiar with the Rationalist movement which appeared
in the church of Islam .... Conservative thinkers regarded this movement as a force of distintegration, and considered it a danger to the
stability of Islam as a social polity. Their main purpose, therefore,
was to preserve the social integrity of Islam and to realise that the
only course open to them was to utilise the binding force of Shari'at,
and to make the structure of their legal system as rigorous as possible.

"2. The rise and growth of ascetic Sufism, which gradually developed
under influences of a non-Islamic character, a purely speculative
side, is to a large extent responsible for this attitude. ...

"The spirit of total other-worldliness in later Sufism obscured
men's vision of a very important aspect of Islam as a social polity. ...

The Muslim state was thus left generally in the hands of intellectual mediocrities, and the unthinking masses of Islam, having no personalities of a higher calibre to guide them, found their security only in blindly followng the Schools.

"3. On the top of all this came the destruction of Baghdad . . . For fear of further disintegration, which is only natural in such a period of political decay, the conservative thinkers of Islam focused all their efforts on the one point of preserving a uniform social life for the people by a jealous exclusion of all innovations in the law of Shari'at as expounded by the early doctors of Islam. . . . But they did not see, and our modern Ulama do not see, that the ultimate fate of a people does not depend so much on organisation as on the worth and power of individual men.

. . .

"It is not true to say that Church and State are two sides or facets of the same thing. . . . The essence of *Tauhid* as a working idea is equality, solidarity and freedom. . . . The ultimate Reality, according to the Quran, is spiritual, and its life consists in its temporal activity. The spirit finds its opportunities in the natural, the material, the secular. All that is secular is therefore sacred in the roots of its being. . . . As the Prophet so beautifully puts it: 'The whole of this earth is a mosque.' The state according to Islam is only an effort to realize the spiritual in a human organisation. But in this sense all state, not based on mere domination and aiming at the realisation of ideal principles, is theocratic.

". . . Islam is a harmony of idealism and positivism; and, as a unity of the eternal verities of freedom, equality, and solidarity, has no fatherland. 'As there is no English Mathematics, German Astronomy or French Chemistry', says the Grand Vizier [of Turkey], 'so there is no Turkish, Arabian, Persian or Indian Islam. . . . You will see that, following a line of thought more in tune with the spirit of Islam, that is to say, the freedom of *Ijtihad* with a view to rebuild the law of Shari'at in the light of modern thought and experience.

". . . According to Sunni Law the appointment of an Imam or Khalifa is absolutely indispensable. The first question that arises in this connection is this — Should the Caliphate be vested in a single person? Turkey's *Ijtihad* is that according to the spirit of Islam the Caliphate or Imamate can be vested in a body of persons, or an elected Assembly. . . .

"... The question which confronts [us] today, and which is likely to confront other Muslim countries in the near future, is whether the Law of Islam is capable of evolution — a question which will require great intellectual effort, and is sure to be answered in the affirmative; provided the world of Islam approaches it in the spirit of Omar — the first critical and independent mind in Islam who, at the last moments of the Prophet, had the moral courage to utter these remarkable words: 'The Book of God is sufficient for us.'

"... From 800 to 1100, says Horten [a European Orientalist], not less than one hundred systems of theology appeared in Islam, a fact which bears ample testimony to the elasticity of Islamic thought as well as to the ceaseless activity of our early thinkers.

"... The assimilative spirit of Islam is even more manifest in the sphere of law. Says Professor Hurgronje — the Dutch critic of Islam: 'When we read the history of the development of Mohammedan Law we find that, on the one hand, the doctors of every age, on the slightest stimulus, condemn one another to the point of mutual accusations of heresy; and, on the other hand, the very same people, with greater and greater unity of purpose, try to reconcile the similar quarrels of their predecessors.'

. . .

"1. In the first place, we should bear in mind that from the earliest times, practically up to the rise of the Abbasides, there was no written law of Islam apart from the Quran. .

"2. Secondly, it was worthy of note that from about the middle of the first century up to the beginning of the fourth not less than nineteen schools of law and legal opinion appeared in Islam.

"3. Thirdly, when we study the four accepted sources of Mohammedan Law . . . the possibility of a further evolution becomes perfectly clear. . . .

"(*a*) The Quran. The primary source of the Law of Islam is the Quran. The Quran, however, is not a legal code. Its main purpose, as I have said before, is to awaken in man the higher consciousness of his relation with God and the universe. . . . Thus the Quran considers it necessary to unite religion and state, ethics and politics in a single revelation much in the same way that Plato does in his Republic.

"It is obvious that with such an outlook the Holy Book of Islam cannot be inimical to the idea of evolution.

"Turning now to the ground work of legal principles in the Quran, it is perfectly clear that far from leaving no scope of human thought and legislative activity the intensive breadth of these principles virtually acts as an awakener of human thought. ... I know the Ulama of Islam claim finality for the popular schools of Mohammedan Law, though they never found it possible to deny the theoretical possibility of a complete *Ijtihad*. ... Did the founders of our schools ever claim finality for their reasonings and interpretations? Never. ...

"... The Law of Islam, says the great Spanish Jurist Imam Shatibi in his *Al-Muwafiqat*, aims at protecting five things — *Din, Nafs, Aql, Mal and Nasl*. Applying this test I venture to ask: 'Does the working of the rule relating to apostasy, as laid down in the *Hedaya*, tend to protect the interests of the Faith in this country?' In view of the intense conservatism of the Muslims of India, Indian judges cannot but stick to what are called standard works. The result is that while the peoples are moving the law remains stationary. ...

"(b) The Hadis. The second great source of Mohammedan Law is the traditions of the Holy Prophet. ...

"... Abu Hanifa, who had a keen insight into the universal character of Islam, made practically no use of these traditions. The fact that he introduced the principle of *Istihsan*, i.e. juristic preference, which necessitates a careful study of actual conditions in legal thinking, throws further light on the motives which determined his attitude towards this source of Mohammedan Law. ...

"(c) The Ijma. The third source of Mohammedan Law is *Ijma* which is, in my opinion, perhaps the most important legal notion in Islam. It is, however, strange that this important notion ... remained practically a mere idea, and rarely assumed the form of a permanent institution in any Mohammedan country. ... It was, I think, favourable to the interest of the Omayyad and the Abbaside Caliphs to leave the power of *Ijtihad* to individual *Mujtahids* rather than encourage the formation of a permanent assembly which might become too powerful for them. ...

"(d) The Qiyas. The fourth basis of *Fiqh* is *Qiyas*, i.e. the use of analogical reasoning in legislation. ... The school of Abu Hanifa seems to have found, on the whole, little or no guidance from the precedents recorded in the literature of traditions. The only alternative open to them was to resort to speculative reason in their interpretations. . . .

". . . The spirit of the acute criticism of Malik and Shafa'i on Abu Hanifa's principle of *Qiyas*, as a source of law, constitutes really an effective Semitic restraint on the Aryan tendency to seize the abstract in preference to the concrete, to enjoy the idea rather than the event. . . . *Qiyas*, as Shafa'i rightly says, is only another name for *Ijtihad* which, within the limits of the revealed tests, is absolutely free; and its importance as a principle can be seen from the fact that, according to most of the doctors, as Qazi Shoukani tells us, it was permitted even in the lifetime of the Holy Prophet. The closing of the door of *Ijtihad* is pure fiction suggested partly by the crystallization of legal thought in Islam, and partly by that intellectual laziness which, especially in the period of spiritual decay, turns great thinkers into idols. If some of the later doctors have upheld this fiction, modern Islam is not bound by this voluntary surrender of intellectual independence. . . .

"This brief discussion, I hope, will make it clear to you that neither in the foundational principles nor in the structure of our systems, as we find them today, is there anything to justify the present attitude. Equipped with penetrative thought and fresh experiences the world of Islam should courageously proceed to the work of reconstruction before them. . . . Humanity needs three things today — a spiritual interpretation of the universe, spiritual emancipation of the individual, and basic principles of a universal import directing the evolution of human society on a spiritual basis. Modern Europe has, no doubt, built idealistic systems on these lines, but experience shows that truth revealed through pure reason is incapable of bringing that fire of living conviction which personal revelation alone can bring. This is the reason why pure thought has so little influenced men, while religion has always elevated individuals, and transformed whole societies. . . . Believe me, Europe today is the greatest hindrance in the way of man's ethical advancement. The Muslim, on the other hand, is in possession of these ultimate ideas on the basis of a revelation, which, speaking from the inmost depths of life, internalizes its own apparent externality. With him the spiritual basis of life is a matter of conviction for which even the last enlightened man among us can easily lay down his life. . . . Let the Muslim of today appreciate his position, reconstruct his social life in the light of ultimate principles, and evolve, out of the hitherto partially revealed purposes of Islam, that spiritual democracy which is the ultimate aim of Islam."

The Future of Science in Islamic Countries (August 1986)

*Prepared by Abdus Salam for inclusion in the volume **Islam and the Future** presented to the Islamic Summit held in Kuwait in January 1987.*

أَشْهَدُ أَنْ لاَإِلَهَإِلاَّ اللهُ وَأَشْهَدُ أَنَّ مُحَمَّدًاعَبْدُه وَرَسُولُه

أَعُوذُ بِاللهِ مِنَ الشَّيْطَانِ الرَجِيمِ

بِسْمِ اللهِ الرَّحْمَنِ الرَّحِيمِ .

> *"In the conditions of modern life, the rule is absolute: the race which does not value trained intelligence is doomed . . . . Today we maintain ourselves, tomorrow science will have moved over yet one more step and there will be no appeal from the judgement which will be pronounced . . . on the uneducated."*
>
> Alfred North Whitehead

Introduction

1. First and foremost, it is important to re-emphasise that the Muslim Ummah constitutes 1/5th of mankind, larger in population than the USA, Western Europe and Japan combined, and only exceeded by China as a unit. In income terms, it represents 1/15th of global GNP — three times as large as that of the Chinese.

2. So far as the sciences are concerned, the Muslim Ummah has a proud past. For 350 years, from 750 CE to 1100 CE, the Ummah[1] had an absolute world ascendency in sciences. From 1100 CE for another 250 years, we shared this ascendency with the emerging West. From the 15th century onwards — this period paradoxically coinciding with the great empires of Islam (Osmanli in Turkey, Safvi in Iran, Mughal in India) — we progressively lost out. There is no question, but today, of all civilisations on this planet, science is the weakest in the lands of Islam. The dangers of this weakness cannot be overemphasised since honourable survival of a society depends directly on strength in science and technology in the conditions of the present age.

3. Why were the Muslims ascendent in sciences? Three reasons: first, the early Muslims were following the injunctions of the Holy Book and the Holy Prophet. According to Dr. Mohammad Aijazul Khatib of Damascus University, nothing could emphasise the importance of sciences more than the remark that "in contrast to 250 verses which are legislative, some 750 verses of the Holy Quran — almost one-eighth of it — exhort the believers to study Nature[2] — to reflect, to make the best use of reason and to make the scientific enterprise an integral part of the community's life." The Holy Prophet of Islam — Peace be upon Him — said that it was the *"bounden duty of every Muslim — man and woman — to acquire knowledge".*

From these injunctions, followed the second reason for our ascendency. Notwithstanding the customary opposition of traditionalists[3], up to the fifteenth century the scientific enterprise and the scientists in early Islam were supported magnificently by the Muslim principalities and by the Islamic society. Thus, to paraphrase what H.A.R. Gibb has written in the context of literature: "To a greater extent than elsewhere, the flowering of the sciences in Islam was conditional . . . on the liberality and patronage of those in high positions. So long as, in one capital or another, princes and ministers found pleasure, profit or reputation in patronising the sciences, the torch was kept burning.". And some princes — like Ulugh Beg at Samarkand — themselves joined in the scientific quest.

The third reason for our ascendency was connected with the cohesion of the Ummah — the Islamic nations, notwithstanding their political differences, acted as a unified Commonwealth, so far as sciences were concerned.

4. As indices of our present backwardness in sciences, one may cite the following:

(*i*) The absolute size of the Islamic scientific community is incredibly small. According to estimates made at the Islamic Conference (held in Islamabad) in 1983, the Muslim nation as a whole, had a total of around 46,000 research and development scientists and engineers. Although such figures are subject to the compilation criteria, one may contrast these with the corresponding figures for a typical developed country. The numbers should be one hundred times larger for a similar population (see Appendix I). Likewise the spending on science and technology with us, does not begin to approach the 1% of GNP suggested by UNESCO as the minimum for healthy science and technology (as contrasted to 2–3% of GNP spent by USA, USSR, Western Europe and Japan).

(*ii*) The late Professor D. J. de Solla Price of Yale University discovered an empirical law some time back, which states that the scientific output in any country is proportional to the GNP — the higher the product, the higher the quantum of research. This law apparently applies to the USA, to all European countries — to China, India, and to a number of developing countries, but not to the lands of Islam — which are way, way below the norm. According to this law, approximately one out of every 15 research papers should have been written by those belonging to Islamic countries. Likewise, one out of every 15 active scientists should belong there. One would be grateful if in fact these figures turned out to show a ratio of $1-200$.[a]

(*iii*) The third indicator of our backwardness is provided by the fact that, barring one or two exceptions, there are no great university research departments or research institutes of any world-calibre in any Islamic country. To take one example, even for the study of desertification — a problem peculiarly severe in Muslim countries — there simply does not exist anywhere in Islamic countries a world-class research institute.

---

[a]Just to substantiate my remark, examine any issue of the multidisciplinary science weekly *Nature*. Note the overwhelming, overpowering, inexorable march of scientific research in unravelling Allah's design and in creating new knowledge in all fields. Note also the paucity of contributions, from any in the Islamic countries. This is truly frightening for its future implications.

In this essay I shall analyse the causes of our backwardness. I shall suggest that provided remedial measures are taken *today*, one may still hope for a future for science and technology in Islamic lands.

5.   **What is Wrong Today with Science and Science-based Technology in Islamic Countries?**

Three things:

(*i*)   *Lack of Commitment*

There has been no declared commitment to acquiring and enhancing scientific knowledge among us. There has been no practical realisation that sciences can be applied to development or to defence as, for example, there was in Japan, at the time of the Meiji Restoration around 1870, when the Emperor took five oaths as part of Japan's new constitution.   One of these oaths was: "Knowledge will be sought and acquired from any source with all means at our disposal, for the greatness of Imperial Japan".

A more recent example of how crucial such commitment[4] can be is provided by the history of the Academy of Sciences of the USSR. Forty years ago, the Soviet Academy of Sciences — created by Peter the Great — was asked to expand its numbers and was set the ambitious task of excelling in all sciences.   Today it numbers a self-governing community of a quarter of a million scientists working in its institutes, with priorities and privileges accorded to them in the Soviet system that others envy.   According to Academician Malcev, this principally came about in 1945, at a time when the Soviet economy lay shattered by the Second World War.   Stalin decided to increase the emphasis on and the recruitment of bright young men and women into sciences. Acting against bureaucratic advice, he announced one day that the emoluments of all scientists and all technicians connected with the Soviet Academy would increase by a factor of three hundred per cent — "no increases for doctors or engineers", according to Academician Malcev — "only for scientists".   Since then, there may be other problems with Soviet science, but it has never failed to attract the brightest intellects in the Soviet society.

One consequence of this lack of practical commitment on our part is that the number of active scientists in the Islamic countries is sub-sub-critical.   Speaking of physics for my own country,[5] "the total number of physics teachers in all the (nineteen) universities of Pakistan is 86, of which only 46 have Ph.D's". Contrast this with one college (Imperial College of Science and Technology), in one university (London) in the UK — the corresponding numbers are 150. Regarding Ph.D. education — in its 100 years of existence — the Punjab University at Lahore has not produced one single Ph.D. in Mathematics and only three in Physics (1982 figures).

(ii) The Manner in which the Enterprise of Science is Run

Science depends for its advances on towering individuals. An active enterprise of science must be run by working scientists themselves and not by bureaucrats or scientists who may have been active once, but have since ossified. When the late Amos de Shalit (then Director of the Weizmann Institute) was asked in a UN Committee, what was the Israeli[6] policy for science, his reply was: "We have a very simple policy for science growth. An active scientist is always right and the younger he is, the more right he is.". Unfortunately in most of our science departments this is far from the accepted norm.

(iii) No Commitment to Self-reliance in Technology

In technology, by and large, none of our Governments have made it a national goal to strive for self-reliance — even for defence technology. And we have paid scant heed to the scientific base of technology i.e. *to the well-recognised truth that science transfer must always accompany technology transfer, if technology transfer is to take.* Thus, while some of our governments and enterprises may claim that they are encouraging the transfer of technology, often all that this means is the importation of designs, machines, technicians and (sometimes even nearly processed) raw materials.

6. Summary Recommendations

For the last thirteen years, I have written in the same vein as in this article — without making any visible impact. Notwithstanding this and notwithstanding the fact that the science gap between the developed and the developing countries is increasing rapidly with the passage of time, I believe we can change the situation if we take appropriate action *now*.[b]

---

[b]I have continually prayed for this to happen and I am sure Allah will not leave these prayers unanswered.

What gives one hope is that there *are* Muslim scientists working principally (though not[c] exclusively) in developed countries who have registered the highest attainments in sciences. *This implies that it is basically the environmental factors in our societies which need to be corrected.* Once such corrections are made, growth of science at the highest possible level — at least in selected areas — will take place, within one or two generations. This happened in Italy for example, where science was stifled by the Church action at the time of Galileo. As soon as propitious conditions returned during the brief Napoleonic era, research suddenly blossomed. And this is what happened in Japan a hundred years back; in the USSR forty years ago; and is happening now in Brazil, in Argentina, in India, in China, in South Korea.

My recommendations are as follows:

(*i*)   *The Absolute Numbers of our Scientists must Increase*

This means making the scientific profession attractive. This means emphasising scientific literacy from the secondary school stage onwards, as well as through our radio and television programmes. This means making every effort by creating the environment for scientific research within our countries to attract back those who have migrated away. In short, this means taking all steps towards implementing the commitment I have already spoken about.

In this context, I have recently been asking the Ulema in India, Bangladesh and Malaysia: Since 1/8th — some 750 verses — of the Holy Book exhort the believers to "study Nature,[2] to reflect, to make the best use of reason and to make the scientific enterprise an integral part of the community's life", do they devote one out of every eight khutbas, in their Friday sermons to science? The uniform reply I received was — they would like to, but they did not know enough science themselves. Has the time not come when, the curricula for *religious* seminaries should contain established parts of modern sciences — like Newton's laws, like astrophysics of stars and galaxies, like a knowledge of fundamental forces of Nature, of their unification, of the genetic code, and of the structure of the Earth?

---

[c]In the Islamic countries there still are a few scientists who keep producing high-class scientific work under the most adverse and discouraging conditions. It is relevant to remark here that scientific creation, at a high level is something for which one has to eat one's heart out.

Should the Islamic Summit not take it upon itself to implement this recommendation?

(ii) We must spend at least 1% of GNP (several billion dollars) on research and development in science and technology — parts of this within each country, and a part for the whole Ummah in a Commonwealth of Sciences. We must devote a fair fraction of such funds on providing *massive* and *hard* training to our younger generation (like Japan did towards the end of the nineteenth century, see Appendix II) in *all* fields of science at all levels of education and research. To give examples of the massive scale needed for research-training, one may note that the UK Science and Engineering Research Council (SERC) awards five thousand grants for Ph.D. training every year. An equal number is awarded by the other research councils — the Agricultural Research Council, the Medical Research Council, the Environmental Research Council and the like. The number of post Ph.D. grants available within the UK (and outside), numbers one thousand every year. And the UK has only 1/15th of the population of the Islamic countries.

(iii) We must create Foundations for Science, both state endowed as well as supported by private donations so that grants for research are available from many sources. *I do not see why 1/8th of the Auqaf Funds should not be devoted to science, in keeping with the emphasis on it in the Holy Book.*[2] (The US has 22,000 private research foundations.) In this context, I do not see why our Islamic banks do not set up their own funds for science and technology as is the case all over Western Europe with mutual banks.

In 1973, the Pakistan Government, on my suggestion, requested the Islamic Summit in Lahore to sanction at least one Foundation for Science for Islam, equal in size to the Ford Foundation, which has a capital of one billion dollars (see the article on "Islam and Science", paragraph 8.2). Eight years later, in 1981, such a foundation was created but with just 50 million dollars *promised* instead of the one billion requested. It may have been more charitable not to have deceived ourselves by such creation. As contrasted with our rivals, we are not serious about science in Islam.

(iv) We must build up research in our universities and in institutes linked to the universities.[6] Our universities must emphasise that teachers spend half of their time on research and the other half on

teaching, as is the norm for developed countries. *I would suggest that at least one university in every Islamic country should devotedly strive and be helped to attain world stature in research. It can be done.*

How can the universities and industry interact in the creation of science and wealth? To answer this question, I reproduce in Appendix III an excerpt from a guest-editorial from the July 1986 issue of the journal *Biotechnology*. Biotechnology is one of the newer sciences; its applications are expected to dominate life, in agriculture, in energy, in medicine, in the 21st century. This excerpt describes the obstacles which the developing world (including the World of Islam) faces in building up expertise in this subject. "Biotechnology thrives on new knowledge generated by molecular biology, genetics and microbiology, but these disciplines are weak, often nonexistent, in the underdeveloped world. Biotechnology springs from universities and other research institutions, centers that generate the basic knowledge needed to solve practical problems posed by society. But the universities of the under-developed world are not research centers . . . And the few creative research groups operate in a social vacuum; their results might be useful abroad, but are not locally . . . Biotechnology needs dynamic inter-actions among the relevant industries. These interactions, however, are weak in countries in which science is perceived as an ornament, not as a necessity . . . Biotechnology requires many highly skilled professionals, but . . . underdeveloped nations lack sufficient people well trained in the pertinent disciplines . . . Economic scarcity and political discrimination induce professionals and graduate students to emigrate or abandon science altogether.".

The guest editor goes on to ask "What can be done" and his answer is: *"First of all, underdeveloped countries must understand that they need to reform their universities . . .* They must recognise that molecular biology is not just another branch of biology, but the one and only tool available for understanding biological structure and function . . . Success in biotechnology depends on the conquest and consolidation of the moving frontiers of cell biology and medicine."

I am sure India, China, Argentina, Brazil, South Korea, among the developing countries, will take heed. *The question is shall the Muslim nations take heed or shall we lose out in this new race to master and utilise biotechnology?*

(*v*) To go back to my recommendations, *we must ensure that our scientific enterprise maintains living contact with international science.*[7] After all, science is being created *outside* the confines of our countries. At present, very few of us, if living and working in our own country, can travel to scientific institutions and meetings abroad. Such travel, as a rule, is considered a wasteful luxury. In some of our countries, incredibly, it needs authorisation from the highest authority in the land! Our countries must join — individually or collectively — the international programmes like those launched by the International Council of Scientific Unions (ICSU) as well as join as associate members, international science organisations like the Centre for European Nuclear Research (CERN).

(*vi*) It is not just the physical isolation of the individual scientists that we suffer from. According to A. Zahlan, (one-time Professor at the American university in Beirut) "There is also the isolation from the norms of international science, the gulf between the way we run the scientific enterprise and the self-governing manner in which it is run in the West" or Japan, or within the community of scientists in the USSR Academy. "We seem to have no developed system of professional organisations, no internal review committees, no independent studies of state of art or of quality, no science foundations administered by the scientists, no independent sources of grants".

This must change.

(*vii*)Every Islamic country must decide its own priorities in science and develop the needed research, both pure and applied. But in addition, there will remain a modicum of disciplines which must be developed collectively for the Commonwealth of Science in Islam. I personally feel a predeliction for mathematics — one time an Islamic Science par excellence — for high energy theory, astrophysics and genetics — among the pure disciplines — and for microelectronics, biotechnology and for materials sciences among the applied sciences. In addition, we must emphasise physics of communications as well as science related to water, agriculture, energy and defense if we wish to survive as independent, self-respecting entity, creating and operating world-class centres of excellence, linked to our universities.[7]

What is crucial is that we set ourselves an ambitious goal through a Summit declared policy. For example, like China, we may resolve — and act — to emulate and excel the United Kingdom in the sciences by the end of the century.

(*viii*)Earlier, I spoke of patronage for the sciences. One vital aspect of this is the sense of security and continuity that a scientist-scholar must be accorded for his work. Like all humans, a scientist or technologist can only give of his best if he knows he will have security, respect and opportunity for his work, and *that he is shielded from discrimination, sectarian and political.*

(*ix*) I have referred throughout this article to a *Commonwealth of Science for the Islamic countries (Ummat-al-Ilm),* even if there may be no political commonwealth in sight. Such a Commonwealth of Science was a reality in our great days, when central Asians like Ibn Sina and Al Biruni would naturally write in Arabic, or their contemporary and my brother in physics, Ibn-ul-Haitham could migrate from his native Basra in the dominions of the Abbasi Caliph to the Court of his rival, the Fatmi Caliph, sure of receiving respect and homage, inspite of the political and sectarian differences of the two regimes which were no less acute then than they are now.

A new Islamic Commonwealth of Science needs conscious articulation, and recognition once again, both by us, the scientists, as well as our governments. Today we, the scientists from the Islamic countries, constitute a very small community. We need to band together, to pool our resources, to feel and work as a community. *To foster this growth, could we possibly envisage from our governments a moratorium, a compact, conferring of immunity, for say the next twenty years,* during which the scientists within this Commonwealth of Science, *within this Ummat-al-Ilm,* could be shielded, so far as political and sectarian differences are concerned, just as was the case in the Islamic Commonwealth of Sciences in the past?

(*x*) To summarise, the renaissance of the sciences within an Islamic Commonwealth of Science is contingent upon five cardinals preconditions: passionate commitment, generous patronage, provision of security, absence of discrimination, self-governance and internationalisation of our scientific enterprise. *The Islamic Summit must take practical steps to accept these principles for all our futures.*

(*xi*) Finally, since this meeting is being held in Kuwait, let me express the wish and the hope that Kuwait may emulate the City State of Athens and acquire the same role as Athens had in the past, for the future of Science in Islam.

7. The Importance of Science for the Muslim Nations

Why am I so passionately advocating our engaging in the enterprise of science and of creating scientific knowledge? This is not just because Allah has endowed us with the urge to know, this is not just because in the conditions of today this knowledge is power and, science in application, the major instrument of material progress and meaningful defence; it is also that as self-respecting members of the international world community, we must discharge our responsibility towards, and pay back our debt for the benefits we derive from the research stock of world science, thus avoiding that lash of contempt for us — unspoken, but certainly there — of those who create knowledge.

Let me end with the prayer: let no future historian record that in the fifteenth century of the Hijra, *Muslim scientific talent was there but there was a dearth of statesmen to marshal and nurture it.*

NOTES

1. There are some — regrettably even among the Muslims — who dismiss the advances made in the great days of science in Islam — in mathematics, in physics, in biology, in chemistry and in medicine — as a "mere continuation of Greek tradition". Even such men cannot gainsay the undoubted fact that through systematic observation and patient experimentation, the Muslims were the first peoples to bequeath to the world the idea that science is an empirical subject. (Thus, in Briffault's words: "The Greeks systematised, generalised, and theorised, but the patient ways of detailed and prolonged observation and experimental inquiry were altogether alien to the Greek temperament . . . . What we call science arose as a result of new methods of experiment, observation and measurement, which were introduced into Europe by the Arabs. (Modern) science is the most momentous contribution of the Islamic civilisation . . . ."

2. The reason for this emphasis on science has been beautifully spelled out by a Christian writer — Huston Smith — in the *Religions of Man* (Harper, 1958):

> "*In an age charged with supernaturalism, when miracles were accepted as the stock-in-trade of the most ordinary saint, Prophet Muhammed refused to traffic with human weakness and credulity. To miraele-hungry idolators seeking signs and portents he cut the*

issue clean: 'God has not sent me to work wonders; He has sent me to preach to you. My Lord be praised! Am I more than a man sent as an apostle?' From first to last he resisted every impulse to glamorise his own person. 'I never said that Allah's treasures are in my hand, that I knew the hidden things, or that I was an angel . . . I am only a preacher of God's words, the bringer of God's message to mankind'. If signs be sought, let them be not of Muhammed's greatness but of God's, and for these one need only open one's eyes. The heavenly bodies holding their swift silent course in the vault of heaven, the incredible order of the universe, the rain that falls to relieve the parched earth, palms bending with golden fruit, ships that glide across the seas laden with goods for man — can these be the handwork of gods of stone? What fools to cry for signs when creation harbours nothing else! In an age of credulity, Prophet Muhammed taught respect for the world's incontrovertible order which was to awaken Muslim science before Christian. Only one miracle he claimed, that of the Quran itself. That he by his own devices could have produced such truth — this was the one naturalistic hypothesis he could not accept".

3. This opposition did succeed in retarding the introduction of printing into the Islamic World till the Napoleonic occupation of Egypt in 1798 — full three hundred and fifty years after Gutenberg's first printing (and dissemination) of the Bible. In Turkey (apart from a short episodic period between 1729 and 1745) printing could not be introduced till 1839 for secular books and not till 1874 for the Holy Quran (see D. J. Boorstin, *The Discoverers* (Vintage Books, New York, 1983)).

4. The same is true of most other communist countries — China and countries of Eastern Europe which exhibit a veneration for science which borders on the religious. In Sofia recently, I saw in the centre of the city, the most prestigious site in the major square occupied by the Academy of Sciences (of which most research institutes of the country are part).

5. According to Dr. Mujahid Kamran, Punjab University (writing in the journal *Concept*, January 1982).

6. To strengthen university science, and to eliminate possible rivalry between universities and the government research institutes, we must adopt the United States pattern where research institutes, even if they are federally financed, are always linked with universities. Why have

the university linkages with institutes of an applied nature (in agriculture, medicine, health and other such fields) proved such a source of strength for US science? The reasons are not far to seek. Firstly, one of the indirect objectives of all such research institutes is, and should be, a wide dissemination of research skills within the community. There is no surer means of doing this than by linking such institutes with universities and letting postgraduate students go through them. Secondly, and reciprocally, the quantum of basic science every applied research laboratory needs for its health and vigour, does not have to be created ab-initio. The linked university faculties provide this.

7. It was this isolation which prompted me to propose the creation of the International Centre for Theoretical Physics so that physicists from developing countries do not make exiles of themselves in order to keep abreast of newer developments in their subjects. This Centre belongs to two United Nations Agencies — IAEA and UNESCO — and is supported by generous grants from the Government of Italy; during 1985 alone, 269 (other) Muslim and 338 Arab physicists were supported at the Centre by funds principally provided by the Governments of Italy and Sweden (the Kuwait Science Foundation and the Qatar University supported 30 of these scientists). Our research budget borders on six million dollars. If all these funds could be devoted to Muslim scientists, we could have supported six times the numbers in each category. These men produce high-class science for the few months while they are at Trieste. I wish there was some mechanism whereby they could be kept producing when they return to their countries and also that other sciences had similar centres.

8. In this context, I cannot overemphasise the value of science towards building national character through the qualities science engenders — thoroughness, patience, pride in one's work, and, above all, tolerance and respect for opinions other than one's own.

Appendix I

To give an outside observer's assessment regarding science in Islamic countries, writing in the prestigious scientific journal, *Nature*, of 24 March 1983, Francis Giles raises the question "What is wrong with Muslim science?" This is what he says: "At its peak about one thousand years ago, the Muslim world made a remarkable contribution to science, notably mathematics and medicine. Baghdad in its heyday and southern Spain built universities to which thousands flocked: rulers surrounded themselves with scientists and artists. A spirit of freedom allowed Jews, Christians and Muslims to work side by side. Today all this is but a memory.

"Even the recent wealth provided by oil exports makes relatively little difference ... science policy and politics, much to the displeasure of many scientists, are closely linked in the Middle East. The region is dominated by dictatorships, benevolent or otherwise ... further complicating any attempt to allow science to take root indigenously. Not surprisingly the brain drain to industrialised countries continues to debilitate intellectual life throughout the Middle East." Harsh criticism, but much of it factual and deserved.

The same issue of *Nature* contains another article on research manpower in Israel from which I quote: "The need for a substantial increase in the number of academically trained people to work in research and development is widely accepted. The National Council for Research and Development has urged that their country will need 86,700 such people in 1995, compared with 34,800 in 1974 — an increase of 150 percent". Compare this figure of 34,800 with around 46,000 researchers in all Islamic countries[d] (the population ratio is 1 : 200).

فَاعْتَبِرُوا يَا أُولِي الْأَبْصَارِ .

"O, those with vision, take heed"

Appendix II

This is a quotation from *Science and the Making of the Modern World* by John Marks (Heinemann Press, 1983) in respect of how Japan built up its science and technology. This article would repay careful study. It illustrates the steps needed to grow science and technology; we do not need to re-invent the wheel.

---

[d]The figures are taken from the Secretariat Report presented to the first meeting of the *Islamic Conference on Science and Technology*, held in Islamabad in Pakistan in May 1983.

"7.5 Science and technology in Japan since the Meiji restoration

In 1869 Mustsushito, an emperor of the Meiji dynasty, regained supreme power in Japan after centuries of rule by the feudal Shoguns. This could be seen as a retreat into the past but in fact it led to the rapid growth of Japan as a technological power which has continued, almost unchecked, ever since.

For nearly 250 years before 1869 Japan had been a closed society — almost isolated from the rest of the world. This had been a deliberate policy designed to exclude European influence, particularly Christianity. A few Dutch trading posts were all that were permitted although some foreign books were imported after 1720. Then in the 19th century the growth of European power and influence (see Chapter 4.11) across the world began to affect Japan. In the 1850s Japan was virtually forced to conclude trading treaties first with America and then with Britain, Holland, Russia and France. These treaties led to much more trade with Europe and America which gradually undermined the traditional feudal structure of Japanese society.

When the emperor regained power in 1869 he represented those who wanted to reverse the isolation policy and open up Japan to Western influence. The emperor proclaimed that 'Knowledge shall be sought throughout the world so as to strengthen the foundations of imperial rule'.[1]

In this chapter we will describe how that precept was put into practice in the 19th and early 20th centuries and how, in very different circumstances, it is still important in Japan today.

THE MEIJI RESTORATION AND SCIENCE AND TECHNOLOGY — 1869–1900

Once the decision had been taken to import Western science and technology, the Japanese government set about the task with characteristic thoroughness. They made detailed surveys of the engineering industries in Europe and America. Then they acted on a broad front. For the short-term they imported foreign engineers and scientists; for the medium-term they sent students abroad and set up colleges in Japan staffed by foreign lecturers; and for the long-term they set up universities and numerous research institutes. Throughout, the emphasis was on the practical application of existing knowledge. In the words of the Prime Minister Prince Ito in 1886:

> *The only way to maintain the nation's strength and to guarantee the welfare of our people in perpetuity is through the results of science ... Nations will only prosper by applying science ... If we wish to place our own country on a secure foundation, insure its future prosperity, and to make it the equal of the advanced nations, the best way to do it is to increase our knowledge and to waste no time in developing scientific research.*[2]

The importance of new techniques

The foundations for the industrial revolution in Japan were laid by the Engineering Ministry established in 1870. Hundreds of foreign engineers were employed to build railways and establish a telegraph network. Modern technology was imported to develop the mining industry and to establish factories for cotton spinning. Most of these foreign engineers were British, but some were from France and the other European countries. Many of these engineers were paid salaries which were four or five times greater than those paid to Japanese government ministers.

Technical education

Great emphasis was placed on technical education and many foreign lecturers were employed in Japanese schools and colleges. They primarily taught practical subjects like engineering, agriculture, medicine and geology together with supporting basic subjects like mathematics, physics, chemistry and biology. They came primarily from Germany, Britain, France and America (see Fig. 7.5.1) and taught in their native languages. Again they were often paid much more than native Japanese.

One particularly important development was the College of Engineering in Tokyo which was staffed mainly by British engineers and which began teaching in 1873. The aim of the College was 'to train men who would be able to design and superintend the works which were necessary for Japan to carry on if she adopted Western methods.'[3] The Prime Minister Prince Ito later said '... that Japan can boast today of being able to undertake such industrial works as the construction of railways, telegraphs, telephones, shipbuilding, working of mines, and other manufacturing works entirely by the hands of Japanese engineers is mainly attributable to the College ...'[4]

Many students were also sent abroad and later came back to teach the next generation of Japanese students.

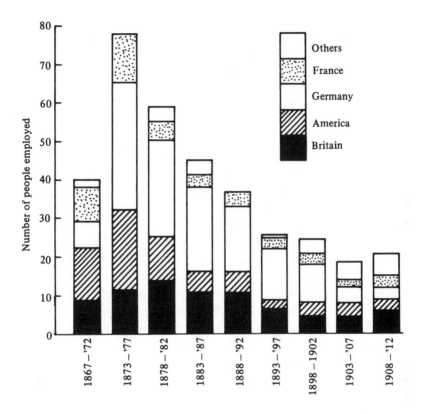

Fig. 7.5.1 Numbers of foreign lecturers in Japan, 1867–1912. *Redrawn from Nakayama, S., 'A Century's Progress in Japan's Science and Technology,' in Technical Japan, vol. 1, part 1, 1968.*

Universities, research institutes and scientific societies

In the period from 1875 to 1900 the Japanese government established many of the same kinds of institutions which make up the scientific community in Europe and North America. The Imperial University of Tokyo was set up in 1877 and, a few years later, it absorbed the Tokyo College of Engineering. Similar universities were established at Kyoto in 1897 and Tohoku in 1911. Again the emphasis was on practical knowledge as can be seen from this extract from the charter of Tokyo University:

> *The aim of the Imperial University shall be to teach and study such sciences and practical arts as meet the demands of the State.*[5]

The government also set up a number of research establishments during the years following the Meiji restoration. Examples include the Naval Hydrographic Division in 1871, the Tokyo Hygenic Laboratory in 1874, the Central Meteorological Observatory in 1875, the Geological Survey Bureau in 1878, the Electro-Technical Laboratory in 1891, the Institute for Research on Infectious Diseases and the Agricultural Experimental Station in 1892 and the Chemical Industrial Research Institute in 1900. Once again there is a clear emphasis on practical research.

Many scientific societies were also established in this period. The Tokyo Mathematical Society was founded in 1877 and later became the Japanese Mathematico-Physical Society. The Tokyo Chemical Society originated in 1878 and in the following year the Tokyo Academy of Sciences was founded, although initially natural scientists were in a minority on this body which was renamed the Imperial Academy of Sciences in 1906. Other societies were established for medicine in 1875, physical geology in 1879, pharmacology in 1881, meteorology and botany in 1882 and zoology in 1888. Societies for heavy engineering tended to be set up a little later — for mining in 1889, construction in 1886, electrical engineering in 1888 and mechanical engineering only in 1897.

Many of these societies have grown extremely rapidly since that time. See, for example, Fig. 7.5.2 which shows how the membership of the Japanese Mathematico-Physical Society increased from 1877 to 1945. From about 1888 onwards the growth was exponential with the number of members doubling roughly every ten years, except for a brief period during World War 1. This growth is much more rapid than the growth in the population of Japan which has increased from about 36 million in 1875 to about 110 million today — a doubling time of about 60 years. And it is even faster than estimates of the growth of Western science where the doubling time is usually estimated as approximately 15 years.

This extremely rapid growth in the numbers of Japanese scientists took place in a number of stages as can be seen from Fig. 7.5.3 which shows the number of physicists active in Japan from 1860 to 1960. At first these were mainly foreign physicists together with a few foreign-trained Japanese (Group I); by about 1910 they had either left Japan, died or retired. These men trained a group of Japanese physicists (Group II) who from the 1890s onwards became the teachers of the first generation of Japanese physicists who were both trained in Japan and taught in

Japanese. After this the physics community in Japan entered on a period of self-sustained growth (Groups III and IIIb) during which both the number of physics graduates (IIIa) and of postgraduate students (IIIb) doubled roughly every seven years.

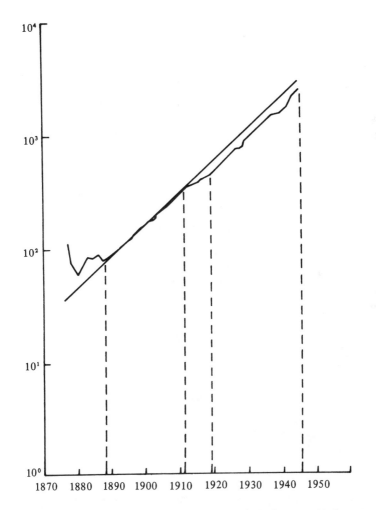

Fig. 7.5.2 Exponential growth in the membership of the Japanese Mathematico-Physics Society, 1877 – 1945 (logarithmic scale); the membership approximately doubled every ten years. *Redrawn from Yagi, E., 'The Statistical Analysis of the Growth of Physics in Japan' in Nakayama, S., Swain, D. L., and Yagi, E. (eds), Science and Society in Modern Japan, Cambridge, Mass: MIT Press, 1974.*

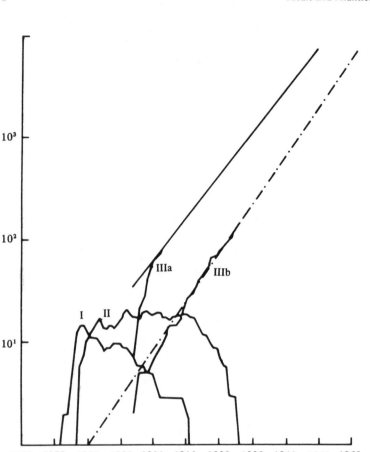

Fig. 7.5.3 Numbers of physicists in Japan, 1860–1960; I — Foreign or foreign-trained physicists, II — Japanese students of group I; III — Japanese physicists taught in Japanese, (*a*) graduates, (*b*) post-graduates continuing for D.Sc. *Redrawn from Nakayama, S., Swain, D. L., and Yagi, E. (eds), Science and Society in Modern Japan, Cambridge, Mass: MIT Press, 1974.*

THE RISE OF JAPAN AS A MILITARY POWER — 1890–1945

The development of science and technology in Japan was one important factor in the growth of Japan as an important military power in the Far East. In 1894–5 Japan defeated China in war and nine years later, in 1904–5, also defeated Russia — the first time in modern times that a European country had been defeated in war by a country outside Europe.

From the 1890s onwards Japan began to develop its heavy industries. Iron and steel production expanded rapidly and much of it was concentrated in the Yawata Irons Works which was financed by the government and operated under military control. The output from this foundry was used mainly in the construction of warships for the Japanese navy. Other heavy industries were also developed at this time such as coal-mining, and the manufacture of rolling stock for the railways and of equipment for the electrical supply industry. From 1897 onwards steam power replaced watermills and by 1914 all heavy industry had switched to electrical power supplies based industries. Probably the most important initiative was the founding of the Institute of Physical and Chemical Research in 1917 which was partly modelled on Germany's Imperial Institute for Physics established in 1887. However, other research institutes were also founded at this time, both by the government and by private industry. Japan was now beginning to develop its own scientific research community, but in contrast with research in Europe, its attention was primarily directed towards Japanese industry and to military technology in particular.

During the 1930s the emphasis on military technology increased and Japanese military power grew rapidly. In 1932 she effectively annexed Manchuria. Then in 1937, taking advantage of the civil war in China, Japan declared war on China and rapidly occupied much of the country, including the capital Peking and the major cities of Shanghai and Nanking. At this time science and technology in Japan became almost completely devoted to military puposes . . .

Then in 1941 Japan made a surprise attack on the American fleet at Pearl Harbor. This attack brought America into World War 2 and it marked the beginning of the war in the Pacific which finally ended in August 1945 with the dropping of the first atomic bombs on Hiroshima and Nagasaki.

SCIENCE AND TECHNOLOGY IN JAPAN SINCE 1945

In 1945 Japan was a defeated nation — her productive capacity had fallen to only 10 per cent of previous levels and there was a threat of food shortages and epidemics. Since then Japan has become one of the most prosperous nations in the world. Science and technology have clearly been important in this transformation but it is much less clear precisely how science and technology have influenced Japanese prosperity and what role the Japanese government has played in the rise of Japan as a major technological power. In this section we will try to illuminate these questions by describing some of the changes which have taken place since 1945.

Post-war reconstruction, 1945–55

In the early years the clear priority was to avert food shortages by the improvement of agriculture. Better strains of rice, more fertilisers and pesticides and improved agricultural machinery all led to greater output. Productivity also increased substantially which meant that more people were available to work in the growing industries of the 1960s.

As in the past, industries like mining and manufacturing were revived by the import of foreign technology. But by contrast with the 1930s, there was virtually no military or defence expenditure on science and technology. One result has been that, since 1945, private companies have provided by far the largest share of the resources devoted to research and development. But the government did control licences for the import of foreign technology and set limits on the foreign ownership of Japanese firms. One important development was the import of quality control technology from the United States.

Economic growth, 1955–73

In this period Japan's industries expanded rapidly. The production of household electrical goods such as TVs, radios and refrigerators grew very fast and major developments also took place in the transport industries — railways, shipbuilding and car manufacture — and in the production of artificial fibres. These changes also led to rapid growth in the production of iron and steel and in the output of the chemical industry. Towards the end of the 1960s the electronics industry also grew very rapidly and a considerable increase took place in expenditure on both research and development and on investment in equipment for the manufacture of semiconductors and integrated circuits.

In all these industries, great emphasis was put on the application of new techniques and many industrial research laboratories were established. The government also became more directly involved in research and development with the establishment of the Science and Technology Agency in 1956 and the Council for Science and Technology in 1959. These agencies have set up a number of research organisations and laboratories such as the Atomic Energy Research Institute in 1956 and the National

Centre for Space Development in 1964. In addition, they have produced a series of reports on the state of science.

Foreign or Japanese technology? 1973 onwards
Since the mid-1970s there has been a significant change in the major aims of science policy in Japan both in industry and in the government agencies. The emphasis is now much more on trying to develop specifically Japanese technology rather than on the efficient application of technology imported from abroad.

The annual reviews published by the Council for Science and Technology identify some weaknesses in Japanese science but also clearly show that Japan is now one of the top six countries involved in large-scale scientific research and developments — the other five are the Soviet Union, the USA, France, Britain and West Germany.

Fig. 7.5.4 shows, for the mid-1970s, the share of these six countries in the world's total GNP and total expenditure on research. The top six countries share about 65 per cent of world GNP but spend nearly 85 per cent of the total expenditure on research. Japan is third, after the USA and the Soviet Union, in both categories with just under 10 per cent of each of the totals. But Japan, like France, only spends about 2 per cent of its GNP on research compared with about 4.5 per cent in the Soviet Union and about 2.5 per cent for the USA, Britain and West

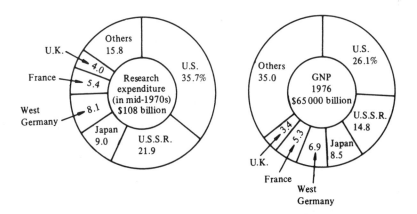

Fig. 7.5.4 Proportions of total world Gross National Product (GNP) and research expenditure in the mid-1970s. *From Science and Technology in Japan, vol. 1, no. 1, January, 1982, Three "I" Publications Ltd.*

Germany. Fig. 7.5.5 shows, for the same six countries, their share of the world's total number of research workers and total population. Together the six countries have about 75 per cent of the research workers but less than 20 per cent of the total population, while Japan is again third after the Soviet Union and the USA in numbers of research workers . . .

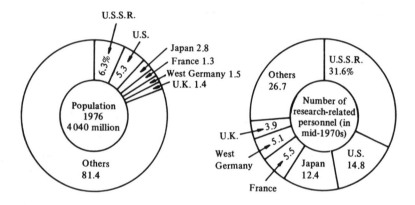

Fig. 7.5.5 Proportions of total world population and research workers in the mid-1970s. *From Science and Technology in Japan, vol. 1, no. 1, January 1982, Three "I" Publications Ltd.*

REFERENCES

1. Fox, G., *Britain and Japan, 1858–1883,* (Oxford: Oxford University Press, 1969), p. 261; quoted in Brock, W. H., *The Japanese Connection,* British Journal for the History of Science, Vol. 14, No. 48, 1981, p. 229.

2. Brock, W. H., op. cit., p. 229.

3. Ibid., pp. 232–3.

4. Ibid., p. 239.

5. Tuge, H. (ed.), *Historical Development of Science and Technology in Japan,* (Tokyo: Kokusai Bunka Shinkokai, 1968), p. 101.

SUMMARY

1. In 1869, after centuries of isolation, Japan adopted a policy of rapid industrialisation.
2. Expert foreign engineers were recruited to introduce new techniques and to staff new colleges and universities; many research organisations and scientific societies were founded.
3. Heavy industries were developed from 1900 onwards and Japanese science and technology became increasingly devoted to military purposes culminating in war with China in 1937 and World War 2 in 1941−5.
4. Since 1945, the efficient application of imported science and technology has played a major part in Japan's emergence as one of the most prosperous nations in the world.
5. The majority of Japan's investment in scientific research and development is made by private industry but the government is now attempting to foster genuine innovation in science and to develop specifically Japanese technology."

Appendix III

This guest-editorial comment from the July 1986 issue of the journal *Biotechnology* outlines the interrelationship of basic and applied research in the newly emerging field of high technology — the science which promises to revolutionise human life in the 21st century.

"Biotechnology in Underdevelopment"

by Daniel J. Goldstein

"The assessment of the viability of biotechnology in any country must be made in the context of its biological sciences and their relationships with the productive sector. A review of these two parameters in the underdeveloped world (with the exception of the special cases of China, Cuba, and Israel) draws a dismal picture.

"Biotechnology thrives on new knowledge generated by molecular biology, genetics and microbiology, but these disciplines are weak, often nonexistent, in the underdeveloped world. Biotechnology springs from universities and other research institutions, centers that generate the basic knowledge needed to solve practical problems posed by society. But the universities of the underdeveloped world are not research centers. Centuries of dependency could hardly produce such institutions. And the few creative research groups operate in a social vacuum; their results might be useful abroad, but are not locally. At the same time, biotechnological opportunities can only be detected and assimilated by innovative industries, and there are few of these in the underdeveloped world. Biotechnology needs dynamic interactions among the relevant industries. These interactions, however, are weak in countries in which science is perceived as an ornament, not as a necessity. Biotechnology is structurally fluid: driving forces oscillate between academy and industry. Rigidity, poor accountability, and conservatism make underdevelopment. Biotechnology requires many highly skilled professionals, because its raw materials are knowledge and skilled intelligence. Underdeveloped nations lack sufficient people well trained in the pertinent disciplines. Economic scarcity and political discrimination induce professionals and graduate students to emigrate or abandon science altogether.

"What can be done? First of all, underdeveloped countries must understand that they need to reform their universities, so that they can turn out people trained to solve problems, seek breakthroughs, and invent. They must recognise that molecular biology is not just another branch of biology, but the one and only tool available for understanding biological structure and function, the first step towards the appropriation and transformation of Nature. Success in biotechnology depends on the conquest and consolidation of the moving frontiers of cell biology and medicine. The history of biotechnology shows how intimate must be the interplay between academy and industry to maintain the competitive edge, to generate new products, and to expand its scope, its profitability, and its social impact. Training some people in the (by now routine) technologies of recombinant DNA could eventually lead to the substitution of certain imports. But we know that the import-substitution policies are self-limiting and but a weak palliative to the real problems. The new strategic products of the agrobiomedical market will continue to be imported.

"The international agencies should be careful when allocating their scant resources for biotechnology. Training programs in routine technologies and short courses on general topics do not constitute acceptable remedies for the lack of research-oriented, high-quality universities. An exaggerated stress on rapid applicability and industrial development projects, in societies that are not used to generating innovative technology and lack a critical mass of creative scientists, may be self-defeating. Noncompetitive enterprises are condemned to rapid obsolescence. Those involved (the industrialised world, which pays a big chunk of the agencies' bills, the agency administrators, and the underdeveloped countries that also pay their share to support the system) cannot absorb many more experiments of this kind. There is now a definite requirement for the real thing, and it will become increasingly difficult to make anyone swallow the usual reports on questionable achievements.

"Why should the industrialised nations care about the biotechnology of underdeveloped nations? After all, until now they have grown richer by extracting from and selling to the periphery. But now the real modernisation of the underdeveloped world is vital to the economic stability of the whole world. The conventional strategies of development have failed, and a realistic solution of the foreign debt problem depends on the economic growth of debtor countries. The only way to achieve this is by applying science and technology to their exportable commodities — leading to world-competitive, high-value-added products. A significant, original science in the debtor countries could generate opportunities for all. Take as an index the quality and economic impact of the work of expatriate scientists of just one such nation, Argentina, in the USA, UK and France.

"It is rather obvious that developing the scientific capabilities of the debtor countries could make joint ventures attractive and profitable. Investments in research, technology, and in their industrial spin-offs would be natural. This would lead to a new type of relationship between the center and the periphery. Instead of the present explosive situation, which closely resembles forced labor, a community of partners could emerge. Relations among partners can become strained, but, as a rule, the potential conflicts are vastly less dangerous and the possibilities of accommodation much greater. Science and high technology, for once, could be constructive tools for reducing tensions and contribute to peace. The alternatives are gory."

PERSPECTIVES ON PHYSICS

Speech at the Nobel Banquet

Delivered on 10 December 1979.

Your Majesties, Excellencies, Ladies and Gentlemen,

On behalf of my colleagues, Professors Glashow and Weinberg, I thank the Nobel Foundation and the Royal Academy of Sciences for the great honour and the courtesies extended to us, including the courtesy to me of being addressed in my language Urdu.

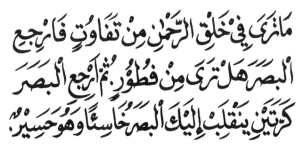

Pakistan is deeply indebted to you for this.

The creation of physics is the shared heritage of all mankind. East and West, North and South have equally participated in it. In the Holy Book of Islam, Allah says

مَا تَرَىٰ فِى خَلْقِ الرَّحْمٰنِ مِنْ تَفَاوُتٍ فَارْجِعِ
الْبَصَرَ هَلْ تَرَىٰ مِنْ فُطُورٍ ثُمَّ ارْجِعِ الْبَصَرَ
كَرَّتَيْنِ يَنْقَلِبْ إِلَيْكَ الْبَصَرُ خَاسِئًا وَهُوَ حَسِيرٌ

"Thou sees not, in the creation of the All-merciful any imperfection. Return thy gaze, seest thou any fissure. Then Return thy gaze, again and again. Thy gaze, Comes back to thee dazzled, aweary."

This in effect is, the faith of all physicists; the deeper we seek, the more is our wonder excited, the more is the dazzlement for our gaze.

I am saying this, not only to remind those here tonight of this, but also for those in the Third World, who feel they have lost out in the pursuit of scientific knowledge, for lack of opportunity and resource.

Alfred Nobel stipulated that no distinction of race or colour will determine who received of his generosity. On this occasion, let me say this to those, whom God has given His Bounty. Let us strive to provide equal opportunities to *all* so that they can engage in the creation of physics and science for the benefit of all mankind. This would exactly be in the spirit of Alfred Nobel and the ideas which permeated his life. Bless You!

Gauge Unification of Fundamental Forces

Address delivered by Abdus Salam on the occasion of the presentation of the 1979 Nobel Prizes in Physics, 8 December 1979.

In June 1938, Sir George Thomson, then Professor of Physics at Imperial College, London, delivered his 1937 Nobel Lecture. Speaking of Alfred Nobel, he said: "The idealism which permeated his character led him to . . . (being) as much concerned with helping science as a whole, as individual scientists . . . The Swedish people under the leadership of the Royal Family and through the medium of the Royal Academy of Sciences have made Nobel Prizes one of the chief causes of the growth of the prestige of science in the eyes of the world . . . As a recipient of Nobel's generosity, I owe sincerest thanks to them as well as to him."

I am sure I am echoing my colleagues' feelings as well as my own, in reinforcing what Sir George Thomson said — in respect to Nobel's generosity and its influence on the growth of the prestige of science. Nowhere is this more true than in the developing world. And it is in this context that I have been encouraged by the Permanent Secretary of the Academy — Professor Carl Gustaf Bernhard — to say a few words before I turn to the scientific part of my lecture.

Scientific thought and its creation is the common and shared heritage of mankind. In this respect, the history of science, like the history of all civilisation, has gone through cycles. Perhaps I can illustrate this with an actual example.

Seven hundred and sixty years ago, a young Scotsman left his native glens to travel south to Toledo in Spain. His name was Michael, his goal to live and work at the once Arab Universities of Toledo and Cordova, where the greatest of Jewish scholars, Moses bin Maimoun, had taught a generation before.

Michael reached Toledo in 1217 AD. Once in Toledo, Michael formed the ambitious project of introducing Aristotle to Latin Europe, translating not from the original Greek, which he knew not, but from the Arabic translation then taught in Spain. From Toledo, Michael travelled to Sicily, to the Court of Emperor Frederick II.

Visiting the medical school at Salerno, chartered by Frederick in 1231, Michael met the Danish physician, Henrik Harpestraeng — later to become Court Physician of Eric IV Waldemarssön. Henrik had come to Salerno to compose his treatise on blood-letting and surgery. Henrik's sources were the medical canons of the great clinicians of Islam, Al-Razi and Avicenna, which only Michael the Scot could translate for him.

Toledo's and Salerno's schools, representing as they did the finest synthesis of Arabic, Greek, Latin, and Hebrew scholarship, were some of the most memorable of international assays in scientific collaboration. To Toledo and Salerno came scholars not only from the rich countries of the East, like Syria, Egypt, Iran and Afghanistan, but also from developing lands of the West like Scotland and Scandinavia. Then, as now, there were obstacles to this international scientific concourse, with an economic and intellectual disparity between different parts of the world. Men like Michael the Scot or Henrik Harpestraeng were singularities. They did not represent any flourishing schools of research in their own countries. With all the best will in the world their teachers at Toledo and Salerno doubted the wisdom and value of training them for advanced scientific research. At least one of his masters counseled young Michael the Scot to go back to clipping sheep and to the weaving of woollen cloths.

In respect of this cycle of scientific disparity, perhaps I can be more quantitative. George Sarton, in his monumental five-volume *A History of Science*, chose to divide his story of achievement in sciences into ages, each age lasting half a century. With each half century he associated one central figure. Thus 450 BC — 400 BC Sarton calls the Age of Plato; this is followed by half centuries of Aristotle, of Euclid, of Archimedes, and so on. From 600 AD to 650 AD is the Chinese half century of Hsüan Tsang, from 650 to 700 AD that of I-Ching, and then from 750 AD to 1100 AD — 350 years continuously — it is the unbroken succession of the Ages of Jabir, Khwarizmi, Razi, Masudi, Wafa, Biruni, and Avicenna, and then Omar Khayam — Arabs, Turks, Afghans, and Persians. After 1100 appear the first Western names: Gerard of Cremona, Roger Bacon — but the honors are still shared with the names of Ibn-Rushd (Averroes),

Moses Bin Maimoun, Tusi, and Ibn-Nafis — the man who anticipated Harvey's theory of circulation of blood. No Sarton has yet chronicled the history of scientific creativity among the pre-Spanish Mayas and Aztecs, with their re-invention of the zero, of the calendars of the moon and Venus and of their diverse pharmacological discoveries, including quinine, but the outline of the story is the same — one of undoubted superiority to the Western contemporary correlates.

After 1350, however the developing world loses out except for the occasional flash of scientific work, like that of Ulugh Beg — the grandson of Timurlane, in Samarkand in 1400 AD; or of Maharaja Jai Singh of Jaipur in 1720 — who corrected the serious errors of the then Western tables of eclipses of the sun and the moon by as much as six minutes of arc. As it was, Jai Singh's techniques were surpassed soon after with the development of the telescope in Europe. As a contemporary Indian chronicler wrote: "With him on the funeral pyre, expired also all science in the East." And this brings us to this century when the cycle begun by Michael the Scot turns full circle, and it is we in the developing world who turn westward for science. As Al-Kindi wrote 1100 years ago: "It is fitting then for us not to be ashamed to acknowledge truth and to assimilate it from whatever source it comes to us. For him who scales the truth there is nothing of higher value than truth itself; it never cheapens nor abases him."

Ladies and Gentlemen, it is in the spirit of Al-Kindi that I start my lecture with a sincere expression of gratitude to the modern equivalents of the Universities of Toledo and Cordova, which I have been privileged to be associated with — Cambridge, Imperial College, and the Centre at Trieste.

1.   Fundamental Particles, Fundamental Forces, and Gauge Unification

The Nobel lectures this year are concerned with a set of ideas relevant to the gauge unification of the electromagnetic force with the weak nuclear force. These lectures coincide nearly with the 100th anniversary of the death of Maxwell, with whom the first unification of forces (electric with the magnetic) matured and with whom gauge theories originated. They also nearly coincide with the 100th anniversary of the birth of Einstein — the man who gave us the vision of an ultimate unification of *all* forces.

The ideas of today started more than twenty years ago, as gleams in several theoretical eyes. They were brought to predictive maturity over a decade back. And they started to receive experimental confirmation some six years ago.

In some sense then, our story has a fairly long background in the past. In this lecture I wish to examine some of the theoretical gleams of today and ask the question if these may be the ideas to watch for maturity twenty years from now.

From time immemorial, man has desired to comprehend the complexity of nature in terms of as few elementary concepts as possible. Among his quests — in Feynman's words — has been one for "wheels within wheels" — the task of natural philosophy being to discover the innermost wheels if any such exist. A second quest has concerned itself with the fundamental forces which make the wheels go round and enmesh with one another. The greatness of gauge ideas — of gauge field theories — is that they reduce these two quests to just one; elementary particles (described by relativistic quantum fields) are representations of certain charge operators, corresponding to gravitational mass, spin, flavour, colour, electric charge, and the like, while the fundamental forces are the forces of attraction or repulsion between these same *charges*. A third quest seeks for a *unification* between the charges (and thus of the forces) by searching for a single entity, of which the various charges are components in the sense that they can be transformed one into the other.

But are all fundamental forces gauge forces? Can they be understood as such, in terms of charges — and their corresponding currents — only? And if they are, how many charges? What unified entity are the charges components of? What is the nature of charge? Just as Einstein comprehended the nature of gravitational charge in terms of space-time curvature, can we comprehend the nature of the other charges — the nature of the entire unified set, *as a set*, in terms of something equally profound? This briefly is the dream, much reinforced by the verification of gauge theory predictions. But before I examine the new theoretical ideas on offer for the future in this particular context, I would like your indulgence to range over to a one-man, purely subjective, perspective in respect of the developments of the last twenty years themselves. The point I wish to emphasise during this part of my talk was well made by G. P. Thomson in his 1937 Nobel Lecture. G. P. said " . . . The goddess of learning is fabled to have sprung full grown from the brain of Zeus, but it is seldom that a scientific conception is born in its final form, or owns a single parent. More often it is the product of a series of minds, each in turn modifying the ideas of those that came before, and providing material for those that come after."

2. The Emergence of Spontaneously Broken SU(2) × U(1) Gauge Theory

I started physics research thirty years ago as an experimental physicist in the Cavendish, experimenting with tritium-deuterium scattering. Soon I knew the craft of experimental physics was beyond me — it was the sublime quality of patience — patience in accumulating data, patience with recalcitrant equipment — which I sadly lacked. Reluctantly I turned my papers in, and started instead on quantum field theory with Nicholas Kemmer in the exciting department of P.A.M. Dirac.

The year 1949 was the culminating year of the Tomonaga-Schwinger-Dyson reformulation of renormalised Maxwell-Dirac gauge theory, and its triumphant experimental vindication. A field theory must be renormalisable and be capable of being made free of infinities — first discussed by Waller — if perturbative calculations with it are to make any sense. More — a renormalisable theory, with no dimensional parameter in its interaction term, connotes *somehow* that the fields represent "structureless" elementary entities. With Paul Matthews, we started on an exploration of renormalisability of meson theories. Finding that renormalisability held only for spin-zero mesons and that these were the only mesons that empirically existed then, (pseudoscalar pions, invented by Kemmer, following Yukawa) one felt thrillingly euphoric that with the triplet of pions (considered as the carriers of the strong nuclear force between the proton neutron doublet) one might resolve the dilemma of the origin of this particular force. By the same token, the so-called weak nuclear force – the force responsible for β radioactivity (and described then by Fermi's nonrenormalisable theory) had to be mediated by some unknown spin-zero mesons if it was to be renormalisable. If massive charged spin-one mesons were to mediate this interaction, the theory would be nonrenormalisable, according to the ideas then.

Now this agreeably renormalisable spin-zero theory for the pion was a field theory, but not a gauge field theory. There was no conserved charge which determined the pionic interaction. As is well known, shortly after the theory was elaborated, it was found wanting. The ($\frac{3}{2}$, $\frac{3}{2}$)

resonance Δ effectively killed it off as a fundamental theory; we were dealing with a complex dynamical system, not "structureless" in the field-theoretical sense.

For me, personally, the trek to gauge theories as candidates for fundamental physical theories started in earnest in September 1956 — the year I heard at the Seattle Conference, Professor Yang expound his and Professor Lee's ideas[1] on the possibility of the hitherto sacred principle of left-right symmetry, being violated in the realm of the *weak nuclear force.* Lee and Yang had been led to consider abandoning left-right symmetry for weak nuclear interactions as a possible resolution of the (τ, θ) puzzle. I remember travelling back to London on an American Air Force (MATS) transport flight. Although I had been granted, for that night, the status of a Brigadier or a Field Marshal — I don't quite remember which — the plane was very uncomfortable, full of crying servicemen's children — that is, the children were crying, not the servicemen. I could not sleep. I kept reflecting on why Nature should violate left-right symmetry in weak interactions. Now the hallmark of most weak interactions was the involvement in radioactivity phenomena of Pauli's neutrino. While crossing over the Atlantic, came back to me a deeply perceptive question about the neutrino which Professor Rudolf Peierls had asked when he was examining me for a Ph.D. a few years before. Peierls' question was: "The photon mass is zero because of Maxwell's principle of a gauge symmetry for electromagnetism; tell me, why is the neutrino mass zero?" I had then felt somewhat uncomfortable at Peierls, asking for a Ph.D. viva, a question of which he himself said he did not know the answer. But during that comfortless night the answer came. The analogue for the neutrino of the gauge symmetry for the photon existed: it had to do with the masslessness of the neutrino, with symmetry under the γ_5 transformation[2] (later christened "chiral symmetry"). The existence of this symmetry for the massless neutrino must imply a combination $(1 + \gamma_5)$ or $(1 - \gamma_5)$ for the neutrino interactions. Nature had the choice of an aesthetically satisfying but a left-right symmetry violating theory, with a neutrino which travels exactly with the velocity of light; or alternatively a theory where left-right symmetry is preserved, but the neutrino has a tiny mass — some ten thousand times smaller than the mass of the electron.

It appeared at that time clear to me what choice Nature must have made. Surely, left-right symmetry must be sacrificed in all neutrino interactions. I got off the plane the next morning, naturally very elated. I rushed to the Cavendish, worked out the Michel parameter and a few other consequences of γ_5 symmetry, rushed out again, got onto a train to Birmingham where Peierls lived. To Peierls I presented my idea: he had asked the original question; could he approve of the answer? Peierls' reply was kind but firm. He said, "I do not believe left-right symmetry is violated in weak nuclear forces at all." Thus rebuffed in Birmingham, like Zuleika Dobson, I wondered where I could go next and the obvious place was CERN in Geneva, with Pauli — the father of the neutrino — nearby in Zurich. At that time CERN lived in a wooden hut just outside Geneva airport. Besides my friends, Prentki and d'Espagnat, the hut contained a gas ring on which was cooked the staple diet of CERN — Entrecote a la creme. The hut also contained Professor Villars of MIT, who was visiting Pauli the same day in Zurich. I gave him my paper. He returned the next day with a message from the Oracle: "Give my regards to my friend Salam and tell him to think of something better." This was discouraging, but I was compensated by Pauli's excessive kindness a few months later, when Mrs. Wu's[3], Lederman's[4], and Telegdi's[5] experiments were announced showing left-right symmetry was indeed violated and ideas similar to mine about chiral symmetry were expressed independently by Landau[6] and Lee and Yang[7]. I received Pauli's first, somewhat apologetic letter on 24 January 1957. Thinking that Pauli's spirit should by now be suitably crushed, I sent him two short notes[8,a] I had written in the meantime. These contained suggestions to extend chiral symmetry to electrons and muons, assuming that their masses were a consequence of what has come to be known as dynamical spontaneous symmetry breaking. With chiral symmetry for electrons, muons, and neutrinos, the only mesons that could mediate weak decays of muons would have to carry spin one. Reviving thus the notion of charged intermediate *spin-one* bosons, one could then postulate for these a type of gauge invariance which I called the "neutrino gauge." Pauli's reaction was swift and terrible. He wrote on 30 January 1957, then on 18 February and later on 11, 12, and 13 March: "I am reading (along the shores of Lake Zurich) in bright sunshine quietly your paper . . ." "I am very much startled on the title of your paper 'Universal Fermi Interaction' . . . For quite a while I have for myself the rule if a theoretician says *universal*

[a]For reference, see Footnote 7, p. 89, of Marshak, Riazuddin, and Ryan (1969), and W. Pauli's letters (CERN Archives).

it just means pure nonsense. This holds particularly in connection with the Fermi interaction, but otherwise too, and now you too, Brutus, my son, come with this word . . ." Earlier, on 30 January, he had written "There is a similarity between this type of gauge invariance and that which was published by Yang and Mills . . . In the latter, of course, no γ_5 was used in the exponent," and he gave me the full reference of Yang and Mills' paper[9]. I quote from this letter: "However, there are dark points in your paper regarding the vector field β_μ. If the rest mass is infinite (or very large), how can this be compatible with the gauge transformation $B_\mu - B_\mu - \partial_\mu \Lambda$?" and he concludes his letter with the remark: "Every reader will realise that you deliberately conceal here something and will ask you the same questions." Although he signed himself "With friendly regards," Pauli had forgotten his earlier penitence. He was clearly and rightly on the warpath.

Now the fact that I was using gauge ideas similar to the Yang-Mills [non-Abelian $SU(2)$-invariant] gauge theory was no news to me. This was because the Yang-Mills theory[9] (which married gauge ideas of Maxwell with the internal symmetry $SU(2)$ of which the proton-neutron system constituted a doublet) had been independently invented by a Ph.D. pupil of mine, Ronald Shaw[10], at Cambridge at the same time as Yang and Mills had written. Shaw's work is relatively unknown; it remains buried in his Cambridge thesis. I must admit I was taken aback by Pauli's fierce prejudice against universalism — against what we would today call unification of basic forces — but I did not take this too seriously. I felt this was a legacy of the exasperation which Pauli had always felt at Einstein's somewhat formalistic attempts at unifying gravity with electromagnetism — forces which in Pauli's phrase "cannot be joined — for God hath rent them asunder." But Pauli was absolutely right in accusing me of darkness about the problem of the masses of the Yang-Mills fields; one could not obtain a mass without wantonly destroying the gauge symmetry one had started with. And this was particularly serious in this context, because Yang and Mills had conjectured the desirable renormalisability of their theory with a proof which relied heavily and exceptionally on the masslessness of their spin-one intermediate mesons. The problem was to be solved only seven years later with the understanding of what is now known as the Higgs mechanism, but I will come back to this later.

Be that as it may, the point I wish to make from this exchange with Pauli is that already in early 1957, just after the first set of parity experiments, many ideas coming to fruition now, had started to become clear. These are:

(*i*) First was the idea of chiral symmetry leading to a $V - A$ theory. In those early days my humble suggestion[2,8] of this was limited to neutrinos, electrons, and muons only, while shortly after, that year, Marshak and Sudarshan[11,12,b], Feynman and Gell-Mann[16], and Sakurai[17] had the courage to postulate γ_5 symmetry for baryons as well as leptons, making this into a universal principle of physics.[c] *Concomitant with the (V − A) theory was the result that if weak interactions are mediated by intermediate mesons, these mesons must carry spin one.*

(*ii*) Second was the idea of spontaneous breaking of chiral symmetry to generate electron and muon masses, though the price which those latter-day Shylocks, Nambu and Jona-Lasinio[18] and Goldstone[19,20] exacted for this (i.e., the appearance of masses scalars), was not yet appreciated.

(*iii*) And finally, though the use of a Yang-Mills-Shaw (non-Abelian) gauge theory for describing spin-one intermediate charged mesons was suggested already in 1957, the giving of masses to the intermediate bosons through spontaneous symmetry breaking, in such a manner as to preserve the renormalisability of the theory, was to be accomplished only during a long period of theoretical development between 1963 and 1971.

Once the Yang-Mills-Shaw ideas were accepted as relevant to the charged weak currents — to which the charged intermediate mesons were coupled in this theory — during 1957 and 1958 was raised the question of what was the third component of the $SU(2)$ triplet, of which the charged weak currents were the two members. There were the two alternatives: the

[b] The idea of a universal Fermi interaction for (P, N), (ν_e, e), and (ν_μ, μ) doublets goes back to Tiomno and Wheeler[13,14] and Yang and Tiomno.[15] Tiomno (1958) considered γ_5 transformations of Fermi fields linked with mass reversal.

[c] Today we believe protons and neutrons are composites of quarks, so that γ_5 symmetry is not postulated for the elementary entities of today — the quarks. If the neutrino also turn out to be massive, γ_5 symmetry is spontaneously broken for it, as it is for electrons, muons, and quarks.

electroweak unification suggestion, where the electromagnetic current was assumed to be this third component; and the rival suggestion that the third component was a neutral current unconnected with electroweak unification. With hindsight, I shall call these the Klein (1938)[21] and the Kemmer[22] alternatives. The Klein suggestion, made in the context of a Kaluza-Klein five-dimensional space-time, was a real tour-de-force; it combined two hypothetical spin-one charged mesons with the photon in one multiplet, deducing from the compactification of the fifth dimension, a theory which looks like Yang-Mills-Shaw's. Klein intended his charged mesons for *strong* interactions, but if we read charged *weak* mesons for Klein's *strong* ones, one obtains the theory independently suggested by Schwinger,[23] though Schwinger, unlike Klein, did not build in any non-Abelian gauge aspects. With just these non-Abelian Yang-Mills gauge aspects very much to the fore, the idea of uniting weak interactions with electromagnetism was developed by Glashow[24] and Ward and myself[25] in late 1958. The rival Kemmer suggestion of a global $SU(2)$ — invariant triplet of weak charged and neutral currents was independently suggested by Bludman (1958) in gauge context and this is how matters stood till 1960.

To give you the flavour of, for example, the year 1960, there was a paper written that year by Ward and myself[26] with the statement "Our basic postulate is that it should be possible to generate strong, weak and electromagnetic interaction terms with all their correct symmetry properties (as well as with clues regarding their relative strengths) by making local gauge transformations on the kinetic energy terms on the free Lagrangian for all particles. This is the statement of an ideal which, in this paper at least, is only very partially realised." I am not laying a claim that we were the only ones who were saying this, but I just wish to convey to you the temper of the physics of twenty years ago — qualitatively no different today from then. But what a quantitative difference the next twenty years made, first with new and far-reaching developments in theory — and then, thanks to CERN, Fermilab, Brookhaven, Argonne, Serpukhov, and SLAC, in testing it!

So far as theory itself is concerned, it was the next seven years between 1961-67 which were the crucial years of quantitative comprehension of the phenomenon of spontaneous symmetry breaking and the emergence of the $SU(2) \times U(1)$ theory in a form capable of being tested. The story is well known and Steve Weinberg has already spoken about it. So I will give the barest outline. First there was the realisation that the two alternatives mentioned above, a pure electromagnetic current versus a pure neutral current — Klein-Schwinger versus Kemmer-Bludman — were not alternatives; they were complementary. As was noted by Glashow (1961) and independently by Ward and myself[27] both types of currents and the corresponding gauge particles ($W^{\pm}$, Z^0, and γ) were needed in order to build a theory that could simultaneously accommodate parity violation for weak and parity conservation for the electromagnetic phenomena. Second, there was the influential paper of Goldstone in 1961 which, utilising a non-gauge self-interaction between scalar particles, showed that the price of spontaneous breaking of a continuous internal symmetry was the appearance of zero mass scalars — a result foreshadowed earlier by Nambu. In giving a proof of this theorem[28] with Goldstone, I collaborated with Steve Weinberg, who spent a year at Imperial College in London. I would like to pay here a most sincerely felt tribute to him and to Sheldon Glashow for their warm and personal friendship.

I shall not dwell on the now well-known contributions of Anderson,[29] Higgs,[30,31,32] Brout and Englert,[33,34] Guralnik, Hagen, and Kibble[35,36] starting from 1963, which showed how spontaneous symmetry breaking using spin-zero fields could generate vector-meson masses, defeating Goldstone at the same time. This is the so-called Higgs mechanism.

The final steps towards the electroweak theory were taken by Weinberg[37] and by myself[38] (with Kibble at Imperial College tutoring me about the Higgs phenomena). We were able to complete the present formulation of the spontaneously broken $SU(2) \times U(1)$ theory so far as leptonic weak interactions were concerned — with one parameter $\sin^2 \theta$ describing all weak and electromagnetic phenomena and with one isodoublet Higgs multiplet. An account of this development was given during the contribution[38] to the Nobel Symposium (organized by Nils Svartholm and chaired by Lamek Hulthen held at Gothenburg after some postponements, in early 1968). As is well known, we did not have then, and still do not have, a prediction for the scalar Higgs mass.

Both Weinberg and I suspected that this theory was likely to be renormalisable.[d] Regarding spontaneously broken Yang-Mills-Shaw theories in general this had earlier been suggested by Englert, Brout, and Thiry (1966). But this subject was not pursued seriously except

[d]When I was discussing the final version of the $SU(2) \times U(1)$ theory and its possible renormalisability in Autumn 1967 during a postdoctoral course of lectures at

[continued overleaf]

at Veltman's school at Utrecht, where the proof of renormalisability was given by 't Hooft[29,40] in 1971. This was elaborated further by that remarkable physicist, the late Benjamin Lee,[41,42,43] working with Zinn-Justin, and by 't Hooft and Veltman.[44,45,e] This followed on the earlier basic advances in Yang-Mills calculational technology by Feynman,[46] De Witt,[47,48] Faddeev and Popov,[49] Mandelstam,[50,51] Fradkin and Tyutin,[52] Boulware,[53] Taylor,[54] Slavnov,[55] Strathdee and Salam.[56] In Coleman's eloquent phrase " 't Hooft's work turned the Weinberg-Salam frog into an enchanted prince." Just before had come the GIM (Glashow, Iliopoulos, and Maiani) mechanism,[57] emphasising that the existence of the fourth charmed quark (postulated earlier by several authors) was essential to the natural resolution of the dilemma posed by the absence of strangeness-violating currents. This tied in naturally with the understanding of the Steinberger-Schwinger-Rosenberg-Bell-Jackiw-Adler anomaly (see Ref. 58) and its removal for $SU(2) \times U(1)$ by the parallelism of four quarks and four leptons, pointed out by Bouchiat, Iliopoulos and Meyer[59] and independently by Gross and Jackiw.[60]

If one has kept a count, I have so far mentioned around fifty theoreticians. As a failed experimenter, I have always felt envious of the ambience of large experimental teams and it gives me the greatest pleasure to acknowledge the direct or the indirect contributions of the "series of minds" to the spontaneously broken $SU(2) \times U(1)$ gauge theory. My profoundest personal appreciation goes to my collaborators at Imperial College, and Cambridge and the Trieste Centre, John Ward, Paul Matthews, Jogesh Pati, John Strathdee, Robert Delbourgo, Tom Kibble, and to Nicholas Kemmer.

In retrospect, what strikes me most about the early part of this story is how uninformed all of us were, not only of each other's work, but also of work done earlier. For example, only in 1972 did I learn of Kemmer's paper written at Imperial College in 1937. Kemmer's argument essentially

Imperial College, Nino Zichichi from CERN happened to be present. I was delighted because Zichichi had been badgering me since 1958 with persistent questioning as to of what theoretical avail his precise measurements on $(g - 2)$ for the muon as well as those of the muon lifetime were, when not only the magnitude of the electromagnetic corrections to weak decays was uncertain, but also conversely the effect of non-renormalisable weak interactions on "renormalised" electromagnetism was so unclear.

[e]An important development in this context was the invention of the dimensional regularisation technique by Bollini and Giambiagi (1972), Ashmore (1972), and 't Hooft and Veltman.

was that Fermi's weak theory was not globally $SU(2)$ invariant and should be made so — though not for its own sake but as a prototype for strong interactions. Then this year I learnt that earlier, in 1936, Kemmer's Ph.D. supervisor, Gregor Wentzel,[61] had introduced (the yet undiscovered) analogue of lepto-quarks, whose mediation could give rise to neutral currents after a Fierz reshuffle. And only this summer, Cecilia Jarlskog at Bergen rescued Oscar Klein's paper from the anonymity of the International Institute of Intellectual Cooperation of Paris, and we learnt of his anticipation of a theory similar to Yang-Mills-Shaw long before these authors. As I indicated before, the interesting point is that Klein was using this triplet, of two charged mesons plus the photon, not to describe weak interaction but for strong nuclear force unification with the electromagnetic — something our generation started on only in 1972 — and not yet experimentally verified. Even in this recitation I am sure I have inadvertently left off some names of those who have in some way contributed to $SU(2) \times U(1)$. Perhaps the moral is that not unless there is the prospect of quantities verification, does a qualitative idea make its impression in physics.

And this brings me to experiment, and the year of the Gargamelle.[62] I still remember Paul Matthews and I getting off the train at Aix-en-Provence for the 1973 European Conference and foolishly deciding to walk with our rather heavy luggage to the student hostel where we were billeted. A car drove from behind us, stopped, and the driver leaned out. This was Musset whom I did not know well personally then. He peered out of the window and said: "Are you Salam?" I said "Yes." He said: "Get into the car. I have news for you. We have found neutral currents." I will not say whether I was more relieved for being given a lift because of our heavy luggage or for the discovery of neutral currents. At the Aix-en-Provence meeting that great and modest man, Lagarrigue was also present and the atmosphere was that of a carnival — at least this is how it appeared to me. Steve Weinberg gave the rapporteur's talk with T.D. Lee as the chairman. T.D. was kind enough to ask me to comment after Weinberg finished. That summer Jogesh Pati and I had predicted proton decay within the context of what is now called grand unification, and in the flush of this excitement I am afraid I ignored weak neutral currents as a subject which had already come to a successful conclusion, and concentrated on speaking of the possible decays of the proton. I understand now that proton decay experiments are being planned in the United States by the Brookhaven, Irvine and Michigan and the Wisconsin-Harvard groups and also by a European collaboration to be mounted in the Mont Blanc Tunnel Garage No. 17. The latter quantitative work on neutral currents at CERN, Fermilab, Brookhaven, Argonne and Serpukhov is, of course, history, but a special tribute is warranted to the beautiful SLAC-Yale-CERN experiment[63] of 1978 which exhibited the effective Z^0-photon interference in accordance with the predictions of the theory. This was foreshadowed by Barkov *et al.*'s experiments[64] at Novosibirsk in the USSR in their exploration of parity-violation in the atomic potential for bismuth. There is the apocryphal story about Einstein, who was asked what he would have thought if experiment had not confirmed the light deflection predicted by him. Einstein is supposed to have said, "Madam, I would have thought the Lord has missed a most marvellous opportunity." I believe, however, that the following quote from Einstein's Herbert Spencer lecture of 1933 expresses his, my colleagues', and my own views more accurately. "Pure logical thinking cannot yield us any knowledge of the empirical world; all knowledge of reality starts from experience and ends in it." This is exactly how I feel about the Gargamelle-SLAC experience.

References

1. T.D. Lee and C.N. Yang, *Phys. Rev.* **104** (1956) 254.
2. A. Salam, *Nuovo Cimento* **5** (1957a) 299.
3. C.S. Wu *et al.*, *Phys. Rev.* **105** (1957) 1413.
4. R. Garwin, L. Lederman, and M. Weinrich, *Phys. Rev.* **105** (1957) 1415.
5. J.I. Friedman, and V.L. Telegdi, *Phys. Rev.* **105** (1957) 1681.
6. L. Landau, *Nucl. Phys.* **3** (1957) 127.
7. T.D. Lee, and C.N. Yang, *Phys. Rev.* **105** (1957) 1671.
8. A. Salam, preprint, Imperial College, London, 1957b.
9. C.N. Yang, and R.L. Mills, *Phys. Rev.* **96** (1954) 191.
10. R. Shaw, "The problem of particle types and other contributions to the theory of elementary particles," Ph.D. thesis, Cambridge University (unpublished), 1955.

11. R.E. Marshak, and E.C.G. Sudarshan, in *Proceedings of the Padua-Venice Conference on Mesons and Recently Discovered Particles* (Societa Italiana di Fisica, 1957).
12. R.E. Marshak, and E.C.G. Sudarshan, *Phys. Rev.* **109** (1958) 1860.
13. J. Tiomno, and J.A. Wheeler, *Rev. Mod. Phys.* **21** (1949a) 144.
14. J. Tiomno, and J.A. Wheeler, *Rev. Mod. Phys.* **21** (1949b) 153.
15. C.N. Yang, and J. Tiomno, *Phys. Rev.* **75** (1950) 495.
16. R.P. Feynman, and M. Gell-Mann, *Phys. Rev.* **109** (1958) 193.
17. J.J. Sakurai, *Nuovo Cimento* **7** (1958) 1306.
18. Y. Nambu, and G. Jona-Lasinio, *Phys. Rev.* **122** (1961) 345.
19. Y. Nambu, *Phys. Rev. Lett.* **4** (1960) 380.
20. J. Goldstone, *Nuovo Cimento* **19** (1961) 154.
21. O. Klein, "On the theory of charged fields," in *Le Magnetisme*, Proceedings of the Conference organized at the University of Strasbourg by the International Institute of Intellectual Cooperation, Paris, 1939.
22. N. Kemmer, *Phys. Rev.* **52** (1937) 906.
23. J. Schwinger, *Ann. Phys.* (NY) **2** (1957) 407.
24. S.L. Glashow, *Nucl. Phys.* **10** (1959) 107.
25. A. Salam, and J.C. Ward, *Nuovo Cimento* **11** (1959) 568.
26. A. Salam, and J.C. Ward, *Nuovo Cimento* **19** (1961) 165.
27. A. Salam, and J.C. Ward, *Phys. Lett.* **13** (1964) 168.
28. J. Goldstone, A. Salam, and S. Weinberg, *Phys. Rev.* **127** (1962) 965.
29. P.W. Anderson, *Phys. Rev.* **130** (1963) 439.
30. P.W. Higgs, *Phys. Lett.* **12** (1964a) 132.
31. P.W. Higgs, *Phys. Rev. Lett.* **13** (1964b) 508.
32. P.W. Higgs, *Phys. Rev.* **145** (1966) 1156.
33. F. Englert, and R. Brout, *Phys. Rev. Lett.* **13** (1964) 321.
34. F. Englert, R. Brout, and M.F. Thiry, *Nuovo Cimento* **48** (1966) 244.
35. G.S. Guralnik, C.R. Hagen, and T.W.B. Kibble, *Phys. Rev. Lett.* **13** (1964) 585.
36. T.W.K. Kibble, *Phys. Rev.* **155** (1967) 1554.
37. S. Weinberg, *Phys. Rev. Lett.* **27** (1967) 1264.
38. A. Salam, in "Elementary particle theory", *Proceedings of the 8th Nobel Symposium*, ed. N. Svartholm (Almqvist and Wiksell, Stockholm, 1968).
39. G. 't Hooft, *Nucl. Phys.* **B 33** (1971a) 173.

40. G. 't Hooft, *Nucl. Phys.* **B 35** (1971b) 167.

41. B.W. Lee, *Phys. Rev.* **D 5** (1972) 823.

42. B.W. Lee, and J. Zinn-Justin, *Phys. Rev.* **D 5** (1972) 3137.

43. B.W. Lee, and J. Zinn-Justin, *Phys. Rev.* **D 7** (1973) 1049.

44. G. 't Hooft, and M. Veltman, *Nucl. Phys.* **B 44** (1972a) 189.

45. G. 't Hooft, and M. Veltman, *Nucl. Phys.* **B 50** (1972b) 318.

46. R.P. Feynman, *Acta Phys. Pol.* **24** (1963) 297.

47. B.S. DeWitt, *Phys. Rev.* **162** (1967a) 1195.

48. B.S. DeWitt, *Phys. Rev.* **162** (1967b) 1239.

49. L.D. Faddeev, and V.N. Popov, *Phys. Lett.* **B 25** (1967) 29.

50. S. Mandelstam, *Phys. Rev.* **175** (1968a) 1588.

51. S. Mandelstam, *Phys. Rev.* **175** (1968b) 1604.

52. E.S. Fradkin, and I.V. Tyutin, *Phys. Rev.* **D 2** (1970) 2841.

53. D.G. Boulware, *Ann. Phys.* (NY) **56** (1970) 140.

54. J.C. Taylor, *Nucl. Phys.* **B 33** (1971) 436.

55. A. Slavnov, *Theor. Math. Phys.* **10** (1972) 99.

56. A. Salam, and J. Strathdee, *Phys. Rev.* **D 2** (1970) 2869.

57. S. Glashow, J. Iliopoulos, and L. Maiani, *Phys. Rev.* **D 2** (1970) 1285.

58. R. Jackiw, in *Lectures on Current Algebra and Its Applications,* by S.B. Treiman, R. Jackiw, and D.J. Gross (Princeton University, New Jersey, 1972).

59. C. Bouchiat, J. Iliopoulos, and P. Meyer, *Phys. Lett.* **B 38** (1972) 519.

60. D.J. Gross, and R. Jackiw, *Phys. Rev.* **D 6** (1972) 477.

61. G. Wentzel, *Helv. Phys. Acta* **10** (1937) 108.

62. F.J. Hasert *et al., Phys. Lett.* **B 46** (1973) 138.

63. R.E. Taylor, in *Proceedings of the 19th International Conference on High Energy Physics,* eds. S. Homma, M. Kawaguchi, and H. Miyazawa (Physical Society of Japan, Tokyo, 1979), p. 422.

64. L.M. Barkov, in *Proceedings of the 19th International Conference on High Energy Physics,* eds. S. Homma, M. Kawaguchi, and H. Miyazawa (Physical Society of Japan, Tokyo, 1979), p. 425.

Physics and the Excellences of the Life it Brings

Speech delivered by Abdus Salam at the Conference on the History of Particles from Pions to Quarks, Fermi National Accelerator Laboratory, May 1985.

The title of my talk tonight is taken from Robert Oppenheimer. He had three types of excellences in mind. First, for the theoretician, the excellence of new ideas while he attempts to read Allah's design; for the experimenter, the excellence of new discoveries, of the pleasure of the search, of sheerly carrying an experimental technique to its limits and beyond; and finally, of the very human desire to spite the theorist. Oppenheimer had these excellences in mind — but also much more; he emphasised the opportunity physics afforded him to come to know internationally a class of great human beings whom one respected not only for their intellectual eminence but also for their personal human qualities — a true reflection of their greatness in physics. And, in addition, he had in mind the opportunities which physics uniquely affords for involvement with mankind — in the parlance of today, in engaging in problems of development and of enhancing the human ideal.

Tonight I wish to speak on some aspects of Oppenheimer's thoughts from a personal point of view. I shall illustrate these by recalling my induction into research, on renormalisation of meson theories, and the excellent men I was privileged to meet while pursuing it. I also wish to speak on excellence through world development as I was asked by the organisers. More precisely, I shall speak on the International Centre for Theoretical Physics, whose creation, under the auspices of the United Nations, I was privileged to suggest in September 1960. The Centre actually came into being only in October 1964 — beyond the cut-off date of this Conference's coverage. However, the ideas that went into the Centre's creation and the battles that had to be won, as well as the physics milieu of the early sixties — with its desire to keep alive the internationalism of the subject, with its emphasis on science rather than technology, and with its perception of brain drain of high level talent, particularly to the US — do fall within the period covered.

The notion of a Centre that should cater particularly to the needs of physicists from developing countries had lived with me from 1954, when I was forced to leave my own country because I realised that if I stayed there much longer, I would have to leave physics, through sheer intellectual isolation. At the September 1960 Rochester Conference, in his banquet speech, Mr. John McCone, the then Chairman of the US Atomic Energy Commision, made a reference to the desirability of creating international centres in physics. He had principally in mind accelerator establishments, which might be created under joint US/USSR/European auspices. After the banquet, over coffee, I remember a conversation with Hans Bethe, Robert Sachs and Nicholas Kemmer in the beautiful hall of the women's residence at Rochester University. We discussed the practical possibility of such centres being created and came to the conclusion that the simplest would be to think in terms of an international theoretical centre.

The same month I had the privilege of being able to voice, on behalf of the Government of Pakistan, this visionary ideal, in the form of a resolution at the annual conference of the International Atomic Energy Agency (IAEA) at Vienna. We were fortunate to receive co-sponsorship of the resolution from the governments of Afghanistan, the Federal Republic of Germany, Iran, Iraq, Japan, Philippines, Portugal, Thailand, and Turkey. As the list of sponsors indicates, the setting up of such a centre was of interest not only to the less privileged countries but also to some of the developed ones. The hope was that a centre of this type, besides providing a venue for collaborative international research for the East and the West, might also help in resolving one of the most frustrating problems that active scientists in poorer countries face — the problem of isolation. Such men, supported by international funds, would come fairly frequently to the Centre to renew their contacts and engage in active research in their fields.

Right from the beginning we received enthusiastic support from the world physics community. Niels Bohr, before his death, expressed his wholehearted support; scientific panels, convened in 1961 and again in 1963 by the Agency's physicist Director General, Dr. Sigvard Eklund, forcefully recommended its creation. Members of the 1961 panel were

Aage Bohr, Paolo Budinich, Bernard Feld, Leopold Infeld, Maurice Levy, Walter Thirring; of the 1963 "Three Wise Men Panel", the members were Robert Marshak, Leon Van Hove and Jayme Tiomno.

Unfortunately, there was not the same unanimous response from the atomic energy commissions around the world. At the 1962 annual conference of the IAEA (where these commissions represent their governments), even though the creation of a centre was accepted in principle after a divided vote — by and large the industrialised countries voted against, and the developing countries for, the Centre — the IAEA's Board of Governors voted the princely sum of $55,000 to set up an International Centre for Theoretical Physics. The United Nations Educational and Scientific Organization (UNESCO) voted $27,000. Thus, additional offers of financial assistance from interested member states had to be solicited; of the five received (from the Government of Italy for a centre to be located in Trieste, from Austria for Vienna, from Denmark for Copenhagen, from Pakistan for Lahore, and from Turkey for Ankara), the most generous was the Italian Government's offer of around $300,000 plus a prestigious building, with Paolo Budinich, Professor of Physics at the University of Trieste, as the moving spirit behind it. This was accepted in June 1963 and the Centre started functioning on 1 October 1964, with a charter for four years. Oppenheimer served on the first Scientific Council of the Centre. In spite of his terminal illness, he came to Trieste, where he helped draft the Centre's Charter. One admired him and his felicity of phrase — even in such legal drafting. Other members of the first Scientific Council were: Aage Bohr, A. Matveyev, V. G. Soloviev, Sandoval Vallarta and Victor Weisskopf. Dr. Alexander Sanielevici, from Romania, was one of the Scientific Secretaries.

Of the Centre and its functioning, I shall speak little tonight, for surely this will be one of the subjects covered by the future conferences in their series. In 1964, when we assembled in Trieste in a rented building, the whole enterprise seemed like a dream. Once again the world's theoretical community rallied around us — plasma physicists as well as particle physicists. We cared not for frills, only for physics. Our goal was to acquire scientific visibility. In this we succeeded. Thus, one year after the Centre's inception, Oppenheimer could comment at the Centre's Council: "It seems to me that the Centre has been successful in these eight or nine months of operation in three important ways. It has cultivated and produced admirable theoretical physics, making it one of the great foci for the development of fundamental understanding of the nature of matter. The Centre has obviously encouraged, stimulated and helped talented visitors from developing countries who, after rather long periods of silence, have begun to write and publish during their visit to the Centre in Trieste. This is true of physicists whom I know from Latin America, from the Middle East, from Eastern Europe and from Asia. It is doubtless true of others. The Centre has become a focus for the most fruitful and serious collaboration between experts from the United States and those from the Soviet Union on the fundamental problems of the instability of plasmas, and of means for controlling it. Without the Centre in Trieste, it seems to me doubtful that this collaboration would have been initiated or continued. In all the work at the Centre of which I know, very high standards prevail. In less than a year it has become one of the leading institutions in an important, difficult and fundamental field''.

To continue the story briefly, in the twenty years of its existence, the Centre has flourished, with physicists from 100 countries, East and West, North and South, ranging over all disciplines of physics — from fundamental physics to physics on the interface of technology, environment, energy, the living state and applicable mathematics. The Centre welcomes around 1,000 physicists from industrialised and 1,000 physicists from developing countries every year for research courses, workshops and meetings and for conducting research, for periods ranging from a few months to a few years.[a] In addition, from a generous grant from the Italian Government, we provide 100 fellowships in experimental physics, tenable at Italian laboratories. We are federated with approximately 200 institutes, mostly in developing countries. In addition, our Scientific Council selects 300 physicists (whom we call Associates of the Centre) — these men and women are accorded the privilege of coming to the Centre three times in six years for periods up to three months per visit, at times of their own choosing, provided they are living and working in developing countries. The Centre's current budget is of the order of five million dollars — three million come from the Government of Italy, one from IAEA, half a million from UNESCO, and the rest from other Government Agencies. The US Department of Energy gives a special grant of $50,000 for visits of US physicists.

[a]The 1986 figures were 2,160 physicists from developing countries and 1,440 from industrial countries.

Although in the founding and running of the Centre we have depended on the volunteer help of the world's leading physicists, it remains a sad fact that the physics communities of the developed countries have, by and large, rendered little assistance in an *organised* form to the cause of physics in developing countries, including the Centre. I wish to stress the word *organised* lest I should be failing to pay a heartfelt tribute to the continued work of great individuals who have made real sacrifices in this cause.

There is no question but that the real amelioration of the worsening situation for physics research in developing countries lies within the countries themselves and the role of the Centre and any other outside agency can only be to help generate self-reliant communities. But outside help — particularly if it is organised help — can make a crucial difference. This could take various forms: for example, the physical societies could help by donating 200-300 copies of their journals to the deserving institutions and individuals and by waiving publication charges. The American Physical Society in fact does provide its publications at half cost to 34 physicists from 13 least developed countries. IUPAP has been helping the Centre defray postage costs for distribution of old runs of journals donated by generous individuals. These schemes should, however, be extended by other societies and laboratories also to cover equipment, and in fact CERN has recently signified its willingness to donate some of its used equipment to laboratories in developing countries. Most important, the research laboratories and the university departments in developed countries could finance visits of their staffs to the institutions in developing countries in an organised manner and reciprocally by creating schemes like the three month Associateship scheme we run at the Centre — at the least for their ex-alumni now working at the developing country universities. Leon Lederman has initiated a scheme at Fermilab whereby a number of Latin American experimental physicists are regularly brought over to train them in techniques of particle physics and ancillary disciplines. And then there are the excellent cooperative schemes of training like the one which T. D. Lee runs for China. These could perhaps be extended to other developing countries.

May I be forgiven for thinking in the following terms: that the physics insitutions in developed countries may consider contributing in their own ways, according to the norms of the well-known United Nations formula, whereby most developed countries have pledged to spend 1% of their (GNP) resources for world development. In the end, it is a moral issue whether the better-off segments of the physics community are willing to look after their own deserving but deprived colleagues, helping them not only materially to remain good physicists, but also joining them in their battle to obtain recognition within their own communities, as valid professionals who are important to the development of both their countries and of the world.

So much for the excellence of a life of physics for realising the ideals of development. Now, I would like to turn to the second aspect of Oppenheimer's thought; some of the excellent and humanly great physicists I came to know internationally in the *early* part of my research.

During the period covered by this Conference (1950-1964), it appears to me that there have been five major developments in theory. First, the rise and the fall of Yukawa's standard model of pions and nucleons. Connected with this was the rise (and the later fall) of the S-matrix theory. The second major development was the understanding of the role of flavour symmetries, in particular, of flavour SU(3). The third development concerned the emergence of chirality; the fourth, the Nambu-Goldstone spontaneous symmetry breaking phenomena; and the fifth the Yang-Mills-Shaw gauge theory and its application to electroweak unification.

I have told the story — at least of my humble part in it — in respect of the last three developments — the rise of chiral symmetry, of spontaneous symmetry breaking and of the electroweak unification — in the Stockholm Lecture of 1979 — including the story of interactions with Pauli, Peierls, Ward, Weinberg, Glashow and others. I shall not repeat this, except to say that I can take legitimate pride in that both the Yang-Mills theory as well as the flavour eightfold way were independently invented by two of my good pupils within the ethos of my research groups at Cambridge and London.

For tonight, I shall concentrate mainly on the story of the short-lived rise of the pion-nucleon theory as the standard model of 1950-51, in consequence of the proof that this was the only theory which could be renormalised then. The people concerned with my story were P.A.M. Dirac, Nicholas Kemmer and Paul Matthews at Cambridge, besides Freeman Dyson, who was visiting Birmingham, and John Ward at Oxford.

The immediate post-war generation — our generation — was brought up to believe implicitly in the Yukawa model of the nuclear forces. The only open question at that time concerned the spin of the meson and the precise form of the nucleon-meson interaction. After Yukawa, Nicholas Kemmer — at least outside Japan — had made the most crucial contributions towards defining this problem. In a classic paper written at Imperial College, London, 1938, he had classified the Yukawa interactions — according to meson-spins and parities, and whether they were direct or derivative couplings. When I started research in October 1949, Kemmer was at Cambridge. Surprising though it may seem, I had started life as an experimental research student in the Cavendish, with a remit to scatter tritium against deuterium for Sam Devons, now Professor at Columbia. My finding myself as an experimenter was in accordance with the Cambridge tradition — handed down from Rutherford's days; those who fared well in the Physics Tripos became experimentalists; those who got third classes were consigned to theoretical research. Soon after starting, I knew the craft of experimental physics was beyond me. It was the sublime quality of patience — particularly patience with the recalcitrant equipment at the Cavendish — which I sadly lacked. Reluctantly, I turned my papers in, and started instead on quantum field theory with Nicholas Kemmer, in the exciting department of P.A.M. Dirac.

I said I started on theory research, but it was not that easy. Those were the days of renormalisation theory with the papers of Tomonaga, Schwinger, Feynman and Dyson providing feverish excitement.. At Cambridge, Nicholas Kemmer was the only senior person interested in these developments. He had behind him not only the kudos of having tabulated all possible meson interactions, but also the reputation of being a prince among men — of generousness to a fault to his students. So I went to Kemmer and requested him to accept me for research. He said he had eight research students already and could not take any more. He suggested I go to Birmingham to work with Peierls. But I could not bear to leave Cambridge — principally because of the beauty of the rose gardens at the Backs of my College — St. John's. (Incidentally, Dirac was at St. John's College also.) I asked Kemmer, "Would you mind if I worked with you peripherally for the time being?" He graciously assented. In my first interview with Kemmer, he said, "All theoretical problems in Quantum Electrodynamics have already been solved by Schwinger, Feynman and Dyson. Paul Matthews has applied their methods to renormalise meson theories. He is finishing his Ph.D. this year. Ask him if he has any problem left."

This was early 1950. So I went to Matthews and asked him what he was working on and if he had any crumbs left. The first piece of advice Matthews gave me was to forget the papers of Schwinger and Feynman and to concentrate on Dyson's two "classical" papers — particularly his most recent in 1949 where he had shown that quantum electrodynamics was renormalisable to all orders in α. He told me he had spent one and a half years already, trying to renormalise meson theories. He had found that only spin zero may work. He was writing up his one-loop calculations for his Ph.D. and had shown that the theory of spin zero mesons was indeed renormalisable up to the second order.

Matthews had at that time already tabulated which theories may possibly be renormalisable with the techniques then known. He had come to the conclusion that no derivative coupling meson theory could be renormalised at all, and that among the direct coupling theories with nucleons, the only hopefuls were either spin-zero, or the neutral vector meson theories with conserved currents for nucleons. No charged vector theory (with massive mesons) could be renormalisable. He had also shown that the neutral vector meson theory with mass was a replica of electrodynamics and one could take over the work of Dyson more or less intact and show its renormalisability. Regarding the spin-zero theories, he had shown that one would, at the least, need an additional $\lambda\phi^4$ term where ϕ is the meson field. The corresponding term for electrodynamics ($e^4 A^4$) was gauge variant, as had been remarked on by Dyson — with John Ward actually proving that the corresponding infinity did not exist.

The ϕ^4 term for spin-zero mesons would however be a new fundamental interaction term with a new fundamental constant λ. A new fundamental constant appeared just too radical those days, and we agonised over this. But the real question was: could one be sure that even with this new interaction term, all the infinities could be assimilated to a renormalisation of the meson-nucleon coupling constant, the masses of the mesons and the nucleons, plus a renormalisation of their wave functions and of the new constant. Matthews had worked with one-loop diagrams and shown that renormalisability appeared possible. He could not go beyond one-loop because overlapping infinities started to come in for higher loops and one had to solve this basic problem, before progress could be made. This was the situation around March 1950.

Matthews had his Ph.D. viva shortly afterwards. His external examiner was Dyson who was visiting Birmingham at that time. Dyson used to spend a few months at Birmingham and the rest in the US. In the viva, Dyson had asked Matthews about overlapping infinities. Dyson had asked him, "Have you come across these infinities? And if so, how do you resolve the problems posed by these?" And Matthews had replied, "You have claimed in your paper on Quantum Electrodynamics (QED) that these infinities — which occur in the self-energy graphs — can be properly taken care of. I am simply following you." No further question on these infinities was asked; both Dyson and Matthews kept silent after this brief exchange.

Now overlapping infinities had indeed appeared in QED where a general self-energy graph can be viewed as an insertion of a modified vertex at either end of the lowest order self-energy graph. Insertions of modified vertices at both ends would be tantamount to double-counting. But Dyson, in his paper, while discussing these, had recommended precisely this — that one should subtract the vertex-part sub-infinities twice before subtracting the final overall self-energy infinity. Dyson must be right; but why? And what made life awkward was that whereas this trouble-some overlap occurred only for self-energies in QED, for meson theories, the overlaps of the infinities were everywhere.

With characteristic generosity of which I became a life-long recipient, Matthews said to me, "My viva is over. After my degree, I'm going off, to take a few months holiday. And then I'll go to Princeton. You can have these problems of renormalising meson theories till I get back to work in the fall. And if you don't solve it by then, I'll take it back".

That was the sort of gentleman's agreement we parted on. So I had to get to the bottom of the overlapping infinity problem before the fall. I thought that the best thing for me would be to ask Dyson's help. So I rang him. I said: "I am a beginning research student; I would like to talk with you. I am trying to renormalise meson theories, and there is this problem of overlapping divergences which you have solved. Could you give me some time?" He said, "I am afraid I am leaving tomorrow for the US. If you wish to talk, you must come tonight to Birmingham". So I travelled from Cambridge to Birmingham that evening. Dalitz and his gracious wife put me up for the night.

Next morning, Dyson came to the Department — this was the first time I had met him. I said, "What is your solution to the overlapping infinity problem?" Dyson said, "But, I have no solution. I only made a conjecture". For a young student who had just started on research, this was a terrible shock. Dyson was our hero. His papers were classics. For him to say that he had only made a conjecture made me feel that my support of certainty in the subject was slipping away. But he was being characteristically modest about his own work. He explained to me what the basis of his conjecture was. What he told me was enough to build on and show that he was absolutely right. I travelled with him to London that afternoon. He was due to catch his boat from Southampton later that day. I think it was during that train journey, in conversation with Dyson, that I appreciated for the first time how weak the weak forces really were.

At Cambridge, amid the summer roses at the backs of the Colleges, I went back to the overlapping infinity problem, to keep the tryst with Matthews' deadline. Using a generalisation of Dyson's remarks, I was able to show that the spin-zero meson theories were indeed renormalisable to all orders. At that time transatlantic phone-calls for physics research had not been invented. So I had vigorous correspondence with Dyson, with the fullest participation of Kemmer, my supervisor. Exciting days indeed!

The subtraction procedure that I designed worked in momentum space. A crucial element of the proof was to associate with a given graph a set of integration variables in momentum space such that for the entire graph or for any of the sub-graphs contained in it, every possible infinity could be associated — on a one-to-one basis — with a single sub-integration. Assuming that this was possible, the subtraction procedure left behind an absolutely convergent remainder — absolutely convergent in the mathematical sense. To prove this one-to-one relationship, one had to consider the topology of the graphs. I could show, with Res Jost's help, that this result certainly holds for the so-called renormalisable theories. I have always felt very proud of this particular part of the proof (*Phys. Rev.*, **84** (1951) 426), but to my knowledge, the paper embodying this has never been referred to by anybody ever. I can only assume that the result has been taken on trust and that no one has ever re-checked it.

Contemporaneous with mine was the work of John Ward at Oxford, who devised a most ingenious scheme of regularisation. This depended on differentiation with respect to external momenta and this was the technique used later by Gell-Mann and Low in their beautiful work on the renormalisation group. Later still, other regularisation schemes were devised in X-space, notably by Hepp, Speer, Bogoliubov and Parasiuk. My procedure, however, was a straightforward subtraction in momentum space. And it would also permit a count of the wave function renormalisation (Z) factors correctly in all conceivable situations. (P.T. Matthews and Abdus Salam, *Phys. Rev.* **94** (1954), 185). Matthews and I wrote a brief review of these developments for *The Reviews of Modern Physics* of October 1951 in which we stated the following criterion for acceptability of a proof in this subject: "The difficulty . . . is to find a notation which is both concise and intelligible to at least two people, of whom one may be the author.". We left it unsaid that "the other person may be the co-author".

I can here tell a story about this work being considered deep and believed in, but seldom read. I was invited to the Institute for Advanced Study at Princeton in January 1951. I had by then applied my technique to renormalise charged spin-zero mesons interacting with photons. I took a manuscript copy of the new paper to Robert Oppenheimer to read and, if he approved, to send to the *Physical Review*. I then realised that I had given him a copy with no diagrams in it. So I went to his office to retrieve the manuscript. I had to wait for some while because he had visitors, but then he came out of his inner office, saw me and said, "I enjoyed reading your paper. It is a fine paper". I should have kept quiet but like a fool, I said: "I am sorry, I gave you a copy in which there were no diagrams. I don't think you could have understood it." Oppenheimer visibly changed colour. But he only said, "The results are surely true and intelligible even without diagrams".

This proof of the renormalisability of spin-zero meson-nucleon direct-coupling theory had come at an opportune time. With the discovery by Cecil Powell of the pion and the subsequent determination of its spin as zero, theory and experiment seemed to converge to a definitive standard model of nucleons and pseudoscalar pions, with a direct Yukawa plus the Matthews interaction. Our elation, however, was short-lived. The Yukawa coupling which nature seemed to favour was not the direct renormalisable pseudoscalar coupling but the unrenormalisable pseudovector coupling. The two couplings were only equivalent in the lowest order — but with the large coupling diameters $g^2/4\pi \approx 14$ — was order by order perturbation of any practical significance?

Then came the discovery of the $\Delta(\frac{3}{2}, \frac{3}{2})$ resonance, plus the discovery of the form factor for the nucleon by Hofstatder. These coups de grace finally killed the model. Influential in our thinking was also the paper of

Fermi and Yang, which questioned whether the pion was a fundamental entity or merely a nucleon-anti-nucleon composite.

For me personally, the disenchantment with the pion-nucleon theory had started much earlier. One of the post-war texts on nuclear physics was L. Rosenfeld's, which I believe he wrote in a war-cellar during 1944-45 in Belgium. This was a 600-page book which then cost £6 — the equivalent of something like $80 today. As a research student, I had invested in the book, with great reluctance; it had burnt a hole in my meagre pocket. The book consisted of the theory of the deuteron, a complete analysis of meson-theoretic nuclear forces, with Moller-Rosenfeld mixtures and the like; and a description of pion-nucleon scattering phase-shifts analyses below 1 MeV. Then Hans Bethe came to lecture at the Cavendish; during this lecture he made the categorical statement that all known deuteron parameters as well as any phase shift analyses below 1 MeV, could determine no more than two parameters of the nuclear potential the scattering length and the effective range. While listening to the lecture I kept thinking, "Surely this result Bethe has announced makes a book like that of Rosenfeld irrelevant?" The thought crossed my mind that just after the lecture finishes, everyone who has acquired a copy of the book will be trying to dispose of it. So, immediately after Bethe finished, I rushed to my lodgings at St. John's College, retrieved my copy and made a sprint to the Heffers Bookshop from which I had purchased the book. The sharks at Heffers offered me £3 if they were to buy the book back even though it was in a mint condition. I accepted, but of course now I feel very sorry that I sold it, because the book contained marvellous tables on harmonic functions.

I started my remarks with Dirac, who did not believe in the renormalisation ideas which we were pursuing in 1950-51. He listened to us, but always maintained the hope for a finite theory. He was recently proved right by the rise of supersymmetry theories, some of which are completely finite — among them the $N = 2$ and $N = 4$ supersymmetry theories and more recently by the superstring theories. In three decisive years, 1925, 1926 and 1927, with three papers, Dirac laid, first, the foundations of quantum physics as we know it; secondly, he laid the foundation of quantum theory of fields; and thirdly, that of the theory of elementary particles, with his famous equation of the electron. No man except Einstein has had such a decisive influence in such a short time on the course of physics in this century. But additionally for me, Dirac, whom I later came to know better, at the Trieste Centre, represented the highest reaches of personal integrity of any human being I have ever met. Knowing him has been one of the excellences of my life of physics.

I will conclude with a story of Dirac and Feynman that perhaps will convey to you, in Feynman's words, what we all thought of Dirac. I was a witness of it at the 1961 Solvay Conference. Those of you who have attended the old Solvay Conferences will know that at least then, one sat at long tables that were arranged as if one was sitting to pray. Like a Quaker gathering, there was no fixed agenda; the expectation — seldom belied — was that some one would be moved to start off the discussion spontaneously.

At the 1961 Conference, I was sitting at one of these long tables next to Dirac, waiting for the session to start, when Feynman came and sat down opposite. Feynman extended his hand towards Dirac and said: "I am Feynman." It was clear from his tone that it was the first time they were meeting. Dirac extended his hand and said: "I am Dirac." There was silence, which from Feynman was rather remarkable. Then Feynman, like a schoolboy in the presence of a Master, said to Dirac: "It must have felt good to have invented that equation." And Dirac said: "But that was a long time ago." Silence again. To break this, Dirac asked Feynman: "What are you yourself working on?" Feynman said: "Meson theories" and Dirac said: "Are you trying to invent a similar equation?" Feynman said: "That would be very difficult." And Dirac, in an anxious voice, said: "But one must try." At that point the conversation finished because the meeting had started.

Particle Physics: Will Britain Kill its Own Creation?

by Professor Abdus Salam FRS

This first appeared in **New Scientist,** *London, the weekly review of science and technology, 3 January 1985.*

There is no question but that particle physics is a British invention. Of the four building blocks of matter that nuclear physics and cosmology deal with — electrons, neutrons, protons and neutrinos — two were discovered at Cambridge and one at Manchester. This was before the Second World War, but after the war the tradition continued. Cecil Powell's group at Bristol found the pion, which had been predicted by Hideki Yukawa; and the first of the so-called "strange" particles was discovered by Patrick Blackett, George Rochester and Clifford Butler working at Manchester.

I started research in particle physics 35 years ago at Cambridge, hooked — just as young men and women from all parts of the world are today — by the excitement of this frontier subject. There it was brought to us by Cecil Powell, when he spoke to the undergraduates at Cambridge of his discoveries.

Since then I have spent my life working on two problems; first, to discover the basic building blocks of matter; and secondly, to discover the basic forces among them. It is still not clear if the presently accepted building blocks, known as quarks and leptons, are indeed the final stage in the first of these two searches. However, so far as the second problem is concerned, we appear to have succeeded in a manner much beyond what any one of us ever imagined. We seem not only to have discovered the principle, known as the "gauge principle", that governs the fundamental forces, but also, we appear to have glimpsed a basic unification among them.

Experimental proof exists now that two of the four fundamental forces — the electromagnetic and the weak nuclear — are basically aspects of the same force, the "electroweak force". We hope that the other two forces, the strong nuclear and the gravitational, will also eventually be united with the electroweak, into one single superforce. Over the past decade, therefore, particle physics has been transformed. It has acquired a direction, which it lacked before.

How was this transformation achieved? Even though their existence was predicted before, there is no question but the transformation came about because of the brilliant experimental discoveries of the W and Z particles, announced in 1983 at the European laboratory, CERN. This followed the earlier experimental discovery of the phenomenon of "neutral currents", also made at CERN in 1973. British teams played an important part in both discoveries. And this year the Nobel Prize for physics has been awarded to two people at CERN, for their leading roles in discovering the W and Z particles.

With this background, I must say it comes as a great surprise to me that the British government should have assembled a committee under the chairmanship of Sir John Kendrew to review British participation in high energy particle physics in general, and in the CERN enterprise in particular. The unkindest cut is the committee's second term of reference, which asks it to reflect on re-allocation of the resources released, in whole or in part, to other areas of science. I am reminded of the Galahad story in P. G. Wodehouse where, at a convivial party, one of the Wodehouse characters biffs another with a round of beef. The latter falls unconscious, and all the "undertakers present" start bidding for the body.

In my view, the government should have included a third term of reference, requiring that the review group also consider whether enough funds are being made available to particle physics and even whether they should be increased. This is in view of the support for particle physics dropping from 40 per cent of the science budget to around 20 per cent during the period from 1965 to 1980.

I may be forgiven for saying this because, of the two passions of my life, the second has been to stress the importance of "science transfer" for developing countries. After building up the Theoretical Physics Department at Imperial College, London, I have spent 20 years fighting the battle of stressing the necessity of science transfer for developing countries. I have been fortunate in having been able to create, under United Nations auspices, a Centre for Theoretical Physics at Trieste, in Italy, with large participation of experimental and theoretical physicists from developing countries.

More than a thousand physicists come every year to the international centre at Trieste from developing countries. (Another thousand come from Britain, the US, the USSR, Italy, Japan, France and other industrialised countries.) They and we believe that fundamental physics[a] is important even for developing countries, not only for the technological benefits its spin-off brings, but also because it highlights the excitement of the adventure of science. This is so crucial in persuading developing societies to invest in science in *applicable* areas. Just for these reasons, the more scientifically advanced of the developing countries, China and India, for example, invest fairly heavily, relative to their science expenditures, in areas of accelerator of physics.

Thus to find that Britain, of all countries, should contemplate withdrawing from the international pursuit of a subject that constitutes one of the frontier areas of science, appears to me incredibly destructive for the morale of the scientific enterprise worldwide. And this at a time when, besides recent successes, the same CERN groups that discovered the W and Z have offered hints of further insights to come.

Let me explain this by illustrating the situation in particle physics today. During the past decade, we have registered the third great breakthrough in physics of this century — after relativity and quantum mechanics. The step forward is in the progress achieved towards the understanding of the problem of force. Theory predicted that the "gauge principle" must be shared among all the forces. This theoretical principle manifests itself in the form of the "gauge particles", particles that in a sense "carry" the forces between interacting objects. Early in this century physicists showed that the photons — "particles" of light — convey the electromagnetic force between charged objects. More recently, five years ago, physicists demonstrated the existence of gluons that carry the strong nuclear binding force. And last year, the W and Z particles responsible for the weak nuclear force of radioactivity were found at CERN. We know now, with full experimental confirmation at CERN's high energies, how the W and the Z come together with the photon, signifying the eventual unity of the weak nuclear force with electromagnetism. There is some support for the belief that in this electroweak unification lies the secret of the left-handedness of the molecules of life (*New Scientist*, 19 January 1984, p. 10).

[a]We in Trieste spend 10% of our budget on particle physics.

Spurred by these recent successes, we had been fully expecting the gluon — the gauge particle of the strong nuclear force that binds the atomic nucleus together — in its turn to unite with the W, the Z and the photon. The gluon does exist, as predicted, and it was indeed discovered in 1979 at the DESY laboratory in Hamburg by the Tasso group and others, which include many British members. But on present form, the gluon simply refuses to unify in the manner we thought it would. Finally, we are nowhere near devising a theory — a quantum theory — of the fourth force, gravity, far less to test its eventual possible unification with the other three forces.

But we cannot be content only with discovering the eventual unity of the four fundamental forces. We must also find out what makes these forces look so diverse, in the cold inhospitable $3K$ Universe we inhabit. We have speculated on, but not yet experimentally discovered, the materialisation of what could produce the diversity of these forces. We must discover the beast known as the Higgs-Kibble particle (or surrogates thereof) which may be responsible for the diversity perceived, because with an understanding of the origin of this diversity is linked the under-standing of all those phase transitions which, we now believe, occurred early in the cosmological history of the Universe. Our triumphs are thus today laced with some failures, either because we can today experiment only at energies which are too low, or because of theoretical fog. Quite frankly we have too many theoretical choices. We need experiments to limit these choices further.

Last year the experimental groups at CERN that had discovered the W and Z particles announced a score or so of "events" — particle interactions — that could not be explained by the standard electroweak theory (*New Scientist*, 20 December 1984, p. 5). These events may or may not presage the sorely needed new directions from among the diverse avenues open for the theory. One guess is that we may be on the verge of learning that the basic distinction, which quantum mechanics has taught us to make, between fermions and bosons — and which forms the backbone of all of chemistry and all of the theory of condensed matter — needs supplementation. We may be on the verge of the discovery that every fundamental fermion has a bosonic counterpart — obeying the so-called principle of supersymmetry. I personally believe the answer lies in supersymmetry, in a multi-dimensional space-time of which we have yet apprehended only four dimensions. Just now, it would be madness to stop doing particle physics; it would be tantamount to infanticide.

For three years, CERN will be the only place where this type of sorely needed guidance to the subject can come from. After this the facilities at Fermilab in the US will surpass the energies available at CERN.

What would be the effect of Britain's withdrawal from CERN? One may have understood such a withdrawal by a country such as Spain, Greece or Norway, for the communities of experimental particle physicists there are just beginning to build up. But were Britain to withdraw, the demoralisation inevitably felt among the smaller countries could mean that CERN might have to close down. It could mean the end for this European centre, built over 30 years, with its accelerators designed and constructed principally by British men of science and technology — the late Sir John Adams, Mervyn Hine, John Fox, Michael Crowley-Milling, Roy Billinge and others. Surely that is not the desire of Her Majesty's Government. At the very least CERN would be crippled. And if it were to escape closing down, it would be thanks to the temper of the science administrations in West Germany, France and Italy, core countries that may not forsake CERN. The physicists in Britain would then have to go cap in hand to colleagues in these countries to obtain news of their fresh discoveries.

But why should Britain contemplate leaving particle physics? I am told, to save funds for other basic sciences. Surely Britain's expenditures on basic science as a whole is not so excessive by international standards. The US has increased its spending on fundamental sciences during the past four years by 52 per cent, with a disproportionate increase in particle physics spending. The US Presidential Science Adviser, George A. Keyworth II, told Congress earlier this year: "Government support of basic research is a Federal trust . . . basic research [is] an essential instrument in the nation's long-term welfare . . . because its benefits are so broadly distributed. Quite simply basic research is a vital underpinning for national well being." Clearly, this message must be conveyed to British authorities as well.

At what rate should a basic science subject develop? I believe there is only one criterion: if a subject is worth pursuing — and under any criteria, particle physics is — it will not stand still. It should be allowed to develop at a rate at which it continues to attract the best young people through the excitement it offers. And it does not do any good to have theory, for example, develop faster than experiment or vice versa. The two must develop in tandem. Up to the beginning of this year it seemed that theory had outstripped experiment in particle physics but with the new and tantalising glimpses CERN has offered us of new phenomena it seems that this situation may be reversed. This is all to the good, for physics — and particle physics — is an empirical subject; it would ossify if it remained purely theoretical.

One of the side effects of Britain's leaving the area of high energy experimental particle physics would be that the brightest young men and women who now enter this area for its excitement, would not do so. It is assumed by some that they would take up other sciences. Personally, I feel this is an unwarranted assumption. Many of them may simply opt for stockbroking, as I have gathered from conversations with some of those who may be affected; others will emigrate.

We must not forget one more thing that is fundamental to thinking about CERN and particle physics in Europe. The present site at CERN is going to run out of space for large accelerators. A tunnel of 27 kilometres in circumference is being constructed under the Jura mountains for the newest accelerator, the large electron positron collider (LEP). This is the largest tunnel that can be built around CERN. In this tunnel, the optimum accelerator to be constructed after LEP, within the present *constant* budget (and I hope fervently that it will be built) would be a proton accelerator. This machine would increase 20-fold the present maximum energy in the world. Assuming that such an accelerator, which would rely on the further development of super-conducting magnet technology, is in place by the year 2000, I would give CERN at its present site a life of no more than 25 years, up to the year 2010. No more.

But what would come next in particle physics after the year 2010? In 1928, the late Paul Dirac said: "With quantum mechanics we have solved all of chemistry and most of physics." Some of my colleagues believe that by the year 2000, the grand design of the Universe will have been unravelled, and our discoveries will have changed Dirac's "most of physics" to "all of physics". I do not subscribe to this view. Truthfully, I believe that we shall need higher energies and still higher energy accelerators. To be able to afford them, we must find new means of acceleration.

This may, for example, be a laser-plasma device; for comparable energies, a laser-plasma accelerating device could be as compact as 1 kilometre in length instead of the present 27 kilometres. Such machines could be commissioned 25 years from now, provided Europe sets aside funds to research into a possible design, starting today. Technologically, such accelerators could represent the most advanced synthesis of plasma physics, laser physics and particle physics. The US is already ahead in this, perhaps with laser or particle-beam weapons in mind.

Personally, I would have liked the SERC's Rutherford Appleton Laboratory (RAL) in Oxfordshire to have earnestly started work on this. This is what I meant by saying that the Review Group may consider recommending an increase of funds, in order to give RAL a new role. The laboratory has unique facilities for laser physics, plasma physics and particle physics — all at one place. These should be utilised in wresting technical leadership in an area which, otherwise, is bound to be developed elsewhere. After all, this is the laboratory that invented the superconducting cable now universally used at all accelerators, including Fermilab in the US. This invention has clear industrial potential, but its further development at RAL was curtailed for lack of funds.

I completely and unreservedly agree with the remark frequently made that the fluctuating exchange rate of the pound versus the Swiss franc and the fluctuations this introduces into the rest of British science expenditure is a scandal. Yet such fluctuations do occur for all British contributions to the United Nations bodies, which as a rule are assessed in dollars. The government sees to it that they cause no havoc to any related domestic enterprise. Clearly, the answer to such problems is to let the right agency of the government take care of the British science subscription — that is, Britain should follow Austria, Belgium, France, Italy, Norway and Switzerland, countries which channel their contributions to international science agencies with the involvement of their foreign affairs ministries, and not through a council like SERC, with foreign and domestic expenditures mixed up with each other.

I shall conclude by recalling a past episode in which Britain supported basic science generously and munificently, and which has been related by the science historian D. F. Moyer. On 6 November 1919, the Royal Society held a meeting at which Sir Arthur Eddington reported on the measurement of the deflection of light, confirming Albert Einstein's theory of general relativity. Eddington's expedition had been commissioned for measurements of the solar eclipse observed at Principe, and the British government had made a grant of around £10,000 — during the First World War — which was equivalent, perhaps, to £250,000 today.

According to A. N. Whitehead who was present at the Royal Society's meeting, "The whole atmosphere of tense interest was exactly that of Greek drama: we were the chorus commenting on the decree of destiny as disclosed in the development of a supreme incident. There was dramatic quality in the very staging; the traditional ceremonial, and in the background the picture of Newton to remind us that the greatest of scientific generalisations was now, after more than two centuries, to receive its first modifications."

The meeting was headline news in the *Times* of London the next day. The *Times* reported that the discussion turned from matters of fact to the theoretical bearings of the results. "At this stage [the *Times* reported] Sir Oliver Lodge" — the propounder of the rival aether theory explaining the deflection — "whose contribution to the discussion had been eagerly awaited, was seen leaving the meeting". The next day, the *Times* reported that "the subject was a lively topic of conversation in the House of Commons. Beneath this news, the *Times* published a letter from Lodge "to explain that my having to leave the meeting . . . was due to a long-standing engagement and a 6 o'clock train."

In response to a request from the *Times,* Einstein wrote an essay on his theory which appeared in the issue of 28 November 1919. He graciously thanked his colleagues:

> *"It was in accordance with the high and proud tradition of English science that English scientific men should have given their time and labour, and that English institutions should have provided the material means, to test a theory that had been completed and published in the country of their enemies in the midst of war."*

Clearly, a country that has upheld fundamental science in this manner in the past cannot lightly absolve itself and withdraw from supporting this most exciting adventure of ideas of our times, after having played a leading role in its original creation.

Addendum (23 September 1985)

I would like to add here a remark on the global context within which the Kendrew Report must be viewed. My thesis is that one cannot view expenditures on basic sciences without taking into account the whole picture of public spending on Research and Development (R & D) in its full context.

According to an OECD Report on Public Expenditure on Research and Development, issued in 1984, the UK Treasury spent around four billion dollars on R&D in 1980 — 12.9% of these funds on "Promotion of Knowledge", on Basic Sciences. (Table I). Of this, nearly $\frac{1}{10}$ th — a little more than 1% of the overall public spending on science — went to particle physics. If Kendrew recommendations are fully accepted and particle physics allocation is diminished, the other basic sciences are expected to gain one third of 1% of the overall budget. Now, according to the same report, during 1980, Britain spent no less than 60% (59.4% to be precise) of its total public R&D budget on defence research; the US in contrast spent 47%, France 40%, Germany 15%. The basic sciences budgets in Germany and France were each of the order of 15% of the total.

My thesis is that if the defence research and development budget were to be decreased from 59.4% — not even to the French level of 40% — but just to 59%, all the savings which the Kendrew Report is desirous of allocating to other basic sciences could be found *without*, at the same time, diminishing particle physics.

Table I

Percentage of Public Expenditure on Research & Development
(Selected Items)

	United Kingdom		FR Germany		France		United States		Japan	
	1975	1980	1975	1980	1975	1980	1975	1980	1975	1980
Defense	52.8	59.4	19.2	15.3	32.8	40.1	50.8	47.3	4.7	4.9
Industrial Development	3.4	3.8	9.9	12.4	8.9	7.9	0.3	0.3	14.2	12.2
Health & Related Services	4.1	3.9	15.9	15.3	6.5	7.5	14.8	15.2	12.1	11.2
General Promotion of Knowledge	14.1	12.9	15.7	14.2	17.1	15.0	4.3	3.0	3.5	4.1

(Taken from a report issued by the Organisation for Economic Cooperation and Development (OECD), 1984, entitled "Science and Technology Indicators: Resources devoted to R&D".)

Behind Reality
Professor Abdus Salam FRS
by Paul Andersen

Reproduced from Imagined Worlds by Paul Andersen and Deborah Cadbury with the permission of BBC Enterprises Ltd.

"When I was at school in about 1936 I remember the teacher giving us a lecture on the basic forces in nature. He began with gravity. Of course we had all heard of gravity. Then he went on to say, 'Electricity. Now there is a force called electricity, but it doesn't live in our town Jhang, it lives in the capital town of Lahore, a hundred miles to the east.' He had just heard of the nuclear force and he said, 'That only exists in Europe.'"

This story of what it was like to be taught in an underdeveloped country is told with humour and excitement. For Abdus Salam, the days since have been spent sharing his enthusiasm between two great interests: physics and development in the Third World. In both he has become an international figure.

The 1930s, when Abdus Salam was still at school, was a period of rapid and sustained advance in physics. The discovery of radioactivity at the turn of the century by Madame Curie, and then later the theory of the nuclear atom proposed by Rutherford, had completely changed scientific thinking on what was considered to be the structure of nature. An atom was no longer an indivisible sphere but was made of a tiny nucleus, with electrons surrounding it. How nuclei or electrons interacted with one another was not clear at the time, but by 1946, when Abdus Salam went up to Cambridge University, some progress had been made.

"When I started research we believed that all matter in the Universe consisted of four discrete types — protons, neutrons, electrons and neutrinos — and that there were four fundamental forces between these four types of chunks of matter. Between the proton and the electron there was the electromagnetic force. The second force was the strong nuclear force, a force between a proton and a neutron. The third force, the weak nuclear force, was a force between pairs of protons and neutrons on the one hand and electrons and neutrinos on the other. And finally there was the force of gravity which is an attractive force between all these chunks of matter."

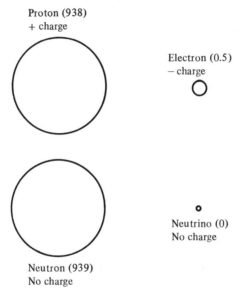

Relative masses of particles in brackets in mass energy units.

"When I started research we believed that all matter in the Universe consisted of four discrete types — protons, neutrons, electrons and neutrinos."

The discovery of these forces and particles had been turning points in modern science. Newton had been the first to observe and analyse the nature of the force of gravity. In the nineteenth century Faraday and Maxwell had described the nature of the electromagnetic force in their work on electricity. Then when Madame Curie made her discovery of radioactivity, what she had in fact been observing was the effects of the weak nuclear force within the atoms of radium. Rutherford's later explanation of the composition of the atom was to set the scene for understanding about the strong nuclear force that held the nucleus together.

It all seemed very tidy. "We felt that the physics of the day had reached the end of the road. The Greeks had wanted to explain all the phenomena of nature through postulating four elements: fire, earth, air and water. We had substituted their elements with four fundamental entities: protons, neutrons, electrons and neutrinos.

"When we look at the world it's incredibly diverse. This diversity seems on the face of it difficult to describe in terms of so few principles. We believe, however, that all these diverse phenomena could be understood in terms of just those four fundamental objects and the four forces. The force of gravity is seen in the fact that we all remain on the Earth and that the planets move around one another. In our language this is a manifestation of the force of attraction between a proton and a neutron and an electron and a neutrino.

"All other phenomena on Earth itself are associated with the second force, the electromagnetic force — the expression of matter in its hardness, its softness and in its colour. It is incredible in its simplicity that all these diverse phenomena are a consequence of a force that exists between the proton and the electron.

"The third force, the strong nuclear force, is something that we are not so conscious of on earth. It is the force that keeps the protons and neutrons together. To see its operation in its purest form you have to go to the stars. The Sun is using this force to produce the vast amount of radiation which comes to us in the form of sunlight.

"Finally, there is the weak nuclear force which manifests itself most clearly through radioactivity. Through the operation of this force, the neutrons are being transformed into protons and electrons and neutrinos." While the mathematics for such a picture is complex, the underlying notion of a few simple particles and their attendant forces being a complete description of the Universe is a startling scientific achievement.

A Unification of the Four Forces

But this image was not to satisfy the physicists for very long. Perhaps there was some way in which these conceptual quartets could be linked together. They might just be the expression of some yet more basic principle.

For example, one of the greatest achievements in scientific thinking in the twentieth century had been the linking of *mass* with *energy*. These two very different concepts were brought together through Einstein's genius in that renowned expression, $E = mc^2$. And there had been earlier examples — the example of unification of electricity and magnetism.

"At the beginning of the century these two were thought to be absolutely distinct forces — the electric force and the magnetic force. However, while they appear very different on the face of it, electricity and magnetism are basically the same force. This was proved by the work of Faraday and Ampere, who showed that if you take an electron that is stationary, then you can sense an electric force in the space around it. Once you begin to move that electron, you will get a manifestation of the magnetic force as well. The magnetic force is not a fundamental thing, but is connected with the electric force through the state of motion of the electron. The realisation that these two forces were of one fundamental origin led to the concept of the electromagnetic force. When I started research, one of the things which we began to look for was the possibility of a further unification between the four known forces."

After his first degree at Cambridge University, Abdus Salam joined the Cavendish Laboratory where Rutherford has carried out his experiments on the structure of the atom. It was an outstanding laboratory for experimental work and a focus for physicists around the world.

As a research student, however, Abdus Salam found his temperament more suited to theoretical rather than practical problems. "I had very little patience with experimental equipment. To be a good experimenter you must have patience towards things which are not always in your control. I think a theoretician has got to be patient, but that is with something of his own creation, his own constructs, his own stupidities. So after three months of experimental work at the Cavendish I just gave up. I went to my supervisor and said, "I'm sorry, I would like to turn back to theory."

It was the right decision. In the following years his work in the field of quantum and particle physics was to achieve many breakthroughs in thinking. While the four fundamental particles were replaced by families of such particles with names like quarks and leptons, the principle of the four forces remained. For many years Einstein had attempted to find a theory that would bring two of these forces, the electromagnetic and gravity, into a unified theory. In his work *General Theory of Relativity*, published in 1915, he had shown that gravity was in fact the expression of the curved geometry of space and time. For the last years of his life Einstein had struggled to achieve this further simplification of nature's principles with a unification of gravity and the electromagnetic force. It was a quest that brought only failure.

Abdus Salam was to attempt the problem from a different perspective and succeed. "Einstein's failure kept physicists away from the question of unification of the forces. Not many people were actively pursuing the idea since it was considered a dead end. We realised that Einstein had been singularly blind about the nuclear forces.

"We consciously chose to ignore gravity because we knew that that was the hardest problem of all. It had beaten a man like Einstein. We started to look for a unification between the electromagnetic force and the weak nuclear force. On the face of it, these two forces are very, very different. You are trying to unite what is a long-range force with a short-range force. The electromagnetic force can be felt at almost any distance. If you put an electron on the table and if you have sensitive enough instruments, you can see the effect of it a hundred metres away, or even one kilometre away, if you wish.

"But not so with the nuclear forces. The nuclear force, and in particular the weak nuclear force, is very short range. The weak force manifests itself only if the proton and the neutron and the electron and neutrino which are participating are 10^{-16} centimetres close to each other. This is an unimaginably short distance, something we never come across in ordinary life. But to me as a physicist, when I speak of distances of 10^{-16} centimetres, the smallness is really irrelevant. My task is to go behind the reality of these short-range and the long-range forces and to search for the unity that may exist between them.

"To give an analogy of what we are trying to do, let us look at ice and water. They can co-exist at zero degrees centigrade, although they are very distinct with different properties. However, if you increase the temperature you find that they represent the same fundamental reality, the same fluid. Similarly, we thought that if you could conceive of a Universe which was very, very hot, something like 10^{16} degrees, unimaginably hot (the present temperature of the Universe which we live in is very low, around $-270°$ centigrade), then it was our contention that the weak nuclear force would exhibit the same long-range character as the electromagnetic force. You would then see the unification of these two forces perfectly clearly.

"To arrive at that sort of temperature, you have to go back into the early history of the Universe. At the time of the Big Bang, the Universe was presumably infinitely hot. And then it started to expand and as it did so it cooled down. When you come to the epoch of the order of 10^{-11}

seconds after the Big Bang, you come to the stage where the low temperature allows the weak nuclear force for the first time to be distinct from the electromagnetic. That is the zero point of the ice and water example. If you are hotter than this "zero" (as you were in the early part of the Universe) then you would see no distinction between the two types of forces. This was our idea for the unification of the forces which of course was arrived at after twenty years of work."

The Big Bang Theory

One of the most striking developments in science over the past decade has been the linking of particle physics with cosmology. The search by physicists for a single theory that will describe the fundamental nature of all matter as we know it has led to an intriguing picture of the early Universe.

The predominant view among cosmologists is that the Universe began with a Big Bang which threw matter out in all directions, and that it has gone on expanding ever since. The first observation to suggest this was made by Edwin Hubble in the 1930s when he discovered that the galaxies were moving away from the Earth and each other. He had found that we inhabit an expanding Universe. An explanation to account for this was that the Universe began with a giant explosion — the Big Bang. However, it was not until the 1960s, when the existence of a microwave background radiation was discovered, that Big Bang theory was finally accepted. This radio signal corresponding to a temperature throughout the Universe of $3°K$ was the lingering remnant of heat from the initial explosion.

In the first moment of time, everything in the Universe was compressed into an unbelievably dense form. So crushed was matter that physicists have to start the clock of time some few moments after zero because the laws of nature will not extend to that point of infinite temperature and mass. It will be a few hundredths of a second later that the undifferentiated soup of matter and radiation begins to resemble the rapid collisions of particles in an accelerator. But here in the early Universe the temperature is at billions of degrees and particles are annihilated and formed in continuous collisions. Matter and radiation are in a constant flux. As this unimaginable fireball of energy expanded in size so it cooled — and has gone on cooling until now, when there is just a trace of heat, at $3°K$.

It was the cooling of the Universe which brought the familiar matter of atoms and molecules into existence — after about one hundred thousand years the temperature had dropped sufficiently for the nuclei to combine with electrons. Physicists have been able to construct a table of time versus the size of the Universe and then create a model of the Big Bang which charts the appearance of the high-energy particles as they are found in the falling temperature of the Universe. Today the Universe is between ten and twenty billion years old and its temperature so low relative to the primordial soup that matter is organised into stars, galaxies and life — frozen into a state of existence.

	TIME	TEMPERATURE	SIZE
Big Bang	10^{-43}	10^{33} °K	10^{-4} cm
	10^{-37}	10^{28} °K	1 km
	10^{-11}	10^{16} °K	1 million km
Today	10^{18}	3° K	10^{22} km

An Asymmetrical Universe and a New Particle

In this view then, the present world is full of interactions that are asymmetrical, but at the moment of creation, in the first nano-seconds of the Big Bang, all interactions were united in one universal force. As the Universe has cooled, so this symmetry has been broken to reveal the distinctive forces of gravity and the weak nuclear, the strong nuclear and the electromagnetic force. The difficulty has been in achieving a satisfactory mathematical formulation that would accurately describe both the symmetrical as well as the asymmetrical situation.

In 1967, Abdus Salam was able, with Steven Weinberg, to provide a unified theory that became the paradigm for much future work. Although they had been working independently, their ideas made use of several important conclusions that had been reached during the previous years. One of these was the idea of spontaneously breaking symmetry. But how does symmetry break spontaneously?

"Suppose you have invited twelve people to a dinner party. They are all waiting to sit down at a large round table. Now if you don't know the rules, you could sit anywhere and take the side plate from your left or right.

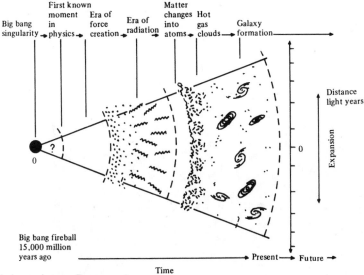

A view of the Big Bang explanation to the origin of the Universe. The unimaginable nature of matter in those early few moments of time is where the present known laws of physics stop.

This is the symmetry situation. But once someone has made the first choice, everyone else has to follow suit. The symmetry of left-right is 'spontaneously broken'."

The acceptance of the new theory of unification was slow in arriving. "In 1967, when our work was completed, there was no notice taken of it at all. I remember lecturing in 1968 at an international conference in Sweden and a very eminent physicist responsible for preparing the conference summary did not even consider the subject worth mentioning. So you can be ignored for the good reason that your ideas are not in the stream of things."

However, the theory had made predictions that could be verified by experiment. The most revealing of these was that a new particle exists at extreme energies. To test this theory they had to convince the experimental physicists working on the great particle accelerators to build new equipment. To create, in principle, conditions that will be similar to those first few moments in the birth of the Universe.

"I have been describing how unification of the weak nuclear and electromagnetic force could occur in terms of the Big Bang and the temperature of the Universe. Now, of course I do not need this concept in

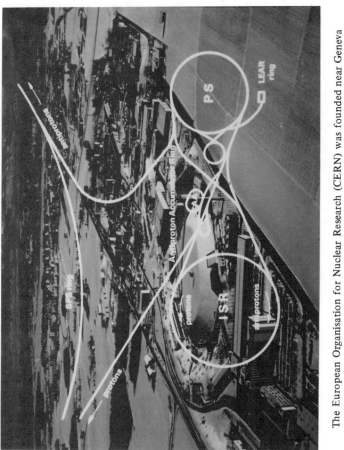

The European Organisation for Nuclear Research (CERN) was founded near Geneva in 1954. The original site in Switzerland was doubled in size by the addition of land across the border (crossed-line) in France. The path of the underground proton synchrotron, SPS, 2.2 km in diameter, is drawn on the photo.

terms of proving the theory because we are going to make that environment in the laboratory. If it occurs in the Universe, fine. If it did not, it would not worry us. We are looking for temperatures of the order of 10^{15} degrees and we can produce them in the particle accelerators. In the European laboratory of CERN, outside Geneva, a 6-kilometre-long track of magnets is used to accelerate particles to enormous velocities. Accelerating protons and antiprotons to these energies is equivalent to giving us that environment of high temperature."

In the late 1970s experiments were conducted at the Stanford Linear Accelerator Laboratory, in California, that confirmed one part of the necessary conclusions of the theory. But it was not until 1983 that final confirmation was obtained with the discovery that the predicted particles — the intermediate vector bosons — did exist. Called W^+, W^- and Z^0, these particles were seen for a few fleeting moments under the cosmic conditions of the CERN accelerator. But this temporary existence was enough to demonstrate that the unification theory was an accurate description of the fundamental nature of matter.

In an unusually bold decision, the 1979 Nobel Prize in Physics was awarded to Abdus Salam and two other theorists, Steven Weinberg and Sheldon L. Glashow, for their work on unification: four years before the confirmatory experiment at CERN. In the intervening years, the work had continued on unification theory and had developed in mathematical terms to show how the third unification with the strong nuclear force may be achieved. But the experimental proof of the third unification was yet further away.

"When we come to the unification of the strong nuclear force with the electromagnetic and the weak nuclear force it is to take the same analogy of ice and water; can we include steam? The transition temperature where these forces unite would be of the order of 10^{28} degrees Kelvin. But if you were going to make an accelerator which could have the type of energy available to produce that kind of environment, then it would need to be so large it would extend from the Earth to the edges of the galaxy." However, this further unification allowed for an indirect experimental proof. The theory had predicted that the fundamental particle, the proton, would decay. But why?

"The proton and the neutron possess the strong force: the two make up one family of particles. The neutrino and the electron possess the weak force, another family. Now if you wish to believe that there is a unification between these two forces, then these two families have in some way to be part of one bigger family. And if they are from the same bigger family, then one member can substitute for another — the particles are inter-convertible under suitable conditions. The implication of that is that eventually the protons (or the neutrons) could disappear and turn into anti-electrons or neutrinos.

"So far in physics, it had been assumed that the proton was a funda-mentally stable particle and that it exists for ever. Indeed if it did not live a long time we would not be here. If the protons and neutrons disappear there is no nucleus, no atom and therefore no matter as we know it. That is why the existence of life was the first proof that the proton lives at least 10^{10} years. The time it has taken for the Universe to produce life.

"Our calculations, however, showed that actually the proton lives much longer. In fact only after 10^{30} years can the proton turn itself into an anti-electron. Or put another way, if we take ten tons of matter, any type of matter, and we look upon the protons contained inside it, we shall find in a year's inspection that one proton out of the ten tons of matter would disappear. This proton turns itself into an anti-electron plus photons. So the task of an experimental proof is to detect that one proton disappearing."

As with Abdus Salam's other theoretical work, it would have to wait for the experimental physicists to be convinced enough to undertake the experiment. The major experiments in particle physics are planned years ahead and employ a great number of specialists. And often, with com-pletely new ideas, the theorists have to wait for new equipment to be built. "I remember there was a large conference in 1974 in London and I spent all of it trying to buttonhole people and persuade them to do this experiment. It was only when the other theory of the unification of electricity with the weak nuclear force was confirmed in 1978, that a stampede began to test this new theory in which the three forces were brought together.

"In 1980 a joint Indian-Japanese team started working in a disused mine in the Kolar gold-fields in southern India. There, 7000 feet below ground, they have this beautifully air-conditioned chamber with 150 tons of iron sheets placed inside. The reason for this depth was to find an environment where you are free from any background of stray events that may ape the results you are looking for. To catch a proton in the act of turning itself into an anti-electron you surround the material, in this case 150 tons of iron, with very sensitive photo multipliers that can detect the minute flashes that occur at the transition. With those 150 tons we expect the counters to click just two or three times in the whole year showing that a few protons have disappeared. Now two or three times is not enough statistically, you must have hundreds of such events before you can really say that the protons do decay. And if they do decay, then three forces will have been united into one single force."

The decay of the proton could offer solutions to some existing problems. One of the puzzles of modern cosmology is the absence of galaxies with any significant amounts of antimatter. We can observe vast amounts of matter throughout the Universe, but what has happened to the antimatter that necessarily must have been in existence in equal amounts at the moment of creation? If there is an asymmetry or an unequal behaviour in the operation of the weak nuclear force between matter and antimatter, then antiprotons may decay faster leaving just protons. So while there may have been the same number of protons and antiprotons in the early Universe, as it expands over time and the antiprotons decay faster, you are left with a largely proton-filled Universe. But this decay must also mean that eventually everything will disappear. In 10^{30} years half the atoms will lose their protons; it will certainly be the end of life as we know it.

Although a number of laboratories began experiments they have yet to provide convincing evidence of proton decay. The Kolar gold-field experiments claim to have seen three events of proton decay. A group in Japan at Kamioka claims two events. A group working at Irvine, Michigan, and Brookhaven, Mississippi, with the largest sample of protons under view, has found no decay candidates. They claim that protons live at least as long as 10^{32} years. Thus proton decay for the moment is uncertain, but this may change. This is science, not dogma.

Science and the Future Generation of Physicists

"It is just not true that when someone like myself gets up and proposes a new theory that I expect I will be right. It may never happen. The proton may not decay. You may be as eminent as you wish, but the youngest member of your audience can get up and contradict you and perhaps be right." It is this perhaps which adds to the excitement Abdus Salam feels about science. For him the challenge of providing new ideas and then putting them forward for others to analyse the test is a constant source of pleasure. He enjoys sharing this enthusiasm with his friends and family.

"I would like to interest my own children, as my father did. Regretfully, they do not listen to me. I tried to bring up my youngest daughter as a physicist. She did do physics at school, and I remember in 1973, when we worked out the theory of proton decay, telling her that I thought the proton was unstable. Well, she went to her A-level teacher telling her this was what her father said. The teacher said, 'My dear girl, whatever nonsense your father teaches you at home, don't put it in the exam paper or you will fail.' And she did fail. And then she took up literature." He enjoys telling this story and although perhaps unable to galvanise his family into scientific endeavour, he has not failed with colleagues and students.

"My whole background from Pakistan was that of mathematics and theory, and I think the emphasis on symmetry is something which I have inherited from the culture of Islam. The belief in unity, in there being one simple cause for all the forces that we see, has a basis in my spiritual background." A devout Moslem, Abdus Salam sees his scientific endeavours as entirely in keeping with the Koran's teaching to 'reflect on the phenomena of nature' and to 'satisfy the sense of wonder at creation'. This shows in his eagerness to engage others. He has worked tirelessly to bring science to the Third World, and to improve the chances for others who come from a background similar to his own.

"For the majority of people like myself there simply was no experimental physics. Experiments cost money, and there are no equipped laboratories. You never get around to doing any experiments and the result is, as it was for me, that you receive a training that is entirely theory. One very stark example of this was told to me by someone who had been visiting Bangladesh. There, one of the schoolteachers was lecturing to the students on a chemical experiment. He said, 'Gentlemen, think of my finger as a test tube, and now I am pouring sulphuric acid into it,' and he showed the movement of pouring the acid onto his finger. 'And now I shall add some iron chips to it — see, this is sulphuric acid with iron chips in it.' This is the situation of experimental work in many of the developing countries. I, fortunately, had the great example of my father who took a vast amount of interest in my school work."

Abdus Salam also had a certain amount of luck. His success at school, in what was then the Punjab region of British India, would normally have led him into a civil service job. "I was very fortunate to get a scholarship to go to Cambridge. The famous Indian civil service examinations had been suspended because of the war and there was a fund of money that had been collected by the Prime Minister of the Punjab. This money had been intended for use during the war, but there was some left and he created five scholarships for study abroad. It was 1946 and I managed to get a place in one of the boats that were full with British families who were leaving before Indian Independence. Of course, if I had not gone that year, in the following year there was the partition between India and Pakistan and the scholarships simply disappeared."

In an effort to make education less of a chance affair for students from the developing countries, Abdus Salam has campaigned for more international aid for educational training. Since the 1950s he has been an active voice in the United Nations as a member of the Advisory Committee on Science and Technology and the Foundation Committee for the UN University.

In 1964 he achieved his ambition with the creation of the International Centre for Theoretical Physics at Trieste. At a meeting of the International Atomic Energy Agency in 1960 he had proposed the setting-up of a centre where scientists from different countries, north and south, east and west, could meet and exchange ideas.

"I certainly feel a responsibility to bring to the notice of the people in power in the developed countries the importance of science and technology for developing countries. I believe that science transfer should precede technology transfer to the poorer countries. But this is often neglected in the aid we receive from the wealthy nations. They forget that unless you have a manpower that is highly proficient in the sciences, the transfer of technology will simply not take place. We set up the Centre at Trieste with the help of the Italian Government and United Nations agencies to enable scientists from developing countries to work at the highest levels of research.

"After I had studied at Cambridge, I had gone back to Lahore in 1951 and was teaching there at the university. But as a physicist, I was completely isolated. It was very difficult even to get the journals and keep in touch with my subject. I had to leave my country to remain a physicist. It is the lack of this contact with others that is the biggest curse of being a scientist in a developing country. You simply do not have the funds, the opportunities, which those from richer countries enjoy as a matter of course. There are not the communities of people thinking and working in the same fields. This is what we have tried to cure by bringing people together at the Centre.

"The Trieste Centre is trying to provide help so that a scientist can remain in his own country for the bulk of the time, but come to one of the Centre's workshops or research sessions for three months or so. There, he will meet the people in his subject, learn new ideas and be able to return charged with a mission to change the image of science and technology in his own country. The majority of the activity at the Centre is concerned with nonparticle physics pursuits. We have begun a course where staff from CERN are providing a course in the use of microprocessors for scientists from all over the developing countries. This is the first such course being done anywhere in the world for these people.

"The most important point is that we in the developing countries have been sold by well-meaning, and perhaps not so well-meaning, people the idea that all we need is technology — borrowed ideas. Lord Blackett, my mentor at Imperial College, used to say, 'There is a world supermarket of technology, go and buy it and take it home.' That is just wrong. Technology simply does not take that way. You need to have in every country a core of people with discrimination at the least, and that is what we are hoping to provide at Trieste."

In its first year, the Centre was host to 150 people. Today, over 2200 visitors from around the world spend time there. All this has meant a greater demand on Abdus Salam's energy and less time for his own research. Despite this, he still manages to pursue the theory of grand unification and to provoke thinking on how it is that matter fits together.

The Final Step

"The next step in this unification is the final step which Einstein wanted to take first. Einstein wanted to unite gravity with electromagnetism. We have replaced electromagnetism by the combined electronuclear force. Can we now achieve the final step and unite this force with the gravitational force?

"Well, if we look at the events after the Big Bang, we believe that possibly the electronuclear force was indistinguishable from gravity when the Universe was 10^{-43} seconds old. There are two distinct ways of looking at this proposal and resolving the difficulties that exist in the mathematical theory.

"One way is to conceive of a Universe having eleven dimensions. This is a new idea. Instead of the four space-time dimensions which we appear to live in and which Einstein worked with, we try to conceive of an eleven-dimensional Universe existing before 10^{-43} seconds. Then around 10^{-43} seconds after the Big Bang there is a transition with seven of those dimensions being 'compactified' into small sizes of the order of 10^{-33} centimetres, and four dimensions remaining with infinite ranges. We cannot discover the seven 'compactified' dimensions by any direct means, but their indirect effect is what we see as the electronuclear charge. This is a very speculative view at this stage."

"A totally different idea is that space-time is indeed no more than four dimensions. But that when we go back to 10^{-43} seconds after the Big Bang, then at these very small distances of the order of 10^{-33} centimetres, we find a granularity in the structure of space and time. That granularity is like cheesiness, worm-holes, or little scoop cuts. And it is this cheesiness which persists to the present epoch as electronuclear charge. Now this idea also is highly speculative. We will need experimentation if we are to find a way that will determine which of these two points of view is correct. It need not be very elaborate experimental work. One suggestion is to look for the force of anti-gravity and if we discover such a force then indeed space-time does have seven extra dimensions compactified, as well as the four of space-time as Einstein conceived them."

Fifty years later, Einstein's problem of bringing gravity into line with our understanding of the other forces of nature remains. To date, the theories of unification of the electromagnetic and nuclear forces are the closest description we have as to how nature behaves. It has enabled the cosmologists' view of the origin of the Universe to be understood in terms of the ultimate building blocks of matter. Our conception of the very beginning of time and space has been brought closer to our grasp and to experimental test. It remains now for the physicists to take us still further to the boundaries of what is reality.

IV.2
SALAM THE MAN

The Lonely Scientists —
Thinking Ahead with Abdus Salam

*Reprinted from **International Science and Technology**, December 1964.*

Here and there among the multitude of people who are concerned about the dreadful dilemmas facing the underdeveloped half of the world are a few men who speak with special authority. They are products of the unindustrialized world, they speak for it, but they have also excelled in the West's own game of physical science. One of these men is Abdus Salam.

Salam is a 38-year-old Pakistani, a graduate of the Punjab University, a Moslem who is likely to include a quotation from the Koran in any public paper, the scientific adviser to the President of Pakistan. He is also a leading student of particle physics and one of the architects of the octet model, a fellow of the Princeton Institute, the youngest fellow of the Royal Society at the time he was appointed, professor of theoretical physics at Imperial College of Science and Technology in London. When I talked to him in London he was just about to take over the direction of a new international institute of theoretical physics.

Salam has a warmly personal manner. His speech conveys the impression of a man with a thorough command of a language not quite native to him — a new subject brings a pause, almost a stammer, while he gathers his thoughts, then an enthusiastic rush of talk. After we had spent a few getting acquainted and had settled down with cups of rich Turkish coffee, I asked my first question:
Is there any opposition between the character of an Asiatic society and the spirit of modern technology?

I would like to say no. Take Japan. But ... let me limit myself for the moment to Pakistan. Now Islamic society was highly technological in the 11th and 12th centuries when the Arabs were strong in the sciences. Even later, in the Turkish days, the technology of the Turks was not poor compared to the emerging European states. After I have said this, of course, I have to admit there are a number of factors in the way life is organized which will have to change if Asia is to become technologically modern.

The question is how.

Up to a level technology is easy. After one starts living in a technological society, one develops a sort of contempt for the thing. It's not hard. It can be acquired easily once the mental attitude towards it changes. It's not like scholarship which needs a long tradition to develop. Now take tradition: I keep telling my boys in Pakistan — Do not despair if you do not produce, for example, mathematicians like Hilbert. You still might produce mathematicians like Ramanujam. Ramanujam was, if you like, relatively untutored . . . .

Intuitive.

An intuitive person, one who could be produced anywhere at any time given a minimum of mathematical training. We could not produce overnight the solid tradition of scholarship which is typified by a Hilbert or a Weirstrass or a Gauss. Luckily, most technology does not need that long tradition of centuries; scholarship does.

But where did you yourself come from?

I come from Pakistan.

I mean — how did Pakistan produce someone like you?

I don't put myself in the category of Hilberts. My subject is theoretical physics, and theoretical physics at the moment is in an intuitive state. It's at a stage when we are sitting on the top of experiments. We are utterly impatient. We don't want to wait from one resonance to the next. As soon as three resonances turn up, we make a complete theory. That theory is upset tomorrow; we don't worry; we start all over again. If you make mistakes, you don't worry. That is the intuitive milieu in theoretical physics. You need different types of gifts; you need good imagination, intuition, perception, seeing a correlation between facts. You do not need that long tradition of erudite knowledge.

This is a temporary condition, of course.

Probably, in another few years things will change; the basic laws will have been established; the thing will beome classical, less exciting. We shall need duller people with deeper scholarship. This just illustrates my point. For in technology, you are not looking for depths of erudition. Apart from a few basic things, the faster technology changes, the better it is.

How do you make it start?

The most important step is breaking the mental barrier. You see, in my country, you preach a thing for five, six, seven years. You go on talking. Nobody listens. And suddenly you find .... For example, take the civil service in Pakistan. The civil service was a legacy of the British empire — men with liberal education, responsible for law and order and revenue collection. Sterling men, first rate administrators. But men with no appreciation of engineering, technology, or science. Not the men best suited for development. I personally do not wish the system to perpetuate. It is the sort of thing we have been crying out against for five, six, seven years. But in the last few years, suddenly we begin to find that a majority of the civil service men are sending their own sons to read physics, chemistry, mathematics, engineering ... for research. You begin to wonder whether the barriers have suddenly begun to fall down.

What are the numbers like? How many young Pakistanis are studying technical subjects?

Let's take the PhD level. And let's concentrate on those being trained in the USA or the UK. Through our Atomic Energy Commission, which does not merely function for the atomic energy program, we have succeeded in training, in the last three years, something like 500 men at the PhD level. Now that's a tremendous number for a country like ours.

Will these men go back to Pakistan?

Oh yes. They are all employees of the Atomic Energy Commission. They will go back to Pakistan. We are trying to throw them into the universities and into other spheres. So I should say we are taking care of the chemists and physicists and partly of the engineers. We are not taking care of biologists — that's a tremendous loss.

Not even agriculturalists?

Not at the moment. We have no organization to do what, for example, the Atomic Energy Commission does.

That seems absurd.

It's absurd — absolutely absurd.

There seems to be an element of old-fashioned intellectual snobbery in the choice of an education.

You are quite right. As a rule it's the glamour subjects which get developed first. That seems to be the pattern all over the world. It's something to be deplored, in the abstract, but you can do nothing about it in a free society. First of all, the boys get more attracted toward the glamour subjects. Second, the government always puts up more money for them. But I do not despair. Once we get the government and the public used to the idea of spending on science, once that tradition develops, then that brings in the second round; the biological sciences, the prospecting sciences, the ones which are economically important get their share too.

And meanwhile, the thing is to encourage the glamour sciences?

I am afraid one can't help it. There is a private enterprise in scientific selection, if you wish. One good man turns up in a country, and he has a bee in his bonnet — he only knows physics, he only knows nuclear engineering. These are the only things he can put his heart into. His force and the energy goes in this direction. What can you do? Stop him giving of his best? Ask him to go back and read medicine?

Often this seems to mean into theoretical physics.

I am glad you said that. For it brings me to the venture, just now closest to my heart. Theoretical physics is one of the few subjects where even a country with very little tradition of science can produce reasonably good people. Japan is a prime example; it wasn't as developed when theoretical physics started there. The Japanese school of physics preceded the heights Japanese technology has reached now. The same thing is happening elsewhere. There are one or two very good physicists in Turkey — one I know is commuting between Columbia and Ankara. A couple of very good Koreans; people from Lebanon; people from India, of course; a number of very good people. Some from Pakistan, some from South America — one or two outstanding men in Brazil, some very fine men in Argentina, and so forth. To my mind, these men are very much worth saving — not only because they are good scientists, but because they are the central . . . .

What do you mean, saving?

Saving for good science, within their own countries. They have the following problem. Theoretical physics is a subject in which — there's a biblical phrase which expresses it — in which speech is the important thing, not the written word. You have got to go around and talk with people and be in contact if only to learn that this particular mess of papers here on my desk is rubbish and these others are the important thing. You can scan the whole blasted lot and never find out what is important. But going to an active place for a day, you can easily get to know the significant from the insignificant. So a man living in isolation or with a small group has every chance of just deteriorating.

He has students but no teachers.

Yes. When I was teaching in Pakistan I had excatly the same problem. In Cambridge, and at the Princeton Institute I had done reasonable work. But at Lahore, I found myself getting out of the subject altogether. So when I was invited to a position at Cambridge, the only choice was to migrate, to make myself an exile. Now if somebody could guarantee those people who are living out there that they can maintain continuity, guaranteed continuity to come out once a year, for three months let us say, and to work in a stimulating atmosphere, they will stay on. They will not have the cruel choice of either giving up physics or their countries.

This is what your new institute will try to do.

This is the project which is dearest to my heart at the moment. You see, in the world of theoretical physics there is the Western group, the United States and Europe, and there are the Eastern European physicists. No one recognizes that they exist, but there is also this third set of people. They may be as good in physics as some in the West or the East, but they have very unequal opportunities.

Do they also represent a different way of thinking?

I definitely believe that every cultural tradition of the human family brings to science a different way of thinking. In theoretical physics I see some of the great Chinese physicists bringing the subject their pragmatism. Or take another example I was recently discussing with Oppenheimer — though he did not necessarily agree. In mathematics or theoretical physics I do not know of any great Jewish complex variable men, or real analysts, but there are great Jewish set theorists, group theorists, and number theorists. This must come from the great Talmudic tradition. We speak now about symmetries in particle physics. When great Negro physicists arise, I wonder, in my lighter moments, if they will introduce the concept of "rhythm" and "harmony" in elementary particles.

So you see a third group in both an intellectual and a political sense.

I do not wish to labor this point about intellectual diversity. But it seemed to me it would be an excellent idea to have an international institute of theoretical physics and one with a special emphasis on this need of underdeveloped countries. The idea started out at the Rochester Conference of High Energy Physics in 1960 with a remark by Mr. McCone. He was then Chairman of the US Atomic Energy Commission, and in his after-dinner speech, he said that it was time now to think of international particle accelerators. A few of us who met after the speech were commenting on it and we said, this is very fine, but let's start at least with a UN-run institute for theoretical physics.

How was the idea received then?

It was opposed at first by the UK. It was opposed by France, by Germany, by Australia, and by Canada. It had only a lukewarm support from the USA and the USSR. We had no friends whatsoever among the great countries. But the idea caught the imagination of the developing countries — then nothing could stop it at the International Atomic Energy Agency meetings in Vienna.

But it wasn't set up at that time, was it?

It was decided that government should make offers of sites for the institute, and the IAEA would choose the most suitable among those offered. A bad way of proceeding ... there was no rational discussion of the ideal place. The governments that made offers were Denmark — a million dollars for a building and about $100,000 towards annual cost; Italy offered buildings and a quarter million dollars for annual costs; we also had an offer from Pakistan and an offer from Turkey. The Italian offer was linked to Trieste. Financially it was the handsomest so the decision was that the institute will be set up in Trieste for four years, and after that the situation will be reviewed, and if necessary it may possibly go to a developing country.

Would that be desirable?

I would like to get the experience of seeing the thing run first. Trieste has some attractions. Somehow Eastern Europe is nearer; it's a semi-international city. Already we have had a tremendous demand for fellowships and senior positions from Eastern Europe — from the Poles, the

Hungarians, the Rumanians, the Yugoslavs; also from South America; also from Asia. The institute will function at the beginning with a senior staff of about fifteen to twenty-five (most PhD) fellows. There is also a new type of fellowship we have instituted. We call it "associateship". The "associates" — several dozen of them — are men from developing countries with the privilege of coming to Trieste at periods of their own choosing from one month to four every year. We shall pay their stay and travel.

It does seem strangely remote from the practical world as a way to help a developing country.

Let us not confuse the full problem with a part of it. I did not suggest this was a panacea for all scientific deficiencies of the poorer countries. If I were an administrator in charge of science in Pakistan, I would certainly do my utmost to stress the basic agricultural and biological sciences. But let me make no bones about it. One needs, in addition, good scientists, first-rate scientists in pure subjects too.

The important thing is to develop a scientific tradition, no matter what science it is?

It's not that. In a free society it is a matter of the example. You must not underestimate what a great physicist can do for the morale of young people in a developing country. They come flocking in to read sciences, rather than literature or law. And another aspect of precept and example, I am pretty certain that this institute, if we can get the word spread that it is functioning the way we want it to, will breed a network of international institutes in other subjects — in practical subjects, like plant breeding or tropical medicine. Idealistically, it's the beginning of a UN university. So I don't despair.

A slow process. A generation.

You don't need a generation. In some ways things are easier in a poor country; they happen in four or five years. That's a generation for us. With us the people you are trying to convince are few — paradoxically perhaps, but the pace is faster. — R.C.

A Man of Science — Abdus Salam
by Nigel Calder

Reprinted from *Science Year: The World Book Science Annual*, (1967).

One summer noon in 1940, Abdus Salam came cycling into Jhang, a country town in the Punjab region of British India. The townspeople had lined the streets to greet him because, at the age of 14, he had just made the highest marks ever recorded in the matriculation examination of Panjab University. The result was a national sensation, but nowhere more than in Jhang, which had so little tradition in schooling.

From that moment, Abdus Salam was public property. Scholarships were to relieve his family of the cost of his further education, first at Panjab University's Government College, Lahore, and later at St. John's College at Cambridge University in England.

Salam was to astonish the ablest men of his time and become a leader in theoretical physics. Today, at 41, he is international property. He directs the new International Centre for Theoretical Physics (ICTP) in Trieste, Italy, on leave of absence from Imperial College of Science and Technology, University of London, an institution similar to the Massachusetts Institute of Technology (MIT), in the United States. He is also chief scientific adviser to the president of Pakistan and one of the "wise men" entrusted by the United Nations (UN) with guiding the application of science and technology to the global war on poverty. But such public recognition says little about the man, or about his role in the world of physics.

Of course, Salam had been a child prodigy, but even his talents could have been stifled by neglect in his corner of the world. The boy had been lucky in his family circle, which has a long tradition of piety and learning. His father was a minor official in the farming community along a tributary of the mighty Indus River that gave India its name. Each day, when the boy came home from school, his father would question him closely on what he had studied. And if any other encouragement was needed, his maternal uncle, a former Moslem missionary to West Africa, supplied it.

As Salam's education proceeded, the traditions of Islam were complemented in his mind by Western studies. He read English literature as well as the *Koran* (sacred book of the Moslems). His prime subject was mathematics, but that would not have been sufficient to save him from the natural destiny of ambitious young men in his country — entry into the civil service. World War II had put a moratorium on new appointments, however, so, in 1946, he went to Cambridge University to continue his studies.

Cambridge captivated him, especially the flower gardens of St. John's. Later he was to turn down a fellowship at neighbouring Trinity College, considered the best college in Great Britain, for aesthetic reasons — the Trinity grounds were not as pleasing as those at St. John's. He became a *wrangler* (Cambridge's traditional term for a first-class mathematician) without much difficulty. Thereafter, Salam followed the advice of Fred Hoyle, the cosmologist, and took a course in advanced physics. "Otherwise," Hoyle told him, "you will never be able to look an experimental physicist in the eye."

Salam did more than take a course, he became a research student in experimental physics in the famous Cavendish Laboratory of Cambridge University. That move could have been a mistake. Salam was no good in the laboratory. He would get bizarre results in his experiments and explain them by inventing a new theory. He importuned the Cambridge theoreticians for something more to his taste. The rare self-confidence and fastidiousness of the young scholar demanded that he question the deepest qualities of nature.

To a Moslem mystic, Allah is to be sought in eternal beauty. And for Salam, beauty comes through finding new, subtle, yet simplifying patterns in the natural world. Anything that threatens to confuse the issue seems to him ugly, filling him with an almost physical revulsion and driving him to clean it away, much as one would remove mud from a shrine.

His first major piece of research, done at Cambridge, completed a vital cleaning operation to get rid of an absurdity in physics. In previous theory, there was nothing to stop an electron from having an infinite electric charge. With great insight, physicists Julian Schwinger, Richard Feynman, and Freeman Dyson had indicated how the difficulty could be overcome, but the complete mathematical proof was lacking. That, Salam supplied.

During the period in which Salam has been active, since the late 1940s, physicists have shredded matter into smaller and smaller pieces, and have proposed new theories to explain them. In all the great advances, Salam has been in the thick of the action. Three of his contributions have been exceptionally important, and illustrate his quest for order.

The first had to do with parity — a theory of physics concerning the symmetry between an event and its mirror image. When a radioactive atom throws out an *electron* (beta particle) it also ejects that most elusive of particles, the neutrino. Both particles spin as they go, and the natural assumption was that the particles were just likely to spin to the left as to the right. At a conference in Seattle, Washington, in 1956, the Chinese-born American physicists Tsung Dao Lee and Chen Ning Yang suggested that such parity of left and right might possibly not be the case.

The startling proposal, which challenged a 30-year-old law on the conservation of parity, nagged Salam as he flew back to England from the Seattle conference. If the ugly, irregular idea of "parity non-conservation" was to be tolerable at all, there had to be a beautiful explanation for it. He recalled that no one had satisfactorily explained why the neutrino had no mass. Any particle will tend to interact with its own field and thereby resist acceleration which is what we mean by mass. Salam saw that nature could dodge this outcome if the neutrino spun only in one direction — in other words, if parity was violated.

More precisely, parity violation had to balance parity conservation exactly. That would mean that of the electrons emerging with neutrinos from radioactive cobalt-60 atoms, an average of three electrons would spin one way to every one that spun the other. By the time his plane had landed in England, Salam had it all clear in his mind. Distinguished colleagues mocked the idea. In 1957, Chien-Shiung Wu of Columbia University in New York City performed the celebrated cobalt-60 experiment that proved the violation of parity — God was left-handed, as Austrian physicist Wolfgang Pauli put it. For every three electrons spinning to the left, one spun to the right, just as Salam had predicted.

But Salam, like many other physicists, was already after bigger game. Could the bewildering variety of particles be elementary? Or, was it, as

Salam asked, that "some are more elementary than others?" The best thing was to seek family groupings, enabling one to say that if one particle exists, then others should exist with familial properties — similar but not identical.

The breakthrough began in 1960, when Yoshio Ohnuki of Nagoya University in Japan introduced the idea of "unitary symmetry" that might exist among particles. It started with the notion that most particles are made up of three entities which are themselves related to one another. Salam was the first non-Japanese physicist, perhaps in a sympathy of oriental minds, to accept the idea. Thus Imperial College, where Salam was professor of theoretical physics, became the centre of development of unitary symmetry.

Salam and John Ward, a visitor to Imperial College, used it in April, 1961, to predict an eightfold family of new particles having twice the spin of the proton, duly discovered some six months later. A research student working with Salam, Yuval Ne'eman from Israel, went on to show that the chief heavy particles, including the proton and neutron, also formed an eightfold family. About the same time, Murray Gell-Mann of the California Institute of Technology came to the same conclusion. He used the symmetry concept to predict a very strange particle — the omega minus — and when that, too, turned up early in 1964, the unitary ideas were established.

The next great advance came from American theorists who extended the unitary symmetry idea to link up separate families of heavy particles into a dynasty of 56 particles. But this theory left out the crucially important ideas of relativity, which omission launched Salam onto his third major contribution to science. This time, working with two associates, Robert Delbourgo and John Strathdee, Salam introduced Einstein's "four dimensions" (three dimensions of space, plus time) to arrive at a still higher pattern. "We are never going to be surprised by the discovery of a new particle again," Salam commented at the time. The earlier theory, leading up to the omega minus, had flaws, and these were carried over into the new theory, as Salam's fellow physicists were quick to point out. The fact remains that the valid portions of the theory represent the highest level of patternmaking in particle physics. As Salam puts it: "We have now run out of indexes."

According to Moslem colleagues, physics, for Salam, is a form of prayer. But he also treats physics as great fun. He holds onto the problem in his mind like a dog with a bone, yet he manages to remain relaxed. He pours out ideas in a continuous stream in discussions with his colleagues. Occasionally, Salam is right — and then his triumphant "I told you so!" might be irritating to anyone who recalled the 99 others, voiced with equal conviction, that were wrong.

The intensity of feeling and humor that goes into his theorizing were illustrated once when he was ill. "I'm sorry," he told a colleague, "I can't do physics now because I can't shout back at you." Generally, Salam talks quietly, thoughtfully, and fluently in a husky voice punctuated by laughter. But he always takes a positive attitude to ideas. "Some theorists are nihilists," he complains. "They are very good at showing where ideas are wrong, but they do not offer anything in their place. I prefer to build." He thinks more or less continuously about the patterns of nature and their mathematical representation, looking for order and beauty. "A broken symmetry breaks your heart." he says. He begins his day at 5 a.m. Like the wise man in the proverb, he goes to bed early, too.

That, then, is the story of the scholarly Punjabi boy who became an outstanding physicist. But there is another Salam: the man of the world in the most modern sense, a man concerned with the politics and organisation of science, and with the terrible problems of poverty and backwardness in his homeland and in half the world.

In 1947, while Salam was finding his place in the unfamiliar world of Cambridge, the British dissolved their Indian empire and the new Moslem nation, Pakistan, came into existence. Four years later, at the age of 25, Salam went back to Lahore. He served as teacher of mathematics at his alma mater, Government College (1951 to 1954), and head of the mathematical department at Panjab University (1952 to 1954). He felt a duty to return home and work among and teach his own people. The move turned out to be unfortunate, although Salam did not give up easily. He spent three troubled years there before professional frustration drove him back to England. Reluctantly, he went down the "brain drain" which robs Asia of much of the talent that it so urgently needs. But he resolved to do all he could to save other young men from the "cruel choice" between homeland and profession.

At Lahore, the lack of facilities was the least of his worries — a theorist, after all, works with plain paper or a blackboard. But, the academic climate in Pakistan was wrong; science was ignored not only by the intellectual leaders of the new nation, but also by the brightest students. Salam, simply, was intellectually lonely. He dabbled fruitlessly in cosmology and the theory of superconductors. "You have to know what other physicists are thinking," he says, "and you have to talk to them. I feared that if I stayed in Lahore my work would deteriorate. Then what use would I be to my country?" Better to be a lecturer in Cambridge than a professor in Lahore.

Salam picked up the threads again with instant success. In 1955, he was asked to serve as a scientific secretary at the first Atoms for Peace Conference convened by the UN in Geneva, Switzerland. Like many others on that famous occasion, Salam was greatly moved and sensed the full strength of world science and its power to work great wonders for the benefit of all men. Two years later, he was chosen to found a department of theoretical physics at Imperial College. He was also elected the youngest fellow of Britain's most select association of scientists, the Royal Society.

Today, Salam is director of his International Centre for Theoretical Physics in Trieste. The possessive "his" is correct. Abdus Salam conceived the centre as a place where men from all countries could work alongside some of the most distinguished minds of physics. As delegate from Pakistan, he proposed its creation to the International Atomic Energy Agency (IAEA) in 1960, and he was himself appointed its first director in 1964. Advanced countries, such as France, Great Britain, the Soviet Union, and the United States, were cool to the idea at first, but they could not resist the enthusiastic support from developing countries that rallied behind Salam. The Italian government provided the greater share of the money for the centre's first four years, donated temporary premises and began work on a fine new building at the coastal resort of Miramare.

The advance that established the centre on the scientific stage and made it a magnet for the world's physicists was Salam's effort, along with Delbourgo and Strathdee, in carrying the unitary symmetry ideas forward. That work was announced within a few months after the centre had opened in October, 1964.

The centre, which Salam envisions as the first department of a UN university, provides a meeting place for leading theoreticians of East and West. In 1965, for instance, Salam arranged a year-long brain-storming session on the attempts to tame the H-bomb — to produce useful power from the hot, heavy gas. Out of the session, presided over by Marshall Rosenbluth, an American, and Raoul Sagdeev, a Russian, came something like an international policy for experimental work aimed at giving mankind access to an unlimited source of power.

Closest of all to Salam's heart is the centre's role in ending the loneliness of men working in the academically underdeveloped countries. Never again should any able theorist suffer the isolation that Salam himself felt when he went back to Lahore. From Africa, Asia and Latin America, professors and students come to spend a few weeks or months at Trieste, where they can "plug in" to the current excitement of physics, sample the latest ideas, and, most important of all, meet informally with the world leaders in the subject. One device pioneered by Salam is already being taken up in other institutions and has attracted special support from the Ford Foundation. It is the "association" plan whereby selected theorists in developing countries are given the privilege of coming to Trieste for three months a year, with the centre paying all the expenses.

The winter at Trieste is the time when many physicists come from the Southern Hemisphere during the summer vacation of their universities. It is for the scientists a time for renewal, an opportunity to communicate with kindred spirits. After teaching for four years at the University of Santiago, Chile, Igor Saavedra felt like "a squeezed lemon." He was tempted to accept a job in London, but the centre at Trieste opened in the nick of time, keeping Saavedra from joining the "brain drain." For East Europeans, Trieste is, above all other considerations, the only place in the world where effective collaboration is possible between the physicists of the East and the West. Salam is also gratified that, through the centre, important contributions to the subject by theoreticians from the developing nations in Africa have begun to appear.

Salam presides benevolently over the centre, aided by his deputy, Paolo Budini of Italy. Few of the visitors can know what battles Salam has fought and goes on fighting to make sure that the centre will survive. In February, 1967, for example, he took the night train to Vienna to talk to the governors of IAEA into extending the centre's life indefinitely. He did not succeed, and he did not conceal his wrath. In former days, a Moslem warrior would draw his sword; Salam unleashes his words. He subscribes to the Islamic tradition that patience is a virtue only up to a certain point, that gentle persuasion can be tried only for so long if you are striving for higher goals.

Abdus Salam's name means, literally, Servant of Peace. The ideal of human brotherhood cultivated in the abstruse mathematics and broken English of the Trieste centre finds a broader and plainer expression in Salam's work for the UN Advisory Committee on Science and Technology. Twice a year, he and 17 other men of learning spend 10 days together at one of the UN's centres — Geneva, Switzerland; New York City; Paris, France; and Rome, Italy. They try to specify ways in which scientific knowledge and technical skills can hasten the advancement of that half of the world now living in poverty.

The UN committee has produced a "World Plan of Action" for building up science and technology within the developing nations and for transferring technical knowledge to countries that desperately need it. The "wise men" have also named particular technologies, such as desalination and the elimination of disease-bearing insects, that need to be developed as fast as possible. Each member has his special preoccupations and enthusiasms. Abdus Salam is particularly interested in involving leading scientists of advanced countries in the problems of world development.

On behalf of his own country, Pakistan, he did just that in a memorable fashion in 1962. The magnificent irrigation system built in the Indus Valley during the British era had deteriorated. Many years of seepage from the great irrigation canals had waterlogged huge areas of farmland, while evaporation from the soil had caused salts to accumulate. When Salam explained the problem, the US government sent leading scientists, agriculturalists, and engineers to West Pakistan. After thorough studies, the team, led by Roger Revelle, then director of the Scripps Institution of Oceanography at La Jolla, California, and science adviser to the Secretary of the Interior, drew up a plan of wells and pumps for draining the land and washing out the salt. Several areas, each about a million acres in size are already successfully under treatment west of Lahore. Over 30,000 farmers have adopted this procedure, greatly increasing agricultural production in West Pakistan.

President Ayub Khan of Pakistan appointed Salam as his personal scientific adviser in 1961, and a close and informal relationship has developed between them. Salam is frank about the human impediments in Pakistan, as in many developing countries, where scientists may proffer constructive suggestions, only to find them ignored by the administrators or dismissed because of the lack of resources to carry them out. Salam's most powerful colleague is Ishrat Usmani, chairman of the Pakistan

Atomic Energy Commission. The commission has gone beyond its basic task of introducing nuclear power. It seeks to encourage general excellence among Pakistani scientists.

In the words of Usmani, "Most of the scientific effort in Pakistan is in a large measure due to Salam's imagination and the weight of his personality. Salam is a symbol of the pride and prestige of our nation in the world of science."

At the same time, Salam confesses that too little attention has been paid to food and agriculture and he is understandably prone to pessimism. In a forecast of the future, he has written: "Twenty years from now the less-developed countries will be as hungry, as relatively undeveloped, and as desperately poor, as today." Yet he recognises slow progress in some directions. In Pakistan, the undue esteem given to the arts, at the expense of science, is being broken down. The president himself has come to share Salam's passionate interest in the publication of better science textbooks. More young people are studying science at the universities.

Since childhood, when he watched the apothecary at Jhang concoct aromatic sherbets from the ancient book of Avicenna, the Persian philosopher-physician, Salam has taken a proud interest in the former glories of Islamic science and literature. He likes to recall the days when Baghdad and Moorish Toledo were, for a time, the world's chief centres of learning. Even today, his vision of the future of Pakistan is not confined to the satisfaction of material needs. "Once a nation starts to think of higher things," he says, "scholars must find a role in that society." During his visits to Pakistan, it is not unusual to find him surrounded by a group of poets reading their verses to him and finding him an appreciative and critical listener.

In keeping with strong Islamic tradition, "Charity begins at home", no young Pakistani seeking help or guidance from Salam is left unaided. His Western students, too, find him generous to a fault in his support of them.

Salam is frequently on the move from continent to continent, yet unlike many of today's jet set scientists he refuses to let public business deflect him from his personal researches. Conversely, in his advisory work in Pakistan and for the UN, he does not allow his scientific sophistication to dampen the simple passion of a man born in a poor community and who knows that he is perhaps the luckiest of all his countrymen.

On the wall of the director's office in Trieste, hangs an inscription of a 16th-century Persian prayer: "He cried: "O Lord, work a miracle!" Salam's strength is that he believes miracles are possible provided one goes out and helps them on their way.

Man of Two Worlds

by Robert Walgate

This first appeared in **New Scientist,** *London, the weekly review of science and technology (26 August 1976).*

Abdus Salam, in a lecture delivered last December to the students of the University of Stockholm, spoke with controlled anger of the exploitation of the Third World by the advanced nations. Piling fact upon fact, finally he burst out passionately with these lines of Omar Khayyam:

> *Ah love! could thou and I with fate conspire*
> *To grasp this sorry scheme of things entire*
> *Would not we shatter it to bits — and then*
> *Remould it nearer to the heart's desire.*

Salam, physicist, FRS, Moslem born by the banks of the Chenab, passionate advocate for the Third World, has the heart of a poet and the mind of a scientist. He loves beauty and looks for it in his science. He is an excellent physicist concerned with deep pattern; he is also a deeply compassionate man. These two threads intertwine through his life.

His work in particle physics has made many important contributions to his subject, not least the unification of two of the forces of nature — the weak and the electromagnetic — in a model which is receiving thorough experimental support. He commutes between Imperial College, London, and his creation, the International Centre for Theoretical Physics in Trieste, a centre where Third World scientists can keep abreast of developments in physics. At 50, Salam is full of energy, travelling all over the world to give lectures, make speeches, and — often successfully — persuade politicians to realise visions. He fell in love with the United Nations when he attended the first Atoms for Peace Conference in 1955, and helped set up the UN Advisory Committee for Science and Technology, of which he was an active member from 1963 until last year. And for eight years he was by personal invitation scientific adviser to President Ayub Khan of Pakistan.

He is direct, disarming, humorous, and deeply serious. He comes of a line of Rajput princelings converted to Islam about the year 1200. His forebears were scholars and physicians: but they were poor. Salam's Moslem upbringing gave him the mores of Islam, the moral code of the Koran, but it is relatively recently that he has come to a spiritual discovery of his religion. "Islam to me is a very personal thing," Salam says, "Every human being needs religion, as Jung has so firmly argued; this deeper religious feeling is one of the primary urges of mankind." But Salam does not consign to eternal hell-fire those outside the fold. "I would like you to become a Moslem, and share the feelings I have, but I wouldn't stick swords into you if you didn't!"

Salam does not believe that there is any conflict between his science and his religion. In physics, he has mostly been involved with symmetries; and "that may come from my Islamic heritage for that is the way we consider the universe created by God, with ideas of beauty and symmetry and harmony, with regularity and without chaos. The Koran places a lot of emphasis on natural law. Thus Islam plays a large role in my view of science; we are trying to discover what the Lord thought; of course we fail miserably most of the time but sometimes there is great satisfaction in seeing a little bit of the truth." Salam also stresses that from 750 – 1200 AD science was almost totally Islamic, and that, "I am simply carrying that tradition on."

"My father had not taken scholarship as a profession, but he was very keen that I should succeed that way. He influenced me very strongly in that respect." The best jobs in Pakistan were civil service jobs; but Salam took a maths degree in Lahore, won a unique scholarship to Cambridge, and while there "drifted into physics".

"There was no question I was very fortunate. If I had not been awarded a scholarship by the then Indian government it would have been totally impossible financially for me to come to Cambridge." The way Salam got the scholarship is to him "something of a miracle". During the Second World War, many Indian politicians wanted to help the British war effort. One of them collected a fund of about £15,000; but the war ended, and he had to decide what to do with the money. He instituted five scholar-ships for foreign education.

Salam and four others were selected. Salam had taken the good care to apply to Cambridge simultaneously; and "the same day I got the scholar-ship, 3 September 1946, I also had a cable saying that an unexpected vacancy had come up at St. John's College — admissions were usually done much earlier — and could I come up that October?" So Salam went to Cambridge, but his four colleagues who were to be offered places next year, never made it. The munificent politician died that year; his successor cancelled the scholarship scheme. "In the end all that effort to collect a War Fund, for buying munitions ended up in one thing alone: to get me to Cambridge!" Salam laughed. "Now one could call it a set of coin-cidences; but my father didn't believe this. He had desired and prayed for this and saw this — I think, rightly — as an answer to his prayers."

Salam emphasises the general moral. "Opportunities are so sporadic in the Third World that the man who is absolutely tops may not even get a chance." There is everything against doing science as a profession; "it is poorly paid, very little endowed. You have to be highly motivated if you take it up; it carries no influence or status in a status-conscious society."

In Cambridge Salam took the part II mathematical tripos and part II physics and came out a Wrangler — a first-class degree. The Cambridge tradition was that those with firsts did experiment, while seconds and thirds did theory. "But for experimental work you need qualities I totally lack — patience, an ability to make things work — I knew I couldn't do it. Impossible. I just hadn't got the patience."

Salam found his way into some problems in quantum electrodynamics, then a subject in the throes of birth (now the most accurate theory known).

"There were a few problems left," said his supervisor, "but all of those have been solved by Matthews." (Paul Matthews, now a professorial colleague of Salam's at Imperial College and shortly to become Vice Chancellor of Bath University. He was then finishing as a Cambridge research student.) "So I went to Matthews and I said — have you got any crumbs left?" Matthews gave him an important problem "for three months". If Salam hadn't solved it in that time Matthews would take it back. Salam solved it, and thereby made an important contribution to "renormalising" (eliminating infinities from) meson theories. It took five months. That was his Ph.D.

Salam returned to what was now Pakistan and to his old university of the Punjab in Lahore as a professor. There was no tradition of doing any postgraduate work; there were no journals; Salam's salary was £700 a year and "I certainly couldn't put the journals on that." There was no possibility of attending any conferences. The nearest physicist to Salam was in Bombay "and that was another country".

The head of Salam's institution told him that though he knew Salam had done some research he could "forget about it". He offered Salam a choice of three jobs: bursar; warden of a hall of residence, or president of the football club. "I chose the football club."

The whole tenor of society was geared against any continuation of research work in physics. Salam was faced with a tragic dilemma; "I had to make a choice; physics or Pakistan." Salam returned to Cambridge. There and subsequently at Imperial College London (where he was appointed professor in 1957 to start the department of theoretical physics) Salam threw himself passionately into physics, inventing the two-component theory of the neutrino, working on particle symmetries and in particular SU(3), and gauge theories with the unification of weak and electromagnetic forces as a goal. But, in addition to this work, his burning concern, fired by his own unhappiness at having to leave his country, was to find ways of making it possible for those like him, to continue working for their own communities while still having opportunities to remain first-rate scientists. "I believe passionately that developing countries need scientists as good as the developed countries do, certainly in the university system." So in 1960 Salam conceived the idea of setting up an International Centre for Theoretical Physics, with funds from the international community — for example, the UN.

To such a centre, those working in the developing countries would come with frequency to renew their contacts with physics, while spending the bulk of their time teaching in their own countries. The centre — rather than the developing country governments — would pay for such visits. Salam, after meeting a lot of indifference in the first world, finally convinced the International Atomic Energy Agency to take up the idea of the centre. Italy, the poor man of Europe, came up with the most generous offer of site and running costs and the ICTP was established in 1964 in Trieste.

After an experience of running the Centre for 12 years there has been a shift in the disciplines the Centre now emphasises, a shift away from fundamental physics to physics which may be more relevant to the needs of the developing countries — for example physics of the condensed matter. "We do post Ph.D. work, not with an eye to industrial laboratories — there are none in most of our countries — but the hope is that if you have teachers in the universities who have worked, for example, in solid state physics, then the next generation at least will have an orientation which is much more industrial.

"Thus we are stressing research in physics of solids, plasma physics, physics of oceans and the Earth, applicable mathematics; physics of technology, of natural resources, together with physics on the frontier. As an example, in solid state physics, Professor John Ziman of Bristol, Norman March of Imperial College and Stig Lundqvist from Sweden, Chiarotti from Italy, Garcia-Molinere from Spain and their colleagues have created (through the work they do at the Centre) a mini-revolution in studies of this subject in the developing countries. This is evidenced in the degree of scientific maturity we now notice in the people coming to ICTP compared to 1964."

Salam emphasises that "it is a most important point to make that — even in a relatively large country like Pakistan — the active physics community numbers no more than 50 persons for a population of some 70 million people. And this is the total sample of men who are responsible for all advanced teaching, for all norms and standards in physics, taught for engineering as well as for all advice to the government on matters concerning technology based on physics.

"Now considering that the active physics community is so small, one can argue whether the teachers we train should be high energy physicists or solid state physicists.

"Many people argue that we shouldn't do any fundamental science at all but concentrate on, say, applied physics of solar energy. Unfortunately things are not so simple. For solar energy research the need is there, but the money is not there, nor are the facilities.

"In the end it will be the US physicists, with the multi-million dollar facilities available to them, who will produce a design that is the epitome of all designs for economic devices in the solar energy field.

"But this does not mean we should not have men trained at the highest possible level in solar energy work, men who know from *inside* what the current work in this discipline is. Perhaps the ideal would be men who commute between fundamentals of solid state physics *as well as* its application to say, solar energy devices. I do not believe this is impossible. To be multidisciplinary in physics is the cross those working in developing countries must be prepared to bear. Another is the philosophy we are trying to live up to at ICTP."

Salam's concern for the Third World has not been confined to ICTP. He has struggled, from inside, with the educational, scientific, and development policies of Pakistan. But his first love has always been physics, with a life that is a tangle of physics and non-physics interests. "It's hard to switch, you find you are in the middle of something very exciting and then you must simply drop it."

Salam gave a current example. At present he is alone with a colleague, Jogesh Pati, in proposing that quarks can be free. It is the right psychological moment to develop the idea, for quark confinement is in theoretical difficulties. But with constant interruption of work through the demands on his time in running the Centre and keeping it alive, Salam bemoans the fact that he cannot spare enough time to develop his ideas.

Does Salam think he's got the balance about right? "Well sometimes I feel I'm being very foolish. I do what is necessary to achieve what I want to, but often less than that." Salam is a man with tremendous energy, and tremendous enthusiasms — but he is one man without time, strung across two worlds and two problems. It is a loss to the world that he cannot have two lives.

Abdus Salam

by John Ziman

Address on the occasion of the Honorary Degree of Doctor of Science by the University of Bristol on 2 July 1981.

Mr. Vice-Chancellor

Only connect! That is the theme that runs through the life and work of Abdus Salam. He has followed the teaching of Islam and has dedicated his life to the principle of unity — the unity of Nature and the unity of Mankind. As a natural philosopher he has seen that the various interactions of the elementary particles must be no more than diverse aspects of a single primary force. As a political and moral leader he has demonstrated that the various interactions of nations and cultures are no obstacle to the brotherhood of Man in science.

In the Faculty of Science, we honour him as one of the finest theoretical physicists in the world. In 1950, he was awarded the Smith's Prize at Cambridge for an outstanding pre-doctoral contribution to physics. Since then he has been continually at the working face of the deepest mine that science has ever pushed down into the bedrock of reality. He has had a major part in every act of the unfolding drama of the discovery and understanding of the primary entities of quantum physics. It is astonishing that a man who is also so active in public affairs should have published some 200 papers on the physics of elementary particles, and is still forging ahead in that intensely competitive and dynamic intellectual enterprise.

In fact, he is still so hard at it that I have not dared to engrave a tablet of his achievements in physics: tomorrow morning, a new experimental observation somewhere might add a whole new theory to the list. Salam has the great scientific gift of suggesting new, physically realistic, theoretical connections that are really worth the effort to confirm. The great theory of the electroweak force, for which he shared the 1979 Nobel Prize in physics, was first put forward 13 years ago. For the next three or four years it was totally neglected, because of apparently insuperable mathematical difficulties. When these had eventually been cleared away, some very delicate experiments were still needed to test the mathematical predictions against physical reality. I remember visiting him in Trieste at that exciting period, with telephone calls from continent to continent to check up on data that seemed, at first, to disconfirm his cherished hypothesis. Salam's personal enthusiasm for physics is delightfully infectious. It was a happy day for us, too, when his persistence was rewarded, and it came out right after all.

What that theory did was to show that certain well-known interactions between elementary particles — for example, the so-called 'weak' force that eventually drives every neutron to decay into a proton and an electron — could be treated as part of the much more familiar electromagnetic force that acts between all charged particles. It was a hard nut to crack. Compared with some of your modern mathematical physicists, Salam's methods are slightly old-fashioned. But he uses such magic sledgehammers as gauge fields and renormalisation theory with a delicate, practised hand. Faraday and Maxwell would have been delighted with his achievement, which is a bit like their unification of magnetism itself with electricity, more than a century ago.

It is good to see science unfolding in the traditional manner. That was a scientific breakthrough in the old style. It has opened the way to yet another revolution in quantum physics, with the goal of a grand unification of all the forces of Nature now in clear sight. Perhaps this is only a mirage — or perhaps another of Abdus Salam's imaginative schemes for the ultimate construction of matter and energy has got it right, and will again be confirmed by the observation of a predicted physical phenomenon that cannot otherwise be explained.

One such prediction of his current theories is that protons themselves, the building blocks of all heavy matter in the universe, should not live forever. Just like neutrons, they should eventually be driven to transform themselves into lighter particles and radiation by a tiny component of a universal force. Fortunately, it is a very small effect. Our present day protons should last a billion, billion times as long as the universe has already existed — which is surely a little longer than it might take for me to master all of Salam's theories fully for myself and then explain them very carefully to this assembly.

Perhaps, Mr. Vice-Chancellor, you will forego that pleasant exercise, and accept this worldwide scientific reputation as evidence of his eminent worthiness for an honorary doctorate in this Faculty. But first, let me present Abdus Salam to you in another aspect, as one of the first citizens of the World. He might be considered as just a leading British scientist, having been Professor of Theoretical Physics at Imperial College, in the University of London, for more than 20 years But in fact he spends a good deal of his time in Trieste, in Italy, and is a frequent visitor to the United Nations in New York. He is a sort of one-man multinational corporation, busily transferring intellectual technology to the less developed countries of the world.

His homeland is Pakistan, a country to which he remains deeply attached. He was born and brought up in the city of Jhang, not far from Lahore, that ancient paradise of Moghul palaces and gardens. From Government College, Lahore, a scholarship transported him to Cambridge, where he showed his mastery of all the mathematics and all the physics that any undergraduate could be permitted to study, and soon had his feet on the swiftly-rising escalator of research. On that brilliant early achievement and promise, he went back to Lahore as a Full Professor at the very tender age of 25. In fact, by ordinary standards of academic success, he was now all set for a comfortable career.

But the next three years must have been the most miserable — and formative — of his life. The old Government College was previously one of the leading academic institutions of British India — but there was little interest in scientific research. As Salam recounts it, the head of the college offered him the choice of three college jobs for any spare time he might have after his teaching duties. He could become warden of the college hostel, or chief treasurer of its accounts, or, if he liked, he could become president of its football club. He says he was fortunate to get the football club — though I suspect that rival clubs didn't feel that way!

The most severe deprivation of those years was the loss of contact with fellow scientists working on the exciting problems of the day. As he analysed it later, this was one of the main reasons for the dispirited research atmosphere in almost all the less developed countries.

Gifted men from countries such as Pakistan, Brazil, Lebanon or Korea work in advanced countries in the West or the Soviet Union. They then go back to build their own indigenous schools. When these men go back to the universities in their home countries, they were perhaps completely alone; the groups of which they formed a part were too small to form a critical mass; there were no good libraries; there was no communication with groups abroad. There was no criticism of what they were doing; new ideas reached them too slowly; their work fell back within the grooves of what they were doing before they left the stimulating environments of the institutions at which they had studied in the West or the Soviet Union. These men were isolated, and isolation in theoretical physics — as in most fields of intellectual work — is death. This was the pattern when I became associated with Lahore University!

Even a thoroughly self-winding genius such as a young Salam could not accept this danger of being slowly buried alive. In 1954 he came back to England, and was soon established in his chair in London. Although he never lost his close personal and professional contacts with his home country, and he takes special pride in being the first Pakistani to win a Nobel Prize, he has not returned to a regular academic appointment in that country.

But Abdus Salam is a man whose heart is as great as his mind. The memory of those anguished years of isolation did not turn sour within him: it became the creative kernel of his greatest achievement. He vowed to provide the means by which other talented young scientists, from less developed countries, could keep themselves from scholarly death by isolation, without having to desert their native lands.

A single line in his biodata records that he has been Director of the International Centre for Theoretical Physics at Trieste since 1964. There is more in that title than the fifty awards he has won from universities and national academies around the world. He created that Centre out of nothing: it is now one of the most successful and respected international institutions of our times. Scientists from developing countries come to Trieste to get the latest scientific news, to learn the latest techniques, and to meet their colleagues from both advanced and developing countries. They come to attend advanced courses, or to work quietly in the library, to argue vehemently with some very bright young chap from Indonesia, or to acquire understanding and insight from a very wise old Professor from Sweden. It is a bustling railway junction of the intellect, bursting out of the handsome buildings it acquired only a few years ago, managed by the brilliant improvisations of a devoted staff, and always short of funds. Yet it lives, and works, and grows, and serves the whole world of the physical sciences.

How was it done? How did that most abstruse of young professors persuade the hard-headed delegates of lumbering international organisations such as the International Atomic Energy Agency, and UNESCO, to put their money into such an out-of-the-way project? How did he so befriend the Italian Government that they gave such support in cash and kind? In these past few years of declining funds and proliferating bureaucracy how has the Centre managed to remain alive and flourishing in the interstices of a system that has brought much grander projects to frustrations?

The Trieste Centre was created, and continues to thrive, through a singular force — the personal willpower of its Director. Let me warn you, Mr. Vice-Chancellor, that Abdus Salam is a manifestation of that imaginary concept of mechanics — the irresistible force. Suppose he asks you to do a little favour for him — say a three-week visit to the University of Vladivostock. You will find you have only three possible responses. The first is, "But Abdus, that's completely forbidden by my religion. I shall be damned to eternity if I go to Vladivostock in August." The second is, "Very sorry, old chap, but all that month I'm absolutely committed to lecturing in Bogota." Most commonly, however, the only remaining degree of freedom is: "Yes — how do I get there?", and off you go. He seems to have that effect on everyone he meets — politicians, government officials, international bureaucrats and fellow scientists. He impresses and persuades by the integrity, purity, and singleness of his purpose, put into the service of his fellow men.

Originally, the Trieste Centre was for the highest of pure science, setting standards of excellence for Third World physics at the most advanced level. But Salam's own experience, both in directing the Centre and as a participant in science policy making for successive governments of Pakistan from about 1960 to 1974, taught him to widen the objectives of science for countries struggling for economic and social development. Over the years, the programme of associateships, advanced courses, seminars, workshops and conferences at Trieste has broadened, to foster and coordinate research in all fields of applicable science. Salam speaks now for the special role of the trained scientist in this development process and for the need for national and international scientific institutions which will make that role attractive and productive. He has thrown his personal charisma, and the immense prestige of his Nobel Prize into a worldwide campaign to establish the essential infrastructure, that can give aid and advice to the smallest and poorest nations in their efforts towards self-development.

In both spheres of philosophy, natural and social, Abdus Salam strives continually to 'connect'. Along that way he has already achieved such a unification of Nature, such a realisation of the ideal of human brother-hood, that it is very proper that we should do him honour, Mr. Vice-Chancellor, to present him to you as one eminently worthy of the degree of Doctor of Science — *honoris causa.*

2nd July 1981   Prof John Ziman, FRS

Science Sublime

The Greeks wished to explain all the phenomena of Nature in terms of four elements: fire, air, earth, and water. Modern science takes the urge to simplify still further. One of the most exciting and romantic endeavours of the twentieth century is the attempt to show that what we now know to be the four fundamental forces of nature — gravity, electromagnetism, and the strong and weak nuclear forces — are aspects of a yet more basic principle. This is the province of the particle physicists, the theoreticians and experimentalists who are concerned with the behaviour of electrons, protons, and the host of other subatomic particles that make up matter. It is a formidable undertaking.

The two forces with which we are most familiar are gravity and electromagnetism. At the beginning of the nineteenth century, electricity and magnetism were thought to be quite distinct, and it was the work of Michael Faraday and, later, James Clerk Maxwell which showed that they were basically two aspects of the same phenomenon. Maxwell's four equations provide a complete description of the production and inter-relation of electricity and magnetism, and subsequently led to the development this century of what are known as the gauge theories. Gravity was, of course, first recognized by Newton, and shown by Einstein in his General Theory of Relativity to be an expression of the curved geometry of space and time. Einstein tried until the end of his life to unify electro-magnetism with gravitation but, like everyone since, failed to do so.

The recognition of the existence of two more fundamental forces — the strong and the weak nuclear forces — came with the discoveries about the structure of the atom in the first decades of this century. It became clear that an atom is not an indivisible sphere like a billiard ball, but consists of a small central region — the nucleus — surrounded by a cloud of particles called electrons. The nucleus itself is composed of particles called protons and neutrons, which are bound together by the strong nuclear force. But atomic nuclei are not always stable. During certain kinds of radioactive decay, for example, the neutrons are transformed into protons, electrons, and elusive particles called neutrinos. Transformations of this kind are governed by the weak nuclear force.

In the late 1950s, Abdus Salam began to address the problem of unifying the weak force and the electromagnetic force. On the face of it, the two forces are very different. One, the electromagnetic force, is a long range force which can be felt at almost any distance. The other, the weak nuclear force, is a short range force acting over unimaginably short distances. But Salam succeeded. In 1979, together with Steven Weinberg and Sheldon Glashow, he was awarded the Nobel Prize for Physics. Their work was mathematical and theoretical, but predicted that certain, as yet undiscovered, particles should exist. It was only in 1983, under the cosmic conditions created in the huge particle accelerator at CERN that these particles were detected, and the theory finally confirmed.

Salam was born in Pakistan in 1926 and went to Cambridge to read mathematics at the end of the second world war. Given his subsequent achievements, he had come to the right place. As an undergraduate, he attended lectures given by the distinguished theoretical physicist Paul Dirac, who had been awarded a Nobel Prize for his work on quantum mechanics and antimatter. After taking his first degree, Salam went on to do research at the Cavendish Laboratory where much of the work on the structure of matter had been done. He is now Professor of Theoretical Physics at Imperial College, London, and Director of the International Centre for Theoretical Physics in Trieste, which he founded to enable young physicists from developing countries to enjoy short, intensive periods of research and international contact. He returns frequently to Pakistan, and is a devout Moslem.

I interviewed Salam at his home in South London. We could hear children playing in the next room and the familiar sounds of domestic life in the background. I could not help but be aware of a contrast between the sheer ordinariness of our surroundings and the intellectual reach of the ideas we were to discuss. Yet in his life, as well as in his research, Salam is able to accommodate seemingly disparate elements with ease:

the familiar and the arcane; a commitment to physics in the developing world and to his own work on unification; to science and to Islam. I wanted to learn something of the route which had brought him from a peasant community in Pakistan to international acclaim in particle physics.

I was trained by my well-wishers, and my father in particular, to think of the Civil Service as a career and it was simply an accident that I became a particle physicist. The accident of the second world war. If the war had not been going on, the Indian Civil Service examination would have been held in the crucial years in which I had to make up my mind for my career and I would by now be a good Civil Servant.

"You had no idea of doing science, then?"

No. It was an accident in the following way. The war had stopped the Civil Service examination. Just after the war the examination was still stopped and having done my MA in mathematics at the University of Lahore, I was given a scholarship to read mathematics further at Cambridge.

"So you had a scientific bent from quite a young age?"

Well, there was a scientific bent all right, but the point was that I was taking mathematics not to do research, but in order to score very high marks in the Civil Service examination. It was a mark spinner if you like.

"So it wasn't that you had a passion for science?"

No. I was certainly *good* in science. In fact, I was recalling the other day that I had written my very first research paper when I was about sixteen years of age and it was published in a mathematics journal. So the research mindedness was there, but there was no motivation for it. But after two years of Cambridge, of course I was hooked on research.

"But I'm not quite sure how you got to Cambridge."

I got to Cambridge by means of a scholarship from a Small Peasants' welfare fund which was set up by the Prime Minister of the State of Punjab at that time.

"Did you come from a peasant background?"

That's right. Although my father was a Civil Servant, he had a small parcel of land and he qualified. So I got one of those scholarships and the interesting thing is that only five scholarships were offered, and the other four people who got them could not get university admission that year. Then came the partition of the country and the scholarships disappeared. So the entire purpose of that fund and those scholarships seemed to be to get me to Cambridge.

"Did you really think that fate was playing a hand? After all, each of these events was very much a matter of chance."

Certainly my father, who was a deeply religious man, always said that this was a result of his prayers. He wanted his son to shine in some field. Of course, in the beginning he was thinking of me as a Civil Servant, but when I decided that I was going to do research, he felt that this was something very appropriate and encouraged me. But the whole sequence of events, my getting a scholarship at the right time, my getting to Cambridge at all at the right time, and then being interested in science, was all, he thought, very much a part of something deeper.

"When you got to Cambridge did you become immediately involved in theoretical physics?"

No, I started in mathematics because I had the mathematical background but, slowly and gradually, during the two years in mathematics I shifted over to theoretical physics. Dirac was lecturing at that time and I attended his lectures. Then I still had a third year free in the sense that I had the scholarship and the choice of whether to go on with higher mathematics — that's Part III of the mathematics tripos — or to do the physics tripos. One of my teachers was the astronomer Fred Hoyle, and I went to ask Fred his opinion about this. He said "If you want to become a physicist, even a theoretical physicist, you must do the experimental course at the Cavendish. Otherwise you will never be able to look an experimental physicist in the eye." And that was very correct advice. But it was a hard year for me doing experimental work at the Cavendish, not having done any for so long. It was the hardest year of my student days.

"What did you find so hard about it?"

The whole attitude towards experiment. It's very interesting. In the Cavendish there used to be a tradition that you were not given any fancy equipment. Just string and sealing wax. You were given every discouragement and you had to overcome this. Now the very first experiment I was asked to do was to measure the difference in wave length of the two sodium D lines, the most prominent lines in the sodium spectrum. I reckoned that if I drew a straight line on the graph paper then its intercept would give me the required quantity I wanted to measure. As you know, mathematically a straight line is defined by two points and if you take one other reading then mathematically that should be enough since you then have three points on that line, two to define the straight line and the third one to confirm. So I spent three days in setting up that equipment. After that I took three readings, and I took them to be marked. In those days the marking of experimental work in the class counted towards your final examination. Sir Denys Wilkinson, who is now Vice Chancellor of Sussex University, was one of the men who supervised our experimental work, and I took it to him. He looked at my straight line, and said "What's your background?" I said "Mathematics". He said "Ah, I thought so. You realize that instead of three readings you should have taken one thousand readings and drawn a straight line through them". I thought "I'll be damned if I can go back and face those three days again". I had by that time dismantled my stuff and I didn't want to go back. So I tried very hard to avoid Sir Denys Wilkinson — he wasn't Sir at that time — during the rest of the year.

I still remember the day the results came out in 1949. I was looking at the results sheets hung in the Cavendish and Wilkinson came up behind me. He looked at me and said "What sort of class have you got?" and I very modestly said "Well, I've got a first class". He turned full circle on his heel, three hundred and sixty degrees, turned completely round and said "Shows you how wrong you can be about people". But going back to Fred Hoyle, I think his advice was absolutely right.

"Now you got the Nobel Prize for unifying certain parts of the theory in particle physics. How did you get the idea?"

It's such an *attractive* idea. You see, the whole history of particle physics, or of physics, is one of getting down the number of concepts to as few as possible. And when you are doing this "getting down" it seems absolutely the natural thing. In fact, it always surprises me that some of my physics friends — and some of them very eminent people, Nobel Prize winners — would not subscribe to the idea. They would find the difficulties in uniting two totally disparate looking phenomena so overwhelming that they would think you stupid to think otherwise.

"Do you think your religious views made you think that they could be unified?"

I think perhaps at the back of my mind. I wouldn't say consciously. But at the back of one's mind the unity implied by religious thought perhaps plays a role in one's thinking.

"Steve Weinberg came to the same theory quite independently. That's surprising isn't it?"

Not at all. The ideas in our subject are common and the diffusion of ideas is astonishingly large. Everybody knows almost everything of what's going on. I think it's a result of the system which we have developed of Summer Schools and symposia, and of course, the pre-print system. It's a very efficient system and we in theoretical physics are probably the best organized for that, for some curious reason. Although, mind you, at the time that Steve and I were working on the theory we were using ideas which, although they had been published, were not highly regarded. In that sense we had the field more to ourselves than would be the case today.

"Did people accept the theory at once?"

No, not at all. The theory was elaborated in 1967 and it was completely ignored. In fact, even before that, when I took another paper — one I wrote in 1964 — to be published in a journal the editor said "The thing that you are predicting has already been tested and not found. Will you add the words that this paper is purely speculative?" And I had to, in order to get the thing published. Those experiments were wrong, the ones which he was alluding to, but we only found that out later.

"So how did the theory come to be accepted?"

As I said, the theory was elaborated in 1967. There was a young physicist called t'Hooft, a Dutchman, who played a very crucial role in showing that the theory was mathematically well established. It was his first piece of work, at the age of 25 or so, and it gave the idea more respectability among the theoreticians. That was in 1971. Then in 1973 the experimenters redid those experiments which had previously shown that our ideas were not correct. They were redone properly at CERN, and that gave the first indication that the theory was on the right lines. But then the experiments were repeated in the United States and they contradicted the Geneva results. And it went on like that for a couple of years, back and forth between the experimenters.

"It's interesting that those experiments turn out to be wrong. As an outsider one thinks that the one thing you have that is reliable in physics is the experimental data. I'm surprised that the facts are so often wrong."

Well, you see, take one experiment which is going on now and which concerns the next stage of unification. I said we have united electromagnetism with the weak nuclear force. But there's a second nuclear force, the so-called strong nuclear force, and that is not yet united with the electroweak force. We hope it will be, and most of us would like to believe that this is happening. Th crucial experiment for this is the decay of the proton. The proton is supposed to be, or was supposed to be, before this theory came along, a fundamentally stable particle. This theory says no, in 10^{32} years all protons will decay. That's a very, very long time. The life of the universe as you know is 10^{10} years. So in 10^{32} years . . . my goodness . . . every proton will decay! Now to see this experimentally, you need 10^{32} protons to be watched for one year before one of them will decay. At the present time the situation is that there is an Indian experiment, 7000 feet deep in the Kolar Gold Field mines, which claims to have seen three events of proton decay. There is the experiment in Japan which claims to have seen one event and there are much more statistically significant experiments in the United States which claim to have seen none. Now what do you believe? The experiments are very difficult. I don't know which way the camel will sit, but this is a crucial experiment. And so it's perfectly possible that some of the experiments are wrong, or that their interpretation is wrong, and that we have to wait for more statistically significant signals.

"Now, you are a theoretician, so you're sitting there, rather grandly, while these experimentalists are testing your theories. But for the people doing the experiments with these big machines life is rather different. The papers they publish have fifty, even a hundred, authors. Do people mind that?"

I think many experimental physicists do not like this situation. Many of them would rather have the old days when just one man or two or three people collaborated and did an experiment and enjoyed it. But the nature of the beast is now such that you cannot help it. You have to have large collaborations because the experiments are costly, and they need vast quantities of equipment. For example, one hundred and fifty experimenters were associated with the two experiments at CERN which finally showed the validity of our theory. And the equipment is incredible. The detecting devices are three storeys high.

"Is your field a very competitive one?"

Oh yes. The practitioners number of the order of about five thousand in theory and about the same number in experiment. And then there is a premium on youth as you know very well.

"Why? Do you just think you're better when you're young?"

No, you are less encumbered. You don't live with your past. You don't live with your failures. You are much more willing to try more ideas in a different way. Older people are also, of course, more encumbered with various types of administrative duties in order to keep the subject running and that sort of thing. But I think most of all it's being unencumbered with the past ideas which you have tried to use and failed. Because then you think "Oh the idea is dead" when it's only the particular approach which you took to the idea which may be dead. I think the younger you are the better it is, if you can take the risk.

"Now were you young when you began to work on your unification theory?"

The idea started about 1957 when I was 31 which is fairly young, but then it took a long time in the execution.

"And did you get up every morning and work on it?"

No, no, not at all. It was intermittent. You worked on that particular set of ideas, then you gave up and took up something else and then you came back to it and so on, publishing little bits as you went along.

"But were you ever wrong? Have you ever been wrong in a major sort of way?"

It's probably just egotism but I can't think of anything where one has been proved completely wrong. There have been many stupid ideas which have led nowhere, of course, but that's the fate of all of us. The majority of our ideas, 99 per cent, lead nowhere. You're lucky if *one* of your ideas is correct in the end.

"And you have no misgivings about that?"

Not at all. But I think in our subject when you look at the successful ideas, you feel there is an inevitability about them. The only word I can use is "sleepwalking". *The Sleepwalkers* is the title of Arthur Koestler's book about Copernicus, Kepler, and Galileo. You just are led more or less from one step to the next.

"Sleepwalking seems a very passive way of doing physics."

It's sleepwalking in the following good sense. The unification ideas needed what we call the gauge theories. The gauge theories were actually first discovered in 1879 by Maxwell — as suggested by the equations for electromagnetism he had written down — then clarified by the German mathematician Hermann Weyl in 1929. They were put into the form in which we use them today by Yang and Mills and my pupil Shaw in 1954. It was the same old set of ideas which had started with Maxwell in 1879 but put in a slightly larger context. Then we, that is Weinberg, Glashow, and myself, said "These gauge ideas are the ideas that we need." That was our contribution. Newton, you remember, was asked why he was so great and he said "I was not so great. I was standing on the shoulders of giants". So my own feeling is that in each generation there is a set of ideas which is more or less common, but people forget and ascribe the entire success of those ideas to the one man who makes the best use of them. In that sense, maybe, physics has always been sleepwalking. When I said that in 1879 Maxwell had a great idea — well, he had inherited a set of ideas from Faraday. He wrote down Faraday's equations and found they were inconsistent, so he supplied one extra term. So in that sense it's inevitable, it's sleepwalking. Take Einstein's ideas which we consider the most revolutionary, the ideas of the curvature of space and time which explain gravitation. They go back to the German mathematician, Gauss, who first, in fact, made the tests to determine the curvature of space. What he didn't do was to put time into it. So there's an inevitability about these ideas. Although it was an act of genius for Maxwell to have found that extra term, and for Einstein to have added time to three-dimensional space, if you trace the history of the ideas they go back by gradations to earlier and earlier generations.

"Do you think if there hadn't been these geniuses these steps would have been made anyhow?"

Yes I do.

"Now, you come from a religious background, was there ever any conflict in doing physics?"

No, why should there be? Because fortunately, and I think I have said this so often in my writings, Islam is one of the three religions, which emphasize the phenomena of nature, and their study. One eighth of the Koran is exhortation to the believers to study nature and to find the signs of God in the phenomena of nature. So Islam has no conflict with science.

"What is the pleasure that you got from physics?"

Well let's put it this way. When you go to sleep and you are exhausted after a day's administrative chores or whatever, what is the thought that gives you maximum relief? I don't know what it is for you, but I get my pleasure from thinking about the problems of physics. It gives me the biggest relaxation.

"You mean thinking about physics is not work for you?"

It's a pleasure. I should qualify this by saying that when you are working something through it can be very, very hard and you eat your heart out. You think this idea should work and it doesn't. Then it can be devastatingly worrisome. But otherwise, most of the time your're thinking about it, it's a pleasure.

"What is that pleasure? Is it the pleasure of reflecting on what you've done that day, or contemplating the beauty of physics?"

Well after you have found something it's marvellous.

"So it's success which gives you pleasure?"

It's not just success. When you are relaxing, as I described, you are not reflecting on the successes of the past. In fact, any paper which you write does not give pleasure for more than a few days. I think for a week at the most. For a week you may be euphoric. "Oh yes, that was a marvellous result." But then it becomes just part of you. Presumably it becomes a part of your pleasure-giving cells, wherever they are, and impels you to further work.

"Do you still feel a sense of awe at the extraordinary nature of particle physics?"

Yes. It is always incredible that what people work out actually does happen.

"But are you more impressed by what people work out, or by what the nature of Nature is?"

Both. As a phenomenon, take Brain Science, for example. It is marvellous. So in that sense physics is not unique. But when I think in terms of the sublime theories that come in physics, I think that is unique.

"Do you like music? I mean, do you get the same sense listening to music?"

I would not say that I find the same sublimity. I find the same sublimity in reading or listening to the Koran, because there I find, for example, after you've been listening to it for half an hour, you suddenly get caught in an elevating fashion.

"But you do see physics as sublime?"

Yes, yes, no question about it. I mean take Einstein's theory — you still, after so many years, you still think "what a sublime, what a marvellous idea it is!"

Abdus Salam

by Rushworth M. Kidder

The International Centre for Theoretical Physics perches on a pine-covered slope overlooking the sapphire-blue Adriatic. But the directors's office looks out the back into the side of a hill. The location seems somehow in character for Abdus Salam, whose purpose lies less with elegant vistas than with little-known corners of the world — and who, despite his Nobel Prize for Physics, typically describes himself as "a humble research physicist from a developing country [Pakistan]."

Asked about his agenda for the next century, he responds without hesitation. "The real issue, to my mind," he says, "is the great divide between the South and the North," referring to those regions of the globe roughly representing the developing and the developed nations.

Although he speaks softly — from a desk chair facing a framed photograph of Albert Einstein and a blackboard chalked with mathematical formulas — his words carry fervor.

This "great divide" between the developing world and the industrial nations, he explains, arises from the fact that each side has a completely different set of problems. The major 21st-century issue facing the North, he says, is the arms race and the threat of nuclear warfare. The problem facing the South is the threat of starvation and utter poverty.

Picking up a copy of *Ideals and Realities*, a collection of his essays, he turns to a piece he wrote about Al Asuli, an 11th-century Islamic physician. Al Asuli, he says, divided the problems of humanity into diseases of the rich and diseases of the poor.

If Al Asuli were alive today, Dr. Salam says, he would make the same distinction. "Half his treatise would speak of the one affliction of rich humanity — the psychosis of nuclear annihilation," he says. "The other half would be concerned with the one affliction of the poor — their hunger and near-starvation. He might perhaps add that the two afflictions spring from a common cause — the excess of science in one case and the lack of science in the other."

For Salam, the operative word here is "science" — which he is careful to distinguish from "technology," or the application of scientific knowledge to human problems. One great difficulty for the developing world, he explains, is the misplaced assumption that sharing the latest in Western machinery, communications, and transportation — technology transfer — will be a panacea for the South.

"Technology transfer is something the South has asked for," Salam says, "and the North is resisting. Quite rightly. I don't blame [the North] for one second for not giving technology as such. Why should you? Why should anybody part with things that nobody else has helped to create?"

"That's where the bread and butter is concerned," he adds, referring to the central role that that the sale of technology plays in the economies of the industrial nations.

Instead of technology transfer, Salam says, the South in the coming century should be asking for a transfer of the basic science out of which technologies can spring. "I wish that the North could decide to give the South as much science as possible." Why this insistence on science?" Because "science is the basis of technology in the present day." He cites the case of Japan. Over the years the Japanese invested heavily in learning "all of science at a very high level. And then they were really successful in their technology."

Similar things are beginning to happen, he says, in five of the developing nations: Argentina, Brazil, China, India, and South Korea. He is especially impressed by the latter country, which he recently visited.

"They took me straight away to the television studios [for a] 2½-hour-long interview," he recalls, "in which they said, 'We have made it a national objective to win Nobel prizes. Can you give us advice?'"

"I told them they were being silly," he says with a chuckle, adding that "they may or may not get Nobel prizes." But he notes with approval that "the very fact that they made it a national objective is a very important thing. That means that they will stock up their libraries, they'll get scientific literature, they'll fund a lot of fellowships, they'll do everything possible to make themselves into a scientifically advanced country."

And that, he suggests, will do more for South Korea than any amount of reliance on Western technology.

As he looks forward into the 21st century, Salam distinguishes several kinds of science that will be practiced. The first he calls "science for science's sake" — the most basic and theoretical research, producing discoveries that sometimes go unappreciated for decades. In general, he says, such research is "probably in a healthy state," despite the never-ending battle to pay for it.

No Science for the Poor

"Then there's science for man's sake," he says, a category he breaks down into three parts: "global science, science for the rich countries, [and] science for the poor countries."

"Science for the poor doesn't exist, simply doesn't exist," he laments — although he notes that the poor countries have plenty of problems that science could help resolve. He cites the current medical concern over AIDS (acquired immune deficiency syndrome): "As long as it remained in Haiti, nobody even bothered about it." Now that it has come to Europe and America, "it will get the attention it deserves."

"It always deserved that attention," he adds wryly.

And what about science for the rich countries? That, he says, gets entangled in defence spending — which, he says, accounts for half of all research spending in the developed world.

For Salam, in fact, the real threat of the nuclear arms race is not that it might ultimate in holocaust. It is that swelling defence costs will sap resources needed to combat the rest of humanity's problems. It's a line of argument, he says, elaborated by President Dwight Eisenhower. "Eisenhower [made] it very clear that every single B-52 bomber that is made in America is depriving not only the poor in the Third World, but also Americans, of sustenance, of shelter, of aid."

"If Eisenhower were alive, he would be just aghast" at current levels of defence spending and the lack of attention to the developing world. Referring to massive defence spending in the North and developing world poverty, Salam says: "Unless you are conscious that the two problems are connected, and that [the developed nations] are squandering the wealth of this world — not only the wealth of this world but also the time and the energies of its scientists and its technologists, which could be used toward bettering humanity — you'll never get to grips with" the basic challenge facing the 21st century.

But what about the peacetime spinoffs that arise from defence-based research? "The statement that defence expenditures have 'fallout' is rubbish, total rubbish," Salam says flatly. "And the statement that since you invest in 'star wars' you will do your toothpaste better is [also] total rubbish."

What's really needed, says Salam, is not the 'fallout' from defence projects but a concentrated effort to study some of the developing world's most pressing problems — starvation, for example. Although Salam supports the idea of food aid for developing nations, he sees it as only a "short-term business." The root of the problem is "food deficiency, drought, and desertification."

The Basic Problem

"This is the basic problem to be solved scientifically," he says. But across much of the developing world, he points out, "there are no scientific studies at all of climate [and] of the underground water situation in the deserts — whether there are underground lakes, and so on." The lack of such studies — which are common enough in the developed countries — supports his contention that "science for the rich" is something quite different from "science for the poor."

One problem in conducting such studies, however, is that they frequently transcend national boundaries.

For that reason, they fall under Salam's third heading of "global science" — the study of the largest interdisciplinary and international issues concerning the global environment.

On this point, he expresses profound pessimism. "There's no such thing as global science as a subject," he complains. Even the disappearance of rain forests, which is commanding increased public attention, is not being considered in global terms.

"People do not take [the rain forest] as a global asset," he laments. "People take it as a problem of Brazil, a problem of Malaysia. How many governments are willing to spend money on that sort of thing? None. Zero." The problem is the lack of "the scientific infrastructure to look at the global problems."

"Everybody seems to be for himself," he says sadly. "There is no global vision at all. It's the lack of global vision that worries me, really. It's the issue of globalism which is missing in science, which is missing in the food problem, which is missing in the health problem." What is needed is "a vision of a sort which I don't see any statesman having."

From his position as an administrator, Salam says he clearly sees the need for sources of funding that would encourage such globalism. He adds that such funds, if they are to come, will have to come from the developed world.

But he again rejects as "rubbish" — one of his favourite words — the idea that "if you save funds from nuclear [arms limitation] you will put them into the welfare of mankind." The temptation, he says, will be for the rich countries simply to funnel the savings back into tax relief — "making the rich richer and the poor hungry man's soul sink lower."

"The whole attitude has to become very different," he notes.

Why Turmoil could Spread

And if the "great divide" between the rich and poor nations is not closed? Salam says that it will be increasingly "hard to ignore [the developing countries'] problems in the 21st century" for two reasons.

First, he says, the North will no longer be able to "insulate itself" politically from the South. If the gap is not narrowed, he says, "What will happen is what is happening already in the Third World" — turmoil, military governments, unrest, and "people on top of each other."

Second, he notes that the world-wide environment "may be affected by lack of attention to the global problems and to scientific globalism."

"In that sense," he says, "no parts of the world are going to be safe from the feeling of turmoil. At the moment, it doesn't [seem to] affect Americans to have starving Africans at their hands. They may very well say, 'Well, if they want to starve, let them starve.'"

"But I dont't think man lives like this," he says. Speaking of rock star Bob Geldof's efforts to raise money for African famine relief, he says, "I think the Geldofs of this world make their point when they show what can be done in a small way."

What, then, does he hope will close the gap? He would like to see industrial nations "specialise" in providing the scientific training to elevate the developing nations. "For example, higher education may be taken up by Britain and the United States. The Russians may take up lower education. The Japanese and the Germans will be asked to do technology."

"That," he concludes, "will be my vision of the future."

IV.3
PERSONAL

Homage to Chaudhri Muhammad Zafrulla Khan

by Muhammad Abdus Salam

Appeared in **Transnational Perspectives**, *Vol. 12, No. 2, 1986.*

Chaudhri Muhammad Zafrulla Khan was one of the greatest human beings I have had the privilege and good fortune of knowing in my life.

I first saw Chaudhri Zafrulla Khan in December 1933. I was then nearly eight years old. I can still see him in my mind's eye — a very handsome figure with a most impressive bearing. I believe the first occasion when he knew of me was when my father wrote to him in 1940 seeking his advice about my future career. He wrote in reply that he would pray for me and he offered three pieces of advice. First, that I should look after my health; health was the basis of all achievement. Second, in respect of studies, he advised me that whatever lectures in the classroom were due for tomorrow, I should prepare for them the day before. And whatever I learned today, I should revise the same day so that it became forever part of me. Third, I should broaden my mind; in particular, whenever I got the opportunity to make an educational journey — or even a journey for pleasure — I should take it, for journeying to new places was conducive to a broadening of one's range of interests.

My first personal contact with Chaudhri Muhammad Zafrulla Khan was in October 1946, when I sailed to the UK to join Cambridge University for studies. Our boat — the P. & O. Franconia — docked at Liverpool. It was a cold and misty morning. Chaudhri Sahib had come to the dockside to meet his nephew who was also travelling on the "Franconia". Chaudhri Zafrulla Khan was at that time Judge at the Supreme Court of India. When we got down from the boat, our heavy cases — my mathematics and physics books which I had packed in them — were lying around in the customs shed. There were few porters due to post war conditions. Chaudhri

Zafrulla Khan said to me, "Take hold of the case from one side and I will take it from the other and we shall carry it to the waiting boat train." This was an amazing reception for a humble student, who had never before encountered such gracious unself-consciousness on the part of a personage so highly placed.

We travelled together to London. During the journey he kept pointing out the beauties of the English countryside, of which he was inordinately fond. The weather was very cold. Seeing me shiver, he kindly gave me one of his (enormously heavy) winter coats. This, in spite of 40 years of use, still survives in the family.

I met him again in 1951 when he was Foreign Minister of Pakistan and came to the Princeton Institute for Advanced Study where I was a Fellow. I spent two days in his company. He was then attending the General Assembly of the United Nations. With him I had the privilege of visiting some of the beautiful historical places on the East Coast of the United States. Fresh from his memorable duels at the United Nations forum, fought with the highest ranking and keenly brilliant adversaries, on behalf of Kashmir, Libya, Tunisia, Algeria, Morocco, and Palestine Arabs, my chief recollection of him is of one who would not suffer fools gladly.

But I really got to know him after 1973 when he came to live at the London Mosque after his retirement from the Presidentship of the International Court of Justice. He was gracious enough to accept to come to my house at Campion Road, nearby, for Sunday breakfasts, whenever I was in London. It could be breakfast only because his day, which he spent working on his translation of the Holy Book or Books of the Hadith and the like, started regularly at 9 a.m. every day including Sundays and could not be interrupted.

These breakfasts were memorable occasions when sometimes we would go over the public episodes in his life which are so beautifully described in his books, particularly his last memoir, *Servant of God.* But this does not convey the lively details which he would narrate to us. For example, he tells the story on pages 67 — 69 of his encounter with Mr. Churchill, but he omits the earlier parts of his story when Mr. Churchill was cross-examined by Sardar-Boota Singh of the Indian Party and the hilarity of that examination. Even so, I shall quote here the story as he himself tells it, to convey some of the wonderful flavour of his narration:

"The public sittings of the Joint committee commenced in the spring of 1933. A large variety of witnesses, Indian and British representing a diversity of interests and views was examined by the Committee. Participation in the proceedings of the Committee proved a very instructive experience. The most outstanding witness who appeared before the Committee was Mr. (later Sir) Winston Churchill. His examination extended over four days. He was firmly opposed to the proposals contained in the White Paper. He condemned them lock, stock and barrel. He looked upon them as a betrayal of its trust by Britain. The barrage of questions directed at him failed to move him a single inch from his stand. His eyes twinkled, he wore a smile, he waved his cigar, he was courtesy and urbanity, but he was utterly unyielding. His questioners could win nothing from him. He held his own against all comers."

"Having watched the drama for a whole day, the Punjab Muslim Delegate felt that it would serve no useful purpose to cross-examine so formidable an adversary as Mr. Churchill. On the morning of the second day the Secretary of State accosted him before the meeting was called to order and enquired: "Do you intend to put any question to Mr. Churchill?"

"No Sir. I consider it would be a profitless exercise."

"Well, he is our cleverest debater in the House, and it is no use trying to catch him out on his previous speeches in which he supported Dominion Status for India. You have seen how he gets out of them. Yes, Dominion *Status*, but status is one thing and function is quite another. India already has Dominion Status. It sent a delegation to the Paris Peace Conference, it is a signatory of the Treaty of Versailles, it is a member of the League of Nations. That is status. But it is not yet ready to function as a Dominion! He thinks India is still what it was when he was serving as a subaltern at Bangalore."

"(The Punjab Muslim delegate) thought over it. His turn came an hour before the close of the afternoon sitting of the Committee. His attitude was deferential, his tone respectful, bordering almost on the apologetic, with a slight touch of deference. Mr. Churchill was cautious, but made a reluctant concession here, a grudging admission there, hedged round with ifs and buts and provideds. When he perceived that he was letting himself be persuaded to yield ground, he began to evade the question put to him, so that it had to be reframed with great care. On one occasion he slipped out of answering the question in one direction, and when the question was carefully rephrased he slipped out in another direction. The questioner's tone became even more bland, almost humble: Mr. Churchill I beg to be forgiven. I am under a disadvantage. English is not my mother tongue. I have twice failed to make my meaning clear. Will you permit me to try once more? The response was gracious: Please, please. The question was put a third time in a shape that did not admit of evasion. Thereafter both the examiner and the witness become more alert. The Committee adjourned. The examination was resumed next morning and continued for another hour. When the questioner concluded with an expression of thanks to the eminent witness, the witness went on record with: My Lord Chairman, may I be permitted to say that I have not noticed that Mr. Zafrulla Khan suffers any disadvantage from lack of knowledge of the English language?"

"When at the end of the fourth day his examination was completed, the cheers of the Committee had the quality of an ovation. (Mr. Churchill) rose from his seat, came across to his Muslim interrogator, shook him by the hand and growled: You have given me the two most difficult hours before the Committee. The questioner acknowledged the growl as if it were an accolade and a token of friendship and so it proved. In subsequent meetings the great Prime Minister would every time present him with a volume of his letters or speeches, and the inscriptions beginning with: Inscribed for Zafrulla Khan, W. S. Churchill; went on mounting the scale: To Zafrulla Khan, from W. S. Churchill; To my friend Zafrulla, from W. S. Churchill; to Zafrulla from his friend W. S. Churchill. Magnanimity was not the least among the many great qualities of the great Prime Minister."

The amazing thing was that Chaudhri Muhammad Zafrulla Khan's memory was faultless, not only about persons, but also about dates and even times of day, for matters which had occurred 50 or 60 years earlier. I also recall with great fondness the narration of his United Nations fights with Big Powers for independence of Libya, Morocco, Tunisia and Algeria (described on pages 179–182 of *Servant of God*); likewise, the heart-warming story of his pilgrimage to Mecca, when he was King Faisal's personal guest, narrated on pages 279-286. In the recounting of all these incidents in his life, what came across strongly was the greatness of his spirit, his intense love — bordering on near-adoration — for the Holy Prophet of Islam, and his own complete reliance on Allah and His divine Will. Also manifest was his love of Persian poetry, of Hafiz and particularly of the mystical verse of Rumi in his Diwan-ee-Shams-Tabriz which he could recite without effort from memory.

Regarding his love for the Holy Prophet, let me tell a story. Chaudhri Muhammad Zafrulla Khan was taken ill with a backache and was confined to bed in a hospital in Wandsworth. I visited him in hospital. I took him *Shamail-i-Tirmizi*, written by Imam Tirmizi, which describes the Holy Prophet's daily life, his looks, what he wore, his daily preoccupations, his family and public life. I said I hoped that sometime in the future, if Allah decrees, I would translate this book into English. I left it with him and went away to Trieste.

I came back a couple of months later and went to see him at his residence. He presented me with a copy of a translation of Tirmizi into English, already completed and *printed*, with a gracious dedication to me. I was astonished at the speed with which he had worked. I mildly protested: I had wished to translate this book myself for my "ghufran". He said "You may not have found time in the immediate future. So I thought while I was confined in the hospital, this would be the most rewarding use of my time."

My last travel with him was when he was invited in 1980 to the inaugural meeting of the Academy of the Kingdom of Morocco, by its Permanent Secretary, the late Dr. Ahmed Taibi Benhima, who knew him from the United Nations days. Chaudhri Sahib was then eight-seven years old — an erect figure, commanding, yet benign. I can vividly recall him sitting next to His Majesty King Hassan at the banquet and at the subsequent function, receiving the King's personal affectionate homage. I saw the same veneration for him in the streets of Fez from students, soldiers — everyone who heard of his name, for what he had done for Morocco's independence.

I have earlier said that he was generous — almost to a fault — to those in need. The story may not be well-known, but after retirement, he dedicated *all* his life's savings to charitable purposes. A large part was spent in rebuilding the living quarters and offices for the Imam next to the London Mosque, as well as towards building the Mahmud Hall (he did not want any mark, commemorating this generosity, on the buildings). The rest — of the order of half a million dollars — he dedicated to setting up a charitable foundation — the Southfields Trust — to help the needy, and for educational purposes.

One Sunday when he honoured us with coming for breakfast, my brother protested to him on generally neglecting his own personal needs. Chaudhri Sahib said he had asked his yearly pension (of around $32,000 a a year) to be deposited straight into the bank account of the foundation he had created. He did not keep any part of his pension for himself. But he had an agreement with the Trust that it would pay him seven pounds a week for life's necessities and once a year the Trust would permit him an economy fare to Pakistan to attend the annual gathering in December. He then added, "I know, through Allah's grace, I am a good advocate, but one judicial case I always lose. That is, whenever I plead to myself for myself."

He has such love for Islam, and such a "ghairat" for his faith, that one could not come away from his company without being fired with his spirit. It is well-known that he spent the last years of his life occupied with his translation of the Holy Quran and the introduction to it, his magnificent biography of the Holy Prophet (*Mohammad, Seal of the Prophets*) published by Routledge & Kegan Paul Ltd. of London, creating single-handedly a veritable one-man Library of Islam in the English language. He said he had left the International Court of Justice, where he could have continued for another term of office, to serve Allah. As he says himself, "Gradually, over the years, the consciousness of God as an experienced reality, rather than as a merely believed phenomenon, was strengthened".

His love for his long-departed mother and the lessons he had learnt from her were often repeated for us. In his book, on page 297, he quotes her as saying, "It is no virtue to be kind to someone we like; virtue is to be kind to those we do not like"; and then, "A friendship is forever, else it is no friendship at all". A mark of this, and a real privilege, was to be included in his prayers at Tahhajad every morning when he said he prayed for 300 persons, naming each one individually. His own oft-repeated saying used to be: "Call to mind when your Lord declared: If you will employ My bounties beneficently, I will surely multiply them unto you; but if you misuse or neglect them, My punishment is severe indeed (14:8)".

I cannot do better to close this note, than by quoting from the last words of his last book *Servant of God*, where he speaks about himself. "His career as a public servant came to an end with the expiry of his second term on the International Court of Justice. He was called to the Bar at the age of twenty one, practised as a lawyer for twenty one years, held executive office in India and Pakistan for fourteen years, was a Judge, national and international, for twenty one years, and a diplomat for three years. He has worn many hats, but the one he now wears is the most honorific of all, and brings him the greatest satisfaction. He is now wholly the servant of God, for which honour all praise is due to God. His one care and concern is that his Gracious Master may be pleased with him, and may continue to afford him, for such time as He may, of His grace and mercy, grant him here below, opportunities of serving Him and His creatures, and bestow upon him the strength and ability to perform that service in a manner acceptable to Him. Of his own he has nothing to devote to His service; life, faculties, capacities, means, relations, friends are all His gifts. He supplicates for wisdom and strength to employ all his gifts in His service, to the winning of His pleasure, to the true service of his fellow beings. For himself he only seeks fulfilment in his Gracious Maker, Creator and Master. He hopes for His mercy, His forgiveness, His forebearance. May He continue to cover up all his numberless faults, shortcomings, vices, sins, disobediences and transgressions under the mantle of His mercy, and safeguard him against humiliation here and hereafter. May He wash him clean of all impurities so that death, when he is pleased to send it, may prove to be a gentle transition from illusion to reality, from faith to fulfilment and utter submission. Amen.

All praise belongs to Allah".

Note

Muhammad Zafrulla Khan, born 6 February 1893; Barrister-at-Law, Lincoln's Inn, 1914; member of Punjab Legislative Council, 1926-1935; Member of Governor-General's Executive Council, India, 1935-1941; Judge Supreme Court of India, 1941-1947; Foreign Minister of Pakistan, 1947-1954; Judge International Court of Justice, 1954-1961 (Vice President, 1958-1961); Permanent Representative of Pakistan at United Nations, 1961-1964; President of General Assembly of United Nations, 1962-1963; Judge International Court of Justice, 1964-1974 (President, 1970-1973). Hon Bencher, Lincoln's Inn, Hony. Fellow, Delegacy of King's College, London, Hony. Fellow, London School of Economics, Hon LL.D. Universities of Cambridge, Columbia, Denver, California (Berkeley), Long Island, Hon D.C.L., Beaver College Penn. Died at Lahore (Pakistan), 1 September 1985.

(From *Servant of God: A Personal Narrative,* The London Mosque, 16 Gressenhall Road, London SW18.)

V

BIODATA

Biodata

1. **ABDUS SALAM**

   Date of birth: 29 January 1926
   Place of birth: Jhang, Pakistan
   Nationality: Pakistani

2. **Educational Career**

   Government College, Jhang and
   Lahore, Pakistan (1938–1946)

   M.A. (Panjab University)
   First place in every
   examination at the Panjab
   University

   Foundation Scholar, St. John's
   College, Cambridge
   (1946–1949)

   B.A. Honours
   Double first in Mathematics
   (Wrangler) and Physics

   Cavendish Laboratory,
   Cambridge (1952)

   Ph.D. in Theoretical Physics

   Awarded Smith's Prize by the
   University of Cambridge for
   the most outstanding *pre-
   doctoral* contribution to
   physics (1950)

3. **Appointments**

   Professor, Government College, Lahore (1951–1954)

   Head of the Mathematics Department
   of Panjab University (Lahore) (1951–1954)

Lecturer, Cambridge University (Cambridge)	(1954 – 1956)
Professor of Theoretical Physics, London University, Imperial College (London)	(1957 –)
Founder and Director, International Centre for Theoretical Physics (Trieste)	(1964 –)
Elected Fellow, St. John's College (Cambridge)	(1951 – 1956)

4. United Nations Assignments

Scientific Secretary, Geneva Conferences on Peaceful Uses of Atomic Energy	(1955 and 1958)
Elected Member of the Board of Governors, IAEA, Vienna	(1962 – 63)
Member, United Nations Advisory Committee on Science and Technology	(1964 – 1975)
Elected Chairman, United Nations Advisory Committee on Science and Technology	(1971 – 1972)
Member, United Nations Panel and Foundation Committee for the United Nations University	(1970 – 1973)
Member, United Nations University Advisory Committee	(1981 – 1983)
Member, Council, University for Peace (Costa Rica)	(1981 – 1986)
Elected Chairman, UNESCO Advisory Panel on Science, Technology and Society	(1981)

5. Other Assignments

Member, Scientific Council, SIPRI (Stockholm International Peace Research Institute)	(1970 –)
Elected Vice President, International Union of Pure and Applied Physics (IUPAP)	(1972 – 1978)
Elected First President of the Third World Academy of Sciences	(1983 –)
Member of the CERN Scientific Policy Committee	(1983 – 1986)
Member of the Board of Directors of the Beijir Institute of the Royal Swedish Academy of Sciences	(1986 – 1988)

Elected first President of the Third World Network of Scientific Organisations	(1988–)

6. Awards for Contributions to Physics

Hopkins Prize (Cambridge University) for the most outstanding contribution to physics during 1957–1958	(1958)
Adams Prize (Cambridge University)	(1958)
First recipient of Maxwell Medal and Award (Physical Society, London)	(1961)
Hughes Medal (Royal Society, London)	(1964)
J. Robert Oppenheimer Memorial Medal and Prize (University of Miami)	(1971)
Guthrie Medal and Prize (Institute of Physics, London)	(1976)
Sir Devaprasad Sarvadhikary Gold Medal (Calcutta University)	(1977)
Matteuci Medal (Accademia Nazionale di XL, Rome)	(1978)
John Torrence Tate Medal (American Institute of Physics)	(1978)
Royal Medal (Royal Society, London)	(1978)
Nobel Prize for Physics (Nobel Foundation)	(1979)
Einstein Medal (UNESCO, Paris)	(1979)
Shri R.D. Birla Award (Indian Physics Association)	(1979)
Josef Stefan Medal (Josef Stefan Institute, Ljublijana)	(1980)
Gold Medal for outstanding contributions to physics (Czechoslovak Academy of Sciences, Prague)	(1981)
Lomonosov Gold Medal (USSR Academy of Sciences)	(1983)

7. Awards for Contributions towards Peace and Promotion of International Scientific Collaboration

Atoms for Peace Medal and Award (Atoms for Peace Foundation)	(1968)
Peace Medal (Charles University, Prague)	(1981)

Premio Umberto Biancamano (Italy)	(1986)
Dayemi International Peace Award (Bangladesh)	(1986)
Genova Sviluppo dei Popoli Prize (Italy)	(1988)

8. **Academies and Societies**

Elected, Fellow, Pakistan Academy of Sciences (Islamabad)	(1954)
Fellow of the Royal Society, London	(1959)
Fellow, Royal Swedish Academy of Sciences (Stockholm)	(1970)
Foreign Member of the American Academy of Arts and Sciences (Boston)	(1971)
Foreign Member, USSR Academy of Sciences (Moscow)	(1971)
Foreign Associate, USA National Academy of Sciences (Washington)	(1979)
Foreign Member, Accademia of Nazionale dei Lincei (Rome)	(1979)
Foreign Member, Accademia Tiberina (Rome)	(1979)
Foreign Member, Iraqi Academy (Baghdad)	(1979)
Honorary Fellow, Tata Institute of Fundamental Research (Bombay)	(1979)
Honorary Member, Korean Physics Society (Seoul)	(1979)
Foreign Member, Academy of the Kingdom of Morocco (Rabat)	(1980)
Foreign Member, Accademia Nazionale delle Scienze (dei XL), (Rome)	(1980)
Member, European Academy of Science, Arts and Humanities (Paris)	(1980)
Associate Member, Josef Stefan Institute (Ljublijana)	(1980)
Foreign Fellow, Indian National Science Academy (New Delhi)	(1980)
Fellow, Bangladesh Academy of Sciences (Dhaka)	(1980)

Member, Pontifical Academy of Sciences (Vatican City)	(1981)
Corresponding Member, Portuguese Academy of Sciences (Lisbon)	(1981)
Founding Member, Third World Academy of Sciences	(1983)
Elected, Corresponding Member, Yugoslav Acedemy of Sciences and Arts (Zagreb)	(1983)
Honorary Fellow, Ghana Academy of Arts and Sciences	(1984)
Honorary Member, Polish Academy of Sciences	(1985)
Corresponding Member, Academia de Ciencias Medicas, Fisicas y Naturales de Guatemala	(1986)
Honorary Life Fellow, London Physical Society	(1986)
Fellow, World Academy of Art and Science (Stockholm)	(1986)
Corresponding Member, Academia de Ciencias Fisicas, Matematicas y Naturales de Venezuela	(1987)
Fellow, Pakistan Academy of Medical Sciences	(1987)
Honorary Fellow, Indian Academy of Sciences (Bangalore)	(1988)
Distinguished International Fellow of Sigma Xi (USA)	(1988)

9. Orders

Order of Nishan-e-Imtiaz (Pakistan)	(1979)
Order of Andres Bello (Venezuela)	(1980)
Order of Istiqlal (Jordan)	(1980)
Cavaliere di Gran Croce dell'Ordine al Merito della Repubblica Italiana	(1980)

10. D.Sc. Honoris Causae

Panjab University, Lahore, Pakistan	(1957)
University of Edinburgh, Edinburgh, UK	(1971)
University of Trieste, Trieste, Italy	(1979)
University of Islamabad, Islamabad, Pakistan	(1979)
Universidad Nacional de Ingenieria, Lima, Peru	(1980)

University of San Marcos, Lima, Peru	(1980)
National University of San Antonio Abad, Cuzco, Peru	(1980)
Universidad Simon Bolivar, Caracas, Venezuela	(1980)
University of Wroclow, Wroclow, Poland	(1980)
Yarmouk University, Yarmouk, Jordan	(1980)
University of Istanbul, Istanbul, Turkey	(1980)
Guru Nanak Dev University, Amritsar, India	(1981)
Muslim University, Aligarh, India	(1981)
Hindu University, Banaras, India	(1981)
University of Chittagong, Bangladesh	(1981)
University of Bristol, Bristol, UK	(1981)
University of Maiduguri, Maiduguri, Nigeria	(1981)
University of the Philippines, Quezon City, Philippines	(1982)
University of Khartoum, Khartoum, Sudan	(1983)
Universidad Complutense de Madrid, Spain	(1983)
The City College, The City University of New York, New York, USA	(1984)
University of Nairobi, Nairobi, Kenya	(1984)
Universidad Nacional de Cuyo, Cuyo, Argentina	(1985)
Universidad Nacional de la Plata, La Plata, Argentina	(1985)
University of Cambridge, Cambridge, UK	(1985)
University of Goteborg, Goteborg, Sweden	(1985)
Kliment Ohridski University of Sofia, Sofia, Bulgaria	(1986)
University of Glasgow, Glasgow, Scotland	(1986)
University of Science and Technology, Heifei, China	(1986)
The City University, London, UK	(1986)
Panjab University, Chandigarh, India	(1987)
Medicina Alternativa, Colombo, Sri Lanka	(1987)
National University of Benin, Cotonou, Benin	(1987)
University of Exeter, UK	(1987)
University of Peking, China P.R.	(1987)
University of Gent, Belgium	(1988)

11. Pakistan Assignments

Member, Atomic Energy Commission, Pakistan	(1958 – 1974)
Elected President, Pakistan Association for Advancement of Science	(1961 – 1962)

Adviser, Education Commission Pakistan	(1959)
Member, Scientific Commission, Pakistan	(1959)
Chief Scientific Adviser to President of Pakistan	(1961 – 1974)
Founder Chairman, Pakistan Space and Upper Atmosphere Committee	(1961 – 1964)
Governor from Pakistan to the International Atomic Energy Agency	(1962 – 1963)
Member, National Science Council, Pakistan	(1963 – 1975)
Member, Board of Pakistan Science Foundation	(1973 – 1977)

12. Pakistani Awards

Sitara-i-Pakistan (S. Pk.)	(1959)
Pride of Performance Medal and Award	(1959)
The Order of Nishan-e-Imtiaz (the highest civilian award)	(1979)

13. As "Servant of Peace"

Member, Scientific Council, SIPRI (Stockholm International Peace Research Institute)	(1970 –)
Awarded the Atoms for Peace Medal and Award (Atoms for Peace Foundation)	(1968)
Peace Medal (Charles University, Prague)	(1981)
Premio Umberto Biancamano, Italy	(1986)
Dayemi International Peace Award (Bangladesh)	(1986)
Member, Council, University for Peace, Costa Rica	(1981 – 1986)

14. Published Papers

Around 250 scientific papers on physics of elementary particles. Papers on scientific and educational policies for developing countries and Pakistan.

15. Scientific Contributions

Research on physics of elementary particles. Particular contributions: (*i*)   two-component neutrino theory and the prediction of the inevitable parity violation in weak interaction;

(*ii*)   gauge unification of weak and electromagnetic interactions — the unified force is called the "electroweak" force — a name given to it by Salam; predicted existence of weak neutral currents and *W, Z* particles before their experimental discovery;

(*iii*)   symmetry properties of elementary particles; unitary symmetry;

(*iv*)   renormalization of meson theories;

(*v*)   gravity theory and its role in particle physics;

(*vi*)   unification of electroweak with strong nuclear forces, grand (electro-nuclear) unification; and

(*vii*) related prediction of proton-decay;

(*viii*) supersymmetry theory, in particular formulation of superspace and formalism of superfields;

(*ix*)   Kaluza-Klein theories;

(*x*)   superstrings.

16. Books

Symmetry Concepts in Modern Physics, Iqbal Memorial Lecture (Atomic Energy Centre, Lahore), 1966.

Edited with E.P. Wigner, *Aspects of Quantum Mechanics* (Cambridge University Press), 1972.

A Man of Science (Research Centre for Cooperation with Developing Countries, Ljubljana, Yugoslavia), 1987.

With Ergin Sezgin, *Supergravity in Diverse Dimensions,* Vols. I and II, to be published by World Scientific Publishing Company in 1988.

Biography, *Abdus Salam,* by Dr. Abdul Ghani (Ma'aref (Printers) Limited, Defence Housing Society, Karachi), 1982.

Ideals and Realities, Selected Essays of Abdus Salam, edited by Z. Hassan and C.H. Lai (World Scientific Publishing Co. Pte. Ltd., Singapore) 1984. Second Edition, 1987. Translated into Chinese, Italian, Arabic, French, Punjabi, Persian, Hindi, Rumanian, Turkish, Bengali and Spanish with Russian, Portuguese, Japanese, and Urdu in progress.

Science and Education in Pakistan (Third World Academy of Sciences, Trieste, Italy), 1987.

Renaissance of Science in Islamic Countries (in press).

Service to the Third World

1. *As a Research Scientist*

Nobel Prize in Physics in 1979 for the prediction of the unification of the electromagnetic with the weak nuclear force. At present Professor Salam is the only living Nobel Laureate in Sciences with a Third World nationality.

2. *As Educator*

(*i*)   Professor, Government College, Lahore (1951 – 1954).

(*ii*)   Head of the Mathematics Department at the Panjab University, Lahore (1951 – 1954).

(*iii*)   Founded and headed the Theoretical Physics Department at Imperial College, London in 1957 (was responsible for producing around 50 Ph.D's in theoretical physics from developing countries like India, Pakistan, Bangladesh, Brazil, Ghana, Nigeria, Lebanon, Iran, Iraq, Greece and others, in addition to U.K. Ph.D's).

(*iv*)   Adviser to the Education Commission of Pakistan in 1959. Author of the Report on the "Structural Changes in Pakistan's Educational System", commissioned by the Government of Pakistan (1970).

(*v*)   Responsible for the Regional Workshop Course in Curriculum Development in Physics, Mathematics and Computer Sciences in Kenya, Africa, 1986. A similar course was organised in Khartoum (Sudan) in 1987 with arrangements to hold it in Addis Ababa Ethiopia, in 1989.

(*vi*)   Responsible for the creation of New Physics Centres and Summer (and Winter) Annually-running Schools at Nathiagali (Pakistan), Cuzco (Peru), Petra (Jordan), Khartoum (Sudan), and Bogota (Colombia). Also responsible for the establishment of new centres for Mathematics and Physics research in Benin, Morocco, Malaysia, and Tanzania.

3. *Experience as Science Administrator*

(*i*)   Member, Pakistan Scientific Commission (1959).

Governor from Pakistan to IAEA Board of Governors (1963).

Member and Chairman (1972), United Nations Advisory Committee on Science and Technology (1964 – 1975).

Member, United Nations Founding Committee for the UN University (1970 – 1973).

Member, UN University Advisory Committee (1981 – 1983).

Chairman, UNESCO Advisory Panel on Science, Technology and Society (1981).

(*ii*) Founded and directed the International Centre for Theoretical Physics (ICTP) since 1964 (set up under the auspices of the IAEA, UNESCO and the Italian Government). The Centre has imparted training for research and research experience to 36,000 visiting scientists since its inception. (During the period 1970 – 87, the Centre received 4,200 high level physicists and mathematicians (mostly for post-doctoral experience) from Africa and the Arab region, 5,000 from Asia, 2,200 from Latin America, 4,000 from the Eastern Europe, 2,600 from North America, plus 10,100 from Western Europe, including 4,200 from Italy.

The Centre started with a budget of $350,000 (in 1964) which has increased by a factor of around 45 to $16 million (1988).

During 1987 alone, the Centre welcomed 3,700 physicists and mathematicians.

(*iii*) Training colleges and Research Workshops have been held at the Centre in Trieste on subjects such as physics of materials and microprocessors, physics of energy, physics of fusion, physics of reactors, physics of solar and other non-conventional energy, geophysics, biophysics, neurophysics, laser physics, physics of oceans and deserts, and systems analysis — this, in addition to high energy physics, condensed matter physics, mathematics, both pure and applied, as well as quantum gravity, cosmology, atomic and nuclear physics.

(*iv*) Since 1981, research workshops and extended research colleges have been organised in developing countries — in China, Ghana, Bangladesh, Colombia, Kenya, Sri Lanka and Sudan: these were 4 – 8 week activities on solid state physics, monsoon dynamics, solar physics, physics of microprocessors, physics of desertification and physics and mathematics teaching.

(*v*)   As Founder President of the Third World Academy of Sciences, Salam has successfully secured a starting annual budget of around $2 million, for stimulating research in basic sciences, in the Third World. The Academy's budget now stands at $4 million.

(*vi*)   Founder President of the Third World Network of Scientific Organisations, (TWNSO) October 1988.

4.   *As Servant of Peace*

Awarded the Atoms for Peace Medal and Award (Atoms for Peace Foundation) (1968).

Member, Scientific Council, SIPRI (Stockholm International Peace Research Institute) (1970 –   ).

Peace Medal (Charles University, Prague) (1981).

Premio Umberto Biancamano, Italy (1986).

Dayemi International Peace Award (Bangladesh) (1986).

Member, Council, University for Peace, Costa Rica (1981 – 1986).

Abdus Salam's name means "Servant of God, Who is Peace".

5.   *Personal Support to Third World Science*

Foundations have been created with all the monetary awards received by Professor Salam to support science in the Third World.

VI
REVIEWS

A Review of the Italian Edition of Ideals and Realities

Review published in **Corriere della Sera**, *6 August 1986, of the Italian translation of the first edition of* **Ideals and Realities** *(Ideali e Realtá, translated by Vincenzo Gatti, published by Edizioni Lint, Trieste, 1986).*

Salam, Physics Nobel Laureate, Inspired by the Quran

It is difficult to find pages as vibrant with human warmth, religiousness, political and moral duty and scientific depth, as in a recent collection of articles (larger than the English edition) of Abdus Salam "Servant of Peace", which is the literal translation of his name, entitled *Ideals and Realities* (Ed. Lint, Trieste, Italy). Perhaps the *Thoughts of the Difficult Years of Einstein* may be able to produce the same emotions in the reader.

Having to say it schematically, one may say that the Physics Nobel Laureate Salam is at the same time two persons who combine only rarely; but when they do, they produce a miracle of humanity. Abdus Salam is a scholar and a sage. As a scholar, that is, as a scientist, he is the last great follower of an ancient tradition of physicists, for whom the intellectual scope of science is the unification of the laws of nature consisting of a few principles, to a grand unification of only one principle. In this search of the "arche", which began in Greece, and continued in Islam (Al Biruni sustained that nature has the same laws everywhere, on Earth as on the Moon), and materialised with the encounter of these two civilisations beginning with modern science, from Galileo to Einstein, Salam has made a fundamental contribution with the electroweak theory for which he received the Nobel Prize in 1979.

As a sage, Salam is again two things: he is a man profoundly religious who finds in the Quran the justification and the best reason for his scientific work, and he is a politician, in the highest and noblest sense of this degraded term, who places all his energy in improving the living conditions of the Third World. Whoever has had the opportunity of meeting him and listening to him, knows that he often refers to poets and to the Holy Book in order to give strength to his ideas.

The following splendid verse of the Quran could then be the seal of the sage:

> *"Though all the trees on earth were Pens*
> *And the Sea was Ink*
> *Seven seas, after, to replenish it,*
> *Yet would the words of the Lord be never spent,*
> *The Lord is Mighty and All Wise."*

Another verse of Omar Khayyam which he often cites gives an idea of the commitment of Salam, the man of action:

> *"Ah love! could thou and I with fate conspire*
> *To grasp this sorry scheme of things entire*
> *Would not we shatter it to bits — and then*
> *Remould it nearer to the heart's desire".*

Excerpt from the Foreword to the Rumanian Translation of Ideals and Realities

by Professor Ioan Ursu,
member of the Rumanian Academy

One may well wonder why several texts in this book underlie an obvious, even though controlled, bitterness, why the world despair surfaces so often, and irresistibly, why does the author see future so rippled with uncertainties. Are we perhaps listening to a pessimist? Why, certainly not. In fact only a man nurturing great confidence in himself, in his fellows, and in future, only an unabated optimist could have endeavoured so many initiatives and campaigns — and all paralleled by an undefatigable scientific effort in one of the most dynamic, demanding and competitive fields of physics. As they say, only those who elude the struggle would not experience the bitter taste of despair.

Abdus Salam's sorrows and disappointments are not, first and foremost, those of a single individual. Spelling them out with all his compelling force and authority, he, in effect, echoes huge communities, and groups of nations among the developing countries, confronted as they are with a world order, bearing the stigma of inequity and injustice and under-development, of the harsh, persistent and exasperating realities of poverty, of hunger, of physical and moral diseases, of illiteracy and intolerance, of confusion and social violence. Bitterly unveiling such realities, distilling his own humane, outraged feelings through an admirable analytical effort, and, by the same token, rejecting the notion that he should ever give up designing projects and hopes for improving the world, the author places himself implicitly on the stands of a fighter for the right of all nations to a creative participation in the making of the World History, one commensurate to the important native resources of their spirituality, for a more equitable access to welfare, for progress and peace in a better, and more right, world.

Review of Ideals and Realities in La Recherche

Translation of the Review of the Second Edition of Ideals and Realities in La Recherche, Vol. 207, February 1989.

What is a believer? Someone who does not despair in the follies of humanity. No definition applies more to Abdus Salam, practising Muslim, physicist of genius, "crossed" by two contradictory causes, that of fundamental science understood in the sense of European rationalism and that of developing countries whose history and heritage did not prepare them to make science the basis of their culture. This second edition of *Ideals and Realities* contains a few new texts to the first: it is a collection of articles and lectures which deal with Abdus Salam's personal experience, belonging to three worlds: Islam on the one hand as a religious source: theoretical physics on the other and the direction of the International Centre for Theoretical Physics at Trieste as a practising professional; finally, the continuous fight, at the heart of the United Nations and other international institutions, against the blockages of the Third World, against those who adhere to the selfishness of the industrialised countries, against those who adhere to internal heaviness and to the choice of development of the poor countries.

The career of the author began in 1940, when, at the age of 14, he entered the University of the Punjab, with the highest marks that any student had ever obtained. The short history recounts that returning from his examinations, the inhabitants of his village were waiting for him in the street to greet him like a hero. At 18 years of age, we find him at St. John's College and soon at the Cavendish Laboratory; the beginning of a brilliant scientific career, and a series of prizes, of elections into numerous academies, till the Nobel Prize in 1979, in homage to his theoretical perceptions, accomplished 13 years before, on the interaction between the electromagnetic and the weak forces, where he could predict the existence of particles (W and Z) which experiments at CERN would confirm much later. Abdus Salam, in Arabic, means "servant of peace": the specialist in quantum theory has never ceased fighting for another international order, insisting on the role that science can play in the development of the Third World. There is something mystical in this scholar; and without doubt the most revealing text of this book is the homage which he renders to his "protector" Chaudri Zafrulla Khan, jurist, also educated in Cambridge, Minister of Foreign Affairs of Pakistan, then President of the International Court of Justice, before retiring to the Mosque in London to become one of the greatest commentators on the Quran. A mystic who gets on well with science but less well with the world. This would give sense, in a less oriental version, to the phrase of "Vicky" Weisskopf, another migrant physicist, who belongs to only two sources, the old Europe of Vienna and the America of MIT, which Salam recounted at CERN: "Humanity is founded on two pillars: compassion and knowledge. Compassion without knowledge is ineffective: knowledge without compassion is inhuman".

Professor Salam at the IAEA General Conference (1977).

Professor Salam receiving his Nobel Medal (December 1979).

Professor Salam with the Secretary-General of the United Nations,
Mr. Javier Perez de Cuellar, July 1985.

At a plenary meeting of the Advisory Committee on Science and Technology (ACAST) at the United Nations Headquarters, New York, Professor Salam talks with Professor Thacker of India.

Professor Salam gives lecture at the Cavendish Laboratory, Cambridge, Scott Lectures (1965). In the front rows are Nobel Laureates Sir George Thompson, Sir John Cockcroft, Sir Nevil Mott and P.A.M. Dirac.

Abdus Salam and Nobel Prize winning physicist P.A.M. Dirac over tea in Cambridge.

Participants at the Symposium on Elementary Particle Interactions (from left to right): M. Tonin, G. Patergnani, G. Furlan, P. Cazzola, N. Dallaporta, G. Peressutti, J. Prentki, G. Andreassi, S. Fubini, L. Bertocchi, C. Villi, B. Bosco, R. Stroffolini, C. Ceolin, P. Budinich, D. Amati, A. Salam, A. Stanghellini, B. Vitale and W. Thirring. Sitting: V. Glaser and L. Taffara.

It was at this Symposium (22–26 June 1960) that Professor Salam proposed the establishment of the International Centre for Theoretical Physics (ICTP).

Professor Salam receiving the Honorary Degree of Doctor of Science at the Aligarh Muslim University, Aligarh, India, January 1981.

Professor Salam with His Holiness, Pope John Paul II.

Professor Salam receiving the Honorary Degree of Doctor of Science at the Universidad Complutense de Madrid, Spain (1983).

Nobel Laureates Professor K. von Klitzing (standing right), Professor K. A. Mueller (left), Professor Abdus Salam (centre), Professor K. Siegbahn (right) and UNIDO Project Leader Mr. G. Rosso-Cicogna (standingl left) met in December 1988 to examine the proposal for the establishment of the International Centre for Science.

The Chinese leader, Mr. Deng Xiaoping, in conversation with
Professor Salam at the Great Hall of the Peoples in Beijing,
September 1986.

Professor Salam in his office with On. Giulio Andreotti,
Prime Minister of Italy.

Professor Salam attending a meeting of the South Commission in Kuwait (1988).

Professor Salam at the Kashgar Teachers College, China (1988).

Abdus Salam as a student at the Government Intermediate College, Jhang, Pakistan (1940). Salam secured 765 marks out of 850 thus creating a new record in the MSLC examination in 1940. (Punjab Kitab Ghar Regd., 19 Mohan Lal Road, Lahore, Pakistan).

Inscription of 16th Century Persian prayer in his office at
Piazza Oberdan in Trieste (1967) reminds Abdus Salam of the
power of miracles, provided one initiates them with hard work.

Professor Abdus Salam with Professor N. N. Bogolubov.